Simulation of ecophysiological processes of growth in several annual crops

Simulation Monographs 29

Simulation Monographs is a series on computer simulation in agriculture and its supporting sciences

Simulation of ecophysiological processes of growth in several annual crops

F.W.T. Penning de Vries, D.M. Jansen,
H.F.M. ten Berge and A. Bakema

Pudoc Wageningen 1989

CIP-data Koninklijke Bibliotheek, Den Haag

ISBN 90-220-0937-8
NUGI 835

Printed in the Netherlands

Contents

Preface

Research has helped many countries to improve agricultural productivity substantially. The scientific approach often used to explore possibilities to increase crop yields is identifying a crucial factor and finding out how to stimulate the positive, or to suppress the negative effects. This type of research often requires a wide range of experiments to establish how different environmental conditions and management factors affect the crop, the soil and the crop pests. The agricultural extension service then brings practical research results to farmers.

This experimental approach yields many results, but also has limitations. For example, in very intensive agriculture a demand develops for management advice that is specific for a certain field, a certain crop and for the weather pattern in a certain year. This aims to use inputs (water, fertilizer, biocides, labour) more efficiently, at less cost to the environment, and, if possible, attaining better yields. Whatever the capacity of a country's agricultural research, it cannot cope with these demands using the experimental approach alone.

Developing countries undertake less agricultural research per crop, per hectare and per person than rich countries. This puts an extra burden on attempts to obtain individual research findings, to integrate them and to develop practical recommendations for farmers. Findings from one place, or from one crop, can only be crudely translated into recommendations for other areas or other crops, unless additional local trials are also undertaken. These are often costly and time-consuming. Increasing the usefulness of experimental results and improving the extrapolability of conclusions from ongoing trials enhances the productivity of agricultural research in developing countries.

Crop growth simulation models are maturing and are increasingly used to support field research and extension in developed and developing countries. Their potential value seems to be substantial. Simulation models draw on knowledge from disciplines such as crop physiology, soil science, agroclimatology and phytopathology. They give quick answers to many 'what if?' questions. Simulation models using proper data can be specific for certain fields and conditions, or can be applied on a regional scale. They can also simulate yields under different weather regimes. This knowledge base has no geographic limitations and can be extrapolated to different conditions, other cultivars, or other cropping schemes. Once objectives have been established, the data required for simulation models can usually be obtained. Progress with models can be much faster than with experimental research alone, inspite of the fact that it is still necessary to evaluate key simulation results with field experiments.

This book explains physiology in a whole-crop context for those who want to apply crop growth models in a broader framework. It provides a theoretical basis, concrete models and concrete data sets for more than a dozen crops. Some data about soils are also presented. The models can be used to simulate production and water use in order to survey and evaluate new sites, crops or cultivars, and husbandry and management techniques before embarking on a large program; they can also establish the impact of year-to-year weather variability on the crop much faster than with more conventional methods. Simulations can sometimes replace field trials, and will usually give field trial results broader and deeper perspectives.

Simulation no longer remains the privilege of specialists. Computers are now widely affordable and usable by organizations or individuals wanting to use them, and software is becoming user-friendly.

We hope that this book will provide many scholars access to simulation models and will help answer specific questions. The models are presented in an open style, rather than as a black-box, so that they can be adapted, modified and further expanded.

Acknowledgements

This monograph uses the results from more than a decade of research and modelling. Some results are explicitly referenced, but many are intertwined in the text and programs. Colleagues at the Department for Theoretical Production Ecology of the Agricultural University and of the Centre for Agrobiological Research (CABO) in Wageningen influenced the work considerably. The result of the synthesis in this Monograph, however, is fully our own responsibility.

Many persons contributed to this monograph in one or several stages of development. J.C. Alagos, H. Drenth, N. van Keulen, J. van der Lek, C.H. Reyes and P. de Vries were explicitly and extensively involved. Many colleagues, visitors, students and course participants also contributed directly or indirectly. A.J. Flinn and C.J.T. Spitters gave valuable editorial advice. The untiring secretarial assistance of H. Rodenburgh and A.A. Silva and drawings by R. Calubiran and C.G. Beekhof is kindly acknowledged.

Drafts of this monograph were used in crop simulation training programs for multidisciplinary teams of scientists from different countries in South East Asia. Their interest in simulation motivated us to complete the text and programs; their scrutiny helped us eliminate many bugs, though an uncounted number must still be there.

Financial support was given by the Dutch government through the Department of Foreign Affairs, Directorate General of International Cooperation.

F.W.T. Penning de Vries
D.M. Jansen
H.F.M. ten Berge
A.H. Bakema

Reader's guide and glossary

The book covers many of the processes of crop growth and water movement and explains how to represent them in simulation models.

The first chapter contains a general introduction to simulation and modelling. It is recommended reading for all those intending to use the simulation models described in this book.

Chapters 2 and 3 introduce and discuss how to simulate the most important crop processes when there is adequate water and nutrients and no pest problems. Modules for simulating crop growth in these situations are presented in Chapter 7, Listings 2-5. By the end of Chapter 3, readers should be able to simulate crop growth in optimal conditions for a range of crop types and weather, and be able to adjust the modules according to individual objectives.

Chapters 4 and 5 discuss simulation of the effects of water stress on crops and introduce simulation of the soil water balance. These chapters are recommended for those persons who will simulate crop growth with periods of water shortage. Modules to simulate the effects of water stress and the soil water balance are presented in Chapter 7, Listings 7-10. By the end of Chapter 5, readers should be able to simulate crop growth with limited water conditions by combining these modules with some of the modules discussed in Chapters 2 and 3.

Chapter 6 discusses weather variables that are inputs for crop models, Listing 11 provides an example. This chapter is recommended to all readers because these data are used in virtually all simulations.

The 11 modules presented can be combined into full simulation programs in several ways. Table 1 shows which combinations are appropriate and in which situations. Full simulation programs are constructed from the modules by the use of a text processor. The final program should have only one initial and one dynamic section. All functions and subroutines can be contained in a library.

A glossary briefly describes frequently used terms and concepts.

Implications for crop growth and specific ways of formulating the modules will not always be evident from the text itself. Simulation practice and using the modules provided is essential. Exercises and answers are found at the end of Chapters 2-5. Exercises are divided into T-exercises which can be answered while reading the text, and S-exercises which require simulation and can be completed at the end of Chapter 3 (for crop growth) or Chapter 5 (for crop growth with water shortage).

Exercises should develop further insights into crop modelling and provide hands-on experience with simulation. Numerical results are given with high precision so that readers can check their simulation results, but this does not

Table 1. Combinations of modules to create full simulation programs. Objectives of the simulation dictate which modules are appropriate.

Module type	< - - MACROS submodels - - >						<- - - - data- - - - >			<subr>
Module name	L1D	L1Q	TIL	L2C	L2SU	L2SS	crop	soil	weather	T12
Listing number	3	4	2	7	8	9	5	10	11	App B
Purpose:										
Potential production	x						x		x	x
Potential production, day-night cycle		x					x		x	x
Potential production, tillers (rice)		x	x				x		x	x
Potential production and transpiration	x			x			x		x	x
Rainfed production, deep water table	x			x	x		x	x	x	x
Rainfed production, shallow ground water	x			x		x	x	x	x	x

suggest absolute accuracy. Do not restrict yourself to obtaining the same answers, but apply the models for your own crop, soil and weather data. Adapt, shorten, or extend programs when desired. Readers are encouraged to examine the varying aspects of model behaviour to explain the results.

Simulation exercises are particularly effective when undertaken by small groups of students or during courses. Running models will help greatly in obtaining a better grasp of simulation and of the entire model. It is only by running, adapting and improving crop models for stated objectives that the knowledge and insights gained can become effective in agricultural research, extension or planning.

Glossary

CSMP The simulation language Continuous System Modelling Program (IBM, 1975).
Function A user defined mathematical, physical or biological relation between one or more inputs and one output (FORTRAN).
Listing A printed version of a module or a program.
MACROS Modules of an Annual CROp Simulator.
Mimick To reproduce the behavior of a small system or subsystem with

equations that are not based on the processes involved.
Module A set of statements in a computer language that together describe a system or a large part of a system; a set of data characterizing a crop, soil or weather; a set of functions and subroutines.
Model A simplified representation of a system. A submodel is a model of a subsystem.
Simulation model A module that represents the relevant processes of a system, usually in the form of a computer program.
Program A complete set of modules and individual statements, and data sets of a particular module and of its driving variables (also called: simulation program).
SAHEL A model for simulating the water balance of free-draining soils with a deep water table.
SAWAH A model for simulating the water balance of soils with impeded drainage, often with a high water table and partially saturated.
Simulate To create a model to study a system and to use a program of the model to reproduce the behavior of the system.
Subroutine A user defined mathematical, physical or biological relation between one or more inputs and one output (FORTRAN).
System A part of the real world consisting of parts that interact and change. The environment of the system exerts influence on it, but the system does not affect its environment. When a system is part of a larger system, it is referred to as a subsystem.

1 Introduction to crop growth modelling

Chapter 1 attempts to answer the question 'Why create a crop growth model?' and to identify methods of determining what factors should be included.

Section 1.1 introduces systems analysis and the dynamic simulation of living systems in their physical environment. Section 1.2 narrows the focus to systems analysis and simulation of annual field crops. Section 1.3 deals with the possibilities for using crop growth models in research and education, and their practical application. Some technical aspects of simulation techniques are discussed in Section 1.4.

1.1 Modelling crop growth

Growing a crop is complex. Some activities, such as planting or seeding, are always needed; others, such as irrigation, fertilization and spraying fungicides are optional. A farmer combines activities effectively because he has a concept or model of how the crop will react to its environment and to husbandry practices. In this sense farmers use multidisciplinary models. However, these mental models are somewhat crude and are difficult to improve or to explain to others.

1.1.1 Descriptive and explanatory models

A crop model is a simple representation of a crop. It is used to study crop growth and to compute growth responses to the environment. Crop models in common use can be distinguished as descriptive and explanatory models.

Descriptive models

A descriptive model defines the behaviour of a system in a simple manner. The model reflects little or none of the mechanisms that are the cause of the behaviour. Creating and using this type of model is relatively straightforward (Figure 1). Descriptive models often consist of one or more mathematical

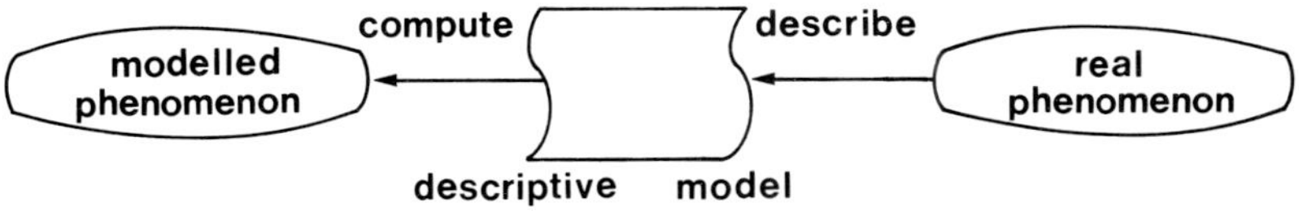

Figure 1. A scheme to indicate how real world observations are brought into a descriptive model.

equations. An example of such an equation is derived from successively measured weights of a crop (Figure 2). This equation is helpful to determine quickly the weight of the crop when no observation was made. However, the growth rate of the crop will not be the same when soil, crop husbandry practices or weather are different. Large deviations can result from differences in weather patterns between years (Figure 3). Adapting the starting point of the regression equation and of the maximum value are possible in hindsight, but predicting these parameters for other fields and in other years is usually too inexact for specific production studies. In theory, it is possible to derive the required constants and equations from many experiments with acceptable accuracy. In practice, however, many variables influence growth patterns. Some, such as soil texture, are constant; others, such as the properties of new cultivars and crop husbandry practices, constantly evolve. Thus, it is impossible to quantify adequately all variables through extensive field experiments. Descriptive models are therefore of value only for situations where interpolation between observations is sought and there is no attempt to quantify the background of the shape of the biomass curve.

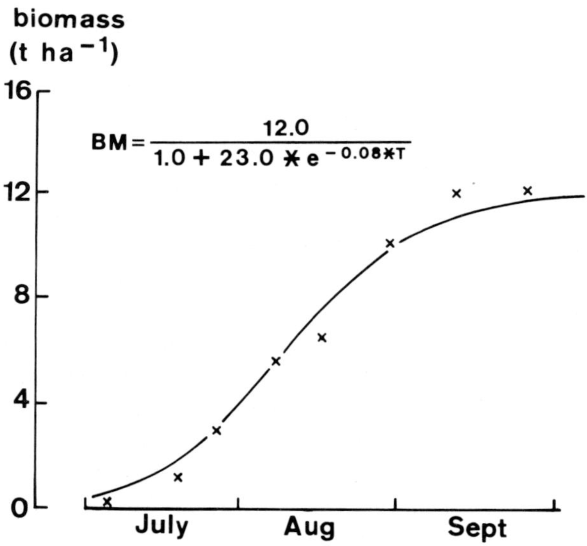

Figure 2. The course of the dry weight of a maize crop in the Netherlands in 1972. Crosses represent observations, the line the regression equation $BM = 12.0 / (1.0 + 23.0 \cdot e^{-0.08 \cdot T})$, where *BM* is the biomass in t ha^{-1}, 12.0 is the maximum value of *BM*, *T* is the time in days since emergence and 1.0, 23.0 and 0.08 are constants.

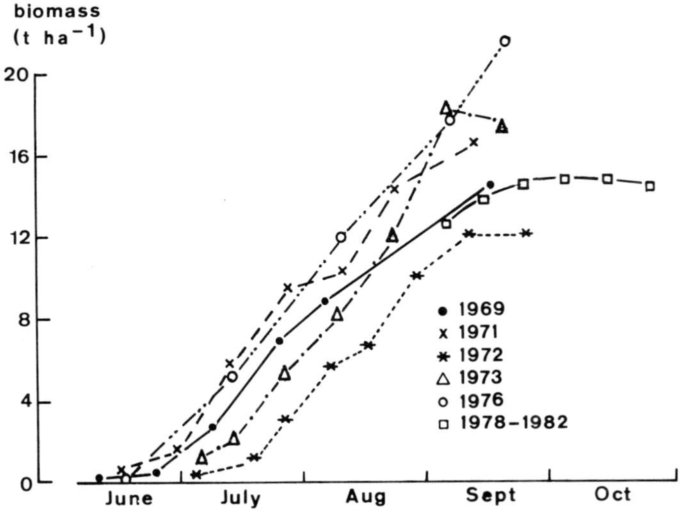

Figure 3. The dry weight of maize crops under optimal conditions in different years in the Netherlands. (Source: Sibma, 1987).

Explanatory models

An explanatory model consists of a quantitative description of the mechanisms and processes that cause the behaviour of a system. These descriptions are explicit statements of the scientific theory and hypotheses. To create an explanatory model, the system is analyzed and its processes and mechanisms are quantified separately. The model is built by integrating these descriptions for the entire system. An explanatory crop growth model contains descriptions of distinct processes such as photosynthesis, leaf area expansion and tiller induction. Crop growth is a consequence of these underlying processes (Figure 4). Each process must be quantified in relation to environmental factors, such as radiation and temperature; and in relation to the crop status, including leaf area, development stage and nitrogen content. Growth rates can then be computed for any stage of the growing season, depending on the actual crop status, the soil and current weather. All important factors can be accounted for in this way, provided there is sufficient theory and data to quantify them. Models considered in this book are mostly of the explanatory type.

The 'behaviour' of the model, i.e., the growth rate for any stage, can be explained by the basic physiological, physical and chemical processes and by the effects of environmental factors on them. De Wit (1970) noted that the hierarchical levels of explanatory processes and of explained behaviour are characterized by time coefficients of different orders of magnitude, and that they are usually the subject of study in different scientific disciplines. The ex-

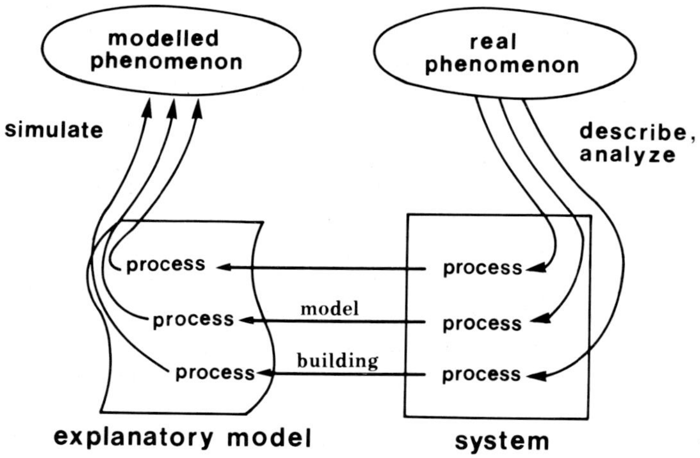

Figure 4. A scheme to indicate how real world observations are analyzed and integrated into an explanatory model to simulate behaviour of the system.

planatory approach to modelling goes deeper by at least one hierarchical level, and sometimes two or more, than the descriptive approach.

1.1.2 Simulation with explanatory models

Simulation models are relatively simple representations of systems in the world around us. A system is defined here as any well delineated part of the real world. The user identifies a system on the basis of objectives and on the intrinsic structure of the world as measured and observed. For an agronomist, a system may be a rice crop; its elements, plant organs (such as leaf, stem and root) and processes (such as growth and transpiration) interact strongly. Weather is a driving variable because it exerts an important driving, or regulating effect on the crop. The crop, on the other hand, has virtually no impact on the weather. In general, driving variables influence the system and its behaviour, but the reverse is not true. Behaviour is the sum of all processes in a system, for example, the growth of a rice crop during a season. A system is 'dynamic' when its states change over time. This may be 'continuous', as when its behaviour and states change relatively slowly, or 'discrete' when changes occur fast or are large (e.g. a tractor changing from the state used to not-used).

A model is dynamic when it simulates the behaviour of a system. State variables in models represent quantities, which may be tangible (such as weight), or abstract (such as development stage). Rate variables represent rates of change of state variables. The photosynthesis rate is an example of a rate variable. In this context, 'simulation' is the study of a system and the computation of its behaviour using a dynamic model.

In explanatory simulation models of dynamic systems, such as those of crops, it is assumed that the rate of change can be closely approximated by considering the rates of processes to be constant during short time periods. This is the state variable approach. In crop simulation time periods must be short compared to the duration of the growing season; often one-day periods are chosen. The biomass formed in such a short time period equals the multiple of growth rate and time period. This is added to the quantity of biomass already present. The growth rate is then recalculated. The new rate is slightly different because environmental conditions or the internal status of the plant will have changed a little. Calculations of rate variables and updating state variables are repeated in sequence until the entire growing season has been covered. Van Keulen & Wolf (1986) give a clear example of this procedure for a growing crop.

Fortunately the number of processes of prime importance for simulating crop growth is limited and detailed calculations to quantify these processes are unnecessary. For example, calculating the efficiency of synthesis of each biochemical compound in a biomass is usually unnecessary; averages for classes of compounds are sufficient. It is also unnecessary, and even counter-productive, to include dynamic aspects of cell physiology in crop growth models. Explanatory models can be of practical use even though knowledge of the processes often does not reach the cellular level. Indeed, it is of limited importance whether processes at the explanatory level (the general level of physiology and soil physics in this book) are descriptive or quantified in more detail (e.g. at the biochemical level), as long as their quantification is valid within the range of conditions for which the model is used. The more detail desired in the results of a model, the more detail the model itself must contain, and hence the more explanatory processes should be included.

1.1.3 Development of explanatory models

Large and complex explanatory models have been developed over the last decade in many places, including Wageningen in the Netherlands. Development has been relatively slow, because among other things, some essential topics were insufficiently understood. These large comprehensive models contain a wealth of information, but are unwieldy. Their use is limited to reassessing hypotheses, for sensitivity analyses and for reference and comparison with other models. These models are seldom used except by the scientists who created them. This not only limits their usefulness, but also undermines the credibility of models and model builders.

In recent years, summary models have been derived from several comprehensive models. These models retain much of the scientific basis and quality of the comprehensive models, but are simpler and much easier to use.

Explanatory models are of three types: preliminary, comprehensive and summary. Explanatory models develop gradually from one type, or phase, to the next. Preliminary models have a simple structure because insights at the explanatory level are still vague. Comprehensive models represent a system in which essential elements are thoroughly understood and incorporate much of this knowledge. Summary models are abstracts of comprehensive models.

Summary and comprehensive models are currently used where weather or soil water limit crop production. Models vary considerably in their value for scientific research, education and applications (Table 2). Examples are given in Chapter 1.3. The content of models considered in Chapters 2-5 is mostly at the summary phase, though parts are still scarcely beyond the preliminary phase.

For more extensive introductory reading see Brockington (1979), Dent & Blackie (1979), Loomis et al. (1979), Penning de Vries & van Laar (1982), Penning de Vries (1983), van Keulen & Wolf (1986), or Rabbinge et al. (1989).

1.2 Crop Production Levels and processes

1.2.1 Levels of crop production

De Wit proposed a classification of systems of crop production based on growth-limiting factors (de Wit & Penning de Vries, 1982; Penning de Vries & van Laar, 1982) and distinguishes four levels of plant production. The crop production systems at any of these levels can be considered as members of a broad class of systems. In order of decreasing yield, these levels are:

Production Level 1

The crop has ample water and nutrients and produces a higher yield than at any other Production Level. Its growth rate depends only on the current state of the crop and on current weather, particularly radiation and temperature.

Table 2. The relative values of important aspects of models in different phases of development.

	Scientific value	Instructive value	Applicability value
Preliminary model	+++	++	+
Comprehensive model	+++	+	++
Summary model	+	+++	+++

With a full canopy, the growth rate of field crops is typically between 150 and 350 kg ha^{-1} d^{-1} of dry matter. This is the 'potential growth rate' and the crop yield 'the potential yield'. These growth conditions are realized on very intensive arable and grassland farms in Western Europe and often in glasshouses.

Production Level 2

The growth rate is limited only by the availability of water for at least part of the growing season. This situation seldom occurs spontaneously, but in semi-arid regions applying fertilizers can result in crop growth at this Production Level. This may also occur in other climates under intensive cropping on light soils.

Production Level 3

The growth rate of the crop is restricted by nitrogen shortage for at least part of the growing season and by water shortage or poor weather for the remainder. This situation occurs frequently in agricultural systems all over the world. Nitrogen shortage occurs particularly in crops when fertilizer is not intensively applied. In the natural environment, even nitrogen-efficient plants cannot always absorb sufficient nitrogen.

Production Level 4

Crop growth is restricted by low phosphorus and other mineral nutrients in the soil for at least part of the growing season. The growth rates are 10-50 kg ha^{-1} d^{-1} and the growing season often lasts less than 100 days. This situation usually occurs in heavily exploited areas where no fertilizers are used.

Rarely do cases fit exactly into one of these Production Levels, but it is practical to reduce specific cases to one of these four categories. This focuses attention on the dynamics of the main environmental factor and on the response of the crop to it. Environmental factors that have no regulatory effect can then be disregarded, because they do not determine the growth rate. The growth rate then sets the absorption rate or efficiency of use of non-limiting factors. If, for example, plant growth is limited by nitrogen, there is little use in studying CO_2 assimilation or transpiration to understand the current growth rate. All emphasis should be placed on nitrogen availability, the nitrogen balance and the plants response to nitrogen.

This analysis of plant production systems allows for considerable narrowing of the subject of study and permits more rapid research progress. Growth-reducing factors, such as diseases, insect pests and weeds, can occur at each of these Production Levels and give them, in a sense, an extra dimension. The fact that actual situations are often more complex does not contradict the general usefulness of this scheme of Production Levels as a basis for distinction between causes and consequences of plant growth.

Note that this use of Production Levels has a crop physiological basis and is not related to descriptions of production systems based on agronomic practice or crop ecology, such as the irrigated, rainfed lowland, deep water and upland production systems in rice growing (IRRI, 1984).

1.2.2 Principal processes of the Production Levels

This systematic analysis of crop production can be taken a step further to formulate simple systems and models at the four levels of plant production.

At Production Level 1 the intensity of radiation, the interception of light and the efficiency of energy use in the plant are key factors for understanding the growth rate. Figure 5, a relational diagram, indicates the essence of models at Production Level 1. Light is a driving variable. Assimilated carbohydrates are stored, usually briefly, in an easily accessible form, such as starch ('reserves'), and are later used for maintenance or growth. Temperature is an external variable that can modify growth rates and photosynthesis. In growth processes, reserves are converted into 'structural biomass' with a specific efficiency. Structural biomass consists of those components that are not mobilized again for maintenance or growth processes elsewhere in the plant. The partitioning of biomass between roots, leaves, stems and storage organs is strongly related to the physiological age of the crop, which itself is a function of temperature.

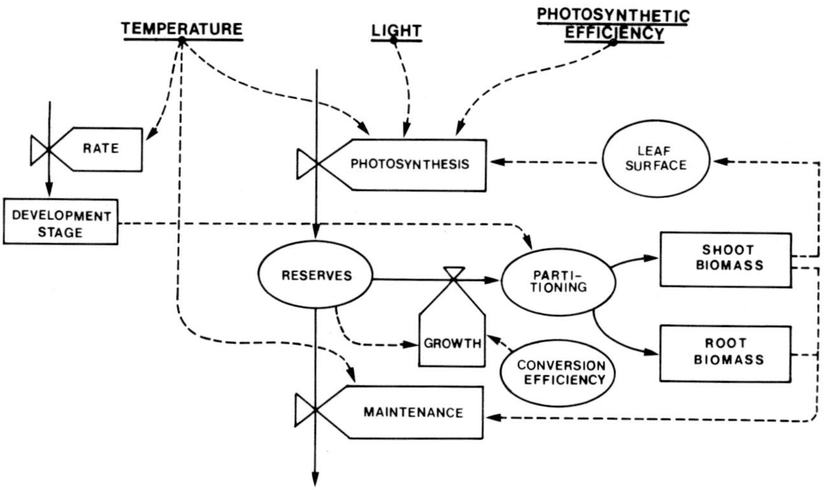

Figure 5. A relational diagram of a system at Production Level 1. Light and temperature are driving variables; the photosynthetic efficiency is a constant. Rectangles represent quantities (state variables); valve symbols, flows (rate variables); circles, auxiliary variables; underlined variables, driving and other external variables; full lines, flows of material; dashed lines, information flow (symbols according to Forrester, 1961).

At Production Level 2, key factors are the degree of exploitation of soil water and the efficiency of its use by the crop (Figure 6). Water shortage leads to stomatal closure and to a simultaneous reduction of CO_2 assimilation and transpiration. Water use efficiency is the ratio of photosynthesis and transpiration rates. The ratio of the actual transpiration rate and the potential rate provides the link between the carbon and water balance. The extent to which the potential transpiration, and consequently the potential photosynthesis rate, is realized, depends on the availability of water. The amount of water stored in the soil is a buffer between rainfall and capillary rise and the processes by which water is lost. This buffering capacity and the simultaneous water loss through transpiration and non-productive processes cause the growth rate to depend only indirectly on rainfall. The relation of plant growth to the principal driving variable of this system is indirect (rather than direct, as at Production Level 1).

At Production Level 3, nitrogen in plant tissues is distinguished by two fractions: mobilizable and immobilizable nitrogen (Figure 7). The amount of nitrogen that can be mobilized for growth of new organs is often considerable.

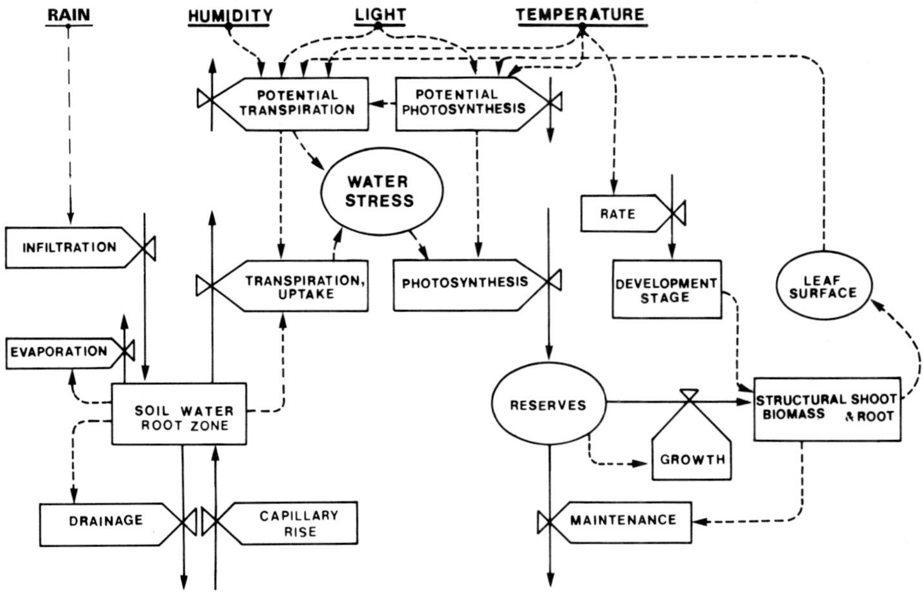

Figure 6. A relational diagram of a system at Production Level 2. Water shortage is the main limiting factor. Rectangles represent quantities (state variables); valve symbols, flows (rate variables); circles, auxiliary variables; underlined variables, driving and other external variables; full lines, flows of material; dashed lines, information flow (symbols according to Forrester, 1961).

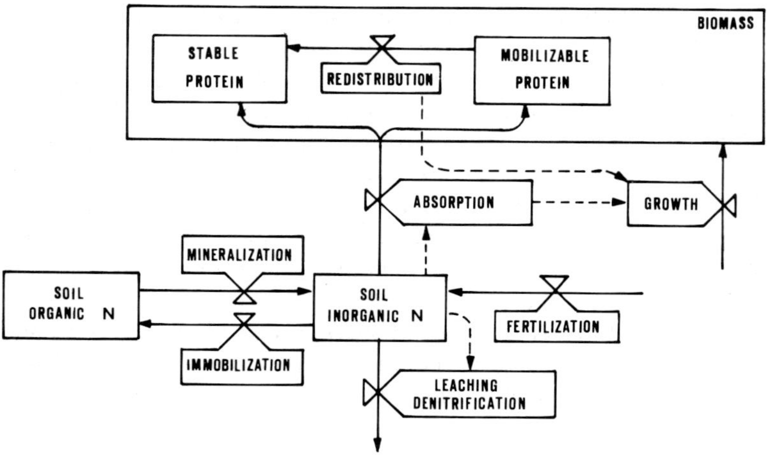

Figure 7. A relational diagram of a system at Production Level 3. Nitrogen shortage is the main limiting factor. Rectangles represent quantities (state variables); valve symbols, flows (rate variables); underlined variables, driving and other external variables; full lines, flows of material; dashed lines, information flow (symbols according to Forrester, 1961).

The concentration of nitrogen in mature tissue may reduce to half or a quarter of its maximum value before the tissue stops functioning. Growth is directly related to the rate of nitrogen absorption only after the internal nitrogen reserve is used. This internal reserve of nitrogen makes the increase in plant dry matter at any moment largely independent of the current absorption of nitrogen.

The relation of nitrogen uptake and growth is, therefore, quite different from that of water uptake and growth. The mobilizable fraction consists of enzymes and membrane proteins that are broken down and exported as amino acids; not all can be considered reserves, because cells cannot function without them. The immobilizable fraction of nitrogen in the tissues is tied up in stable proteins. The growth rate at this Production Level is primarily determined by the availability of nitrogen from the soil and the internal reserve. Hence the rate of CO_2 assimilation is a consequence of the growth rate. The availability of nitrogen from the soil resembles that of water; a variable amount of inorganic nitrogen is present in the soil and most of it is readily available to roots that are sufficiently close. Soil microflora may compete with plants for this nitrogen and other processes may also interfere. Nitrogen in organic matter in the soil is not available to crops. But mineralization, i.e., breakdown of organic matter by microbes, releases nitrogen to the inorganic pool.

Crucial processes of crop growth at Production Level 4 are similar to those at

Production Level 3 (Figure 8). The concentration of phosphorus in ageing tissue decreases in the same way as nitrogen; and, as with nitrogen, plants also have an internal reserve of phosphorus. But the processes that make phosphorus available to roots differ considerably from those for nitrogen. Plants require a much higher root density for adequately exploring the soil for phosphorus; and the quantity of dissolved phosphorus in the soil is so small that the rate of its replenishment determines the phosphorus supply to roots. Mycorrhiza may enhance phosphorus uptake by increasing the explored volume of soil. Both organic and inorganic compounds in the soil may provide and capture dissolved phosphorus.

Chapters 2, 3, 4 and 5 consider models for situations with ample nutrients for crop growth. For models of crops in situations with severe shortages of nitrogen and phosphorus, see van Keulen (1982), Hansen & Aslyng (1984) and van Keulen & Wolf (1986).

1.3 Uses of crop growth models

1.3.1 Determining when to use a simulation model

De Wit et al. (1978) stated: 'In our opinion, simulation models, if they are useful at all, should form a bridge between reductionists, who analyze proc-

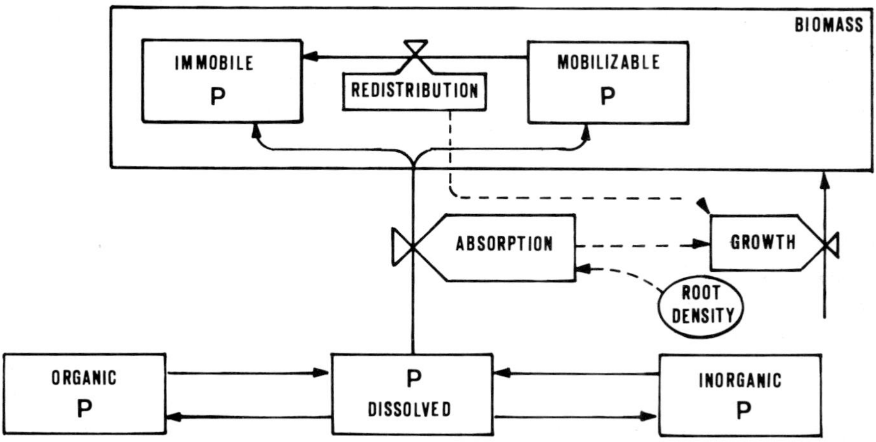

Figure 8. A relational diagram of a system at Production Level 4. Phosphorus shortage is the main limiting factor. Rectangles represent quantities (state variables); valve symbols, flows (rate variables); underlined variables, driving and other external variables; full lines, flows of material; dashed lines, information flow (symbols according to Forrester, 1961).

esses separated from their physical, chemical or biological background, and generalists who are interested in the performance of whole systems in which the individual processes operate in their natural context. Both the reductionist and the generalist should recognize their work in the simulation program. By comparing detailed output, the generalist can independently evaluate how the model operates with field data, and the reductionist can determine whether the treatment of processes that form the basis of the simulation model correspond with his ideas. To the reductionist, simulation can be a guide to areas where research is most promising for further understanding of the system studied. To the generalist, simulation extends his capability to envisage how a whole system functions.'

Scientists who are generalists follow a bottom-up approach to modelling. By using a mathematical model, they can describe their subject more clearly and study the implications more easily. As the generalist's model is improved and elaborated, it gives a broader view and covers more topics. As yet, understanding the real world of agriculture and biology is still far from complete. Further improvement of crop growth models is necessary.

Research administrators, policy makers and industrial leaders may use a top-down approach when seeking the solution to a problem or surveying possibilities. They first identify the problem and then determine specifications to which the answer must comply. One or more techniques from mathematics or information sciences, such as statistics, linear programming, simulation, data base management and expert systems, can be used to provide these answers. Statistics can create clarity about the relations between variables that fluctuate. Linear programming can help determine the optimal combination of many inputs and factors to achieve a certain goal, such as optimal land use (see Kingwell & Pannell, 1987). Simulation techniques can cope with complex relations between state and rate variables in a system and enable computation of a systems behaviour in specific circumstances and new environments. Data base management techniques allow storage of masses of data plus the relations between them; each item remains traceable individually or in groups, making information quickly, efficiently and completely available to many people. Expert systems, an emerging tool, can help select the best choice from many possibilities in a question-and-answer dialogue, the computer playing the role of the expert.

All techniques require the availability of sufficient basic data and knowledge. For simulation, the relations between all principal variables of the system and the values of key constants must be known. This should not be overlooked. If these relations and values are not known, it may be better not to choose simulation for the practical problem; it may be better to devote a major research effort to accumulate the necessary knowledge.

Finding basic data to use in this book was not easy. The amount in open literature is limited and an inventory for a range of crops was difficult to make. This was true even for parameters that are quite commonly used in science and for which good examples are documented, such as the maximum rate of leaf photosynthesis. This is symptomatic of the fact that we are still in the early stages of applying crop growth models.

Potential users of crop growth models frequently ask scientists to produce or adapt a model for a specific crop and a specific problem. Model developers look more towards potential users, not least, because they may sponsor their work. A brief discussion of where modelling stands in terms of usefulness to non-modellers is presented in Subsection 1.3.4. Models can also be appreciated for their value for research and for instruction, see Subsections 1.3.2 and 1.3.3. It is worthwhile to recall that the usefulness of a model changes considerably as it evolves (see Table 2 Subsection 1.1.3).

Any tool can be adequate or inappropriate, depending on the goal for which it is applied. This holds for models just as much as for mechanical tools. A particular crop growth model can be very suitable for achieving a specific goal, but totally inappropriate for another. The user and the developer must carefully define objectives before using, adaptating, or developing a model. Though this may appear obvious, many modelling efforts have suffered from under-specified objectives (IIASA, 1980). Defining objectives can also help prevent excessive optimism about results and unconsciously pushing objectives higher and higher. It also helps resist the temptation to let the model derail and become an encyclopaedia of science.

Models are not yet at a stage (and may never reach the stage) when they can be used or applied without understanding how they work. As yet few persons are trained to work with crop growth simulation models.

1.3.2 Using models to guide research

For the past 15 years, simulation models have been used, among other things, to determine how far crop growth in different situations can be explained from documented theory and data. Objectives for this modelling have been broadly, but not explicitly defined (Ng & Loomis, 1984). In the process of developing models, topics were identified where crucial insights were missing, and hence research was needed. McKinion (1980) documented this research progress of a group in southeastern USA. An example from Wageningen describes briefly how modelling helped guide research.

In the sixties, attempts to compute the photosynthesis rate of crop canopies yielded several explanatory models, e.g. the model by de Wit (1965). These were static models; time was not included as a variable. Their results were used, among others, to estimate potential food production for certain areas of the world and hence to provide perspectives and goals for crop husbandry and breeding (de Wit, 1967; Linneman et al., 1979). These efforts stimulated quantitative research on leaf photosynthesis (e.g., Louwerse & van Oorschot, 1969).

Next, a preliminary dynamic simulation model was constructed. This included an abstract of the static photosynthesis model. Respiration was first taken as a fixed fraction of photosynthesis, and later as a fixed fraction per day of the biomass, plus an amount proportional to the growth rate. Adding a functional equilibrium between root and shoot growth (by which crops attempt to maintain an optimal water content) made this model an ELementary CROp growth Simulator (ELCROS) (de Wit et al., 1970). However, respiration was quantified unsatisfactorily and in subsequent research it was determined that respiration due to growth is directly related to the chemical composition of the new biomass (Penning de Vries et al., 1974). Respiration caused by maintenance could only be partially quantified and is a weak part of crop growth models even today.

Micrometeorology was introduced in the models (Goudriaan, 1977) to simulate, among others, the effect of canopy resistance on heat and gas exchange. This refined and improved the simulation of transpiration. Next, the BAsic CROp growth Simulator (BACROS) was evolved (de Wit et al., 1978). BACROS has been evaluated with field crop experiments with two-week periodic harvests. But as this comprehensive model simulates growth processes with time intervals of hours, it also needed evaluation on a shorter time scale. A mobile laboratory was created (Alberda et al., 1977) to measure assimilation and transpiration continuously in the field. This led to another round of checking, correcting and improving BACROS.

The major conclusions from research about the discrepancies between observed and measured gas exchange were that stomatal resistance is often controlled by the photosynthesis rate (Louwerse, 1980), and that the contribution of diffuse light to canopy photosynthesis had been underestimated (Lantinga, 1985). Wageningen scientists now use BACROS as a reference model and yardstick for developing other models and as a basis for developing summary models, such as SUCROS (a Simple and Universal CROp growth Simulator) (van Keulen et al., 1982). Current research is aimed at further improving BACROS and SUCROS, among others, by research on plant morphology and maintenance respiration and by combining them with models from other scientific fields.

1.3.3 Using models in education

Simulation models can be used for hands-on learning about the behaviour of a system in different situations when the real system is too large, too slow, or too expensive for teaching purposes. Examples are weather models, models of the annual cycle of farm activities and airplane flight simulators. There is no difference to the trainee between descriptive and explanatory models as long as they reproduce reality sufficiently well. Crop growth simulation models have, as yet, seldom been used in this way. Huke's interactive model (1985) of a rice farm in Bangladesh demonstrates several options and hazards involved in growing crops, but is not meant to be a realistic simulation of the real world.

Studying quantitative crop physiology with a model is also not yet common. Educative models should be thoroughly evaluated, and must be lucid, well documented and physically available to students; few models score high on all criteria. Summary type models are best for this purpose and some have recently been published. Learning with crop models has been part of the International Post Graduate Courses in Wageningen for some years (Penning de Vries et al., 1988) and it has been incorporated in the training program of the Wageningen Agricultural University. Some books in the series of Simulation Monographs teach crop physiology in a whole crop context.

To allow more widespread use of models by non-specialists, simplified access is generally needed. This may be achieved either by simplifying the model itself to only a few lines, or by adding a program before the model. The latter program, called an 'interface', can ask a novice-user for the most crucial inputs and suggest defaults for others. Most model simplifications, so far, are of the first group. Technical limitations in simulation techniques have made this unavoidable. However, with the advent of powerful but cheap computers, this is no longer a strong argument. Providing interfaces between the end-user and well-evaluated models may be a better development.

1.3.4 Application of crop growth models

Application refers here to using a model outside the scientific discipline in which it was constructed. In the near future, opportunities for applying simulation models for crop production lie particularly in the domains of potential crop production and crop production with temporary water shortage but ample nutrients (i.e., Production Levels 1 and 2). These domains include predicting short term yield, extrapolating and interpolating crop performance over large regions and simplifying and combining with other models to create links with other sciences. Applying models can lead to more effective use of existing knowledge for extension, agronomic and cropping systems research and breeding, to more efficient experimentation and for further integrating the scientific disciplines involved in crop production. Some examples are given in this book; others, concerning rice, are given elsewhere (Penning de Vries, 1987). A broad overview of crop simulation models and their applications is presented by Whisler et al. (1986).

A survey of the consequences of different crop husbandry measures, such as different planting densities, can be pursued with a model and alternatives can easily be compared (Ng & Loomis, 1984). Results of different timings and dosages of fertilizer applications can be simulated. This opens the door for improving the efficiency of fertilizers and biocides for specific cases, and for reducing the loss of excess inputs to the environment. Guided management, such as this, is already in use on a large scale for optimizing fungicide application in wheat in the Netherlands (Zadoks et al., 1984) and is under development for nitrogen fertilization in arable crops. Crop modelling is also used in irrigation scheduling in the USA.

Crop yields can be predicted some time before harvest by using expected weather data. This is important for crops where trade or planning post-harvest operations starts early. Because long-range weather forecasts are not yet reliable, predicting weather is not yet very exact. However, using a range of reasonable weather patterns, a 'fork' of yield expectations can be determined. Figure 9 shows this for a tulip bulb crop in the Netherlands. The further the season progresses, the shorter and narrower the fork becomes and the better commercial options can be considered. Models are not yet being used for this type of prediction.

Crop performance can be predicted for climates where the crop has not been grown before, or not grown under optimal conditions. This has been used for wheat in Zambia (van Keulen & de Milliano, 1984) and Southeast Asia (Aggarwal & Penning de Vries, 1988). Using simulation models in this manner has been successful in several cases. Even though it may not be accurate, it is probably as good as an experts opinion, and is easier to get! However, it is not

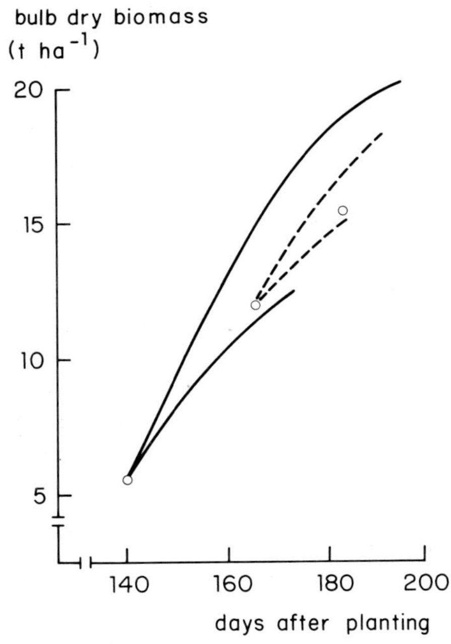

Figure 9. The course of biomass of a tulip bulb crop in the Netherlands, planted on November l. Each pair of lines begins from observed values. The lower branch of each pair represents expected growth in a very poor season, the upper branch in a very good season. Repeating the simulation after a month narrows the range of expected values (Source: Benschop, 1986).

used very often, possibly because of the lack of essential data (Versteeg, 1985) and possibly because few agronomists realize how modelling can help to establish what yield is possible. Van Keulen & Wolf (1986) have used a crop growth model to estimate yield levels of various food crops on a regional scale where there were no or insufficient external inputs to ensure potential growth. It is expected that this type of simulation can help strengthen regional development and agricultural planning in developing countries (de Wit et al., 1988).

A particular form of extrapolation is that in which the physiological or morphological characteristics used in the model are modified. Some crop characteristics are known or expected to vary between cultivars. With simulation techniques, a breeder can survey the impact that breeding for specific characteristics may have (Landivar, 1979, de Wit et al. 1979, Ng & Loomis, 1984). Few documented examples of this type of simulation exist. It is also possible to explore the effects of the increasing ambient CO_2 concentration on crop yield, harvest index and water use, and can help breeders to anticipate future requirements (Goudriaan et al., 1984).

An almost infinite number of combinations of soil type, weather and agricultural practices exist. Experimenting in all desirable situations is impossible, but using models increases the human capacity. With sequential years of weather data, estimates can be provided for weather-related variability in yield and water use. A helpful technique is generating a long series of daily values for precipitation from historic records of only a few years (Subsection 6.2.5). Variability in yields of sensitive crops or cropping sequences due to variations in weather can be tested with large sets of weather data, which speeds crop assessment by many years. This has been undertaken for Faba bean crops in Western Europe (Grashoff et al., 1987) and for rice in the Philippines (Morris, 1987).

Combining a crop growth model with a model for a related biological or physical system, but with a similar time coefficient, can yield an extremely powerful means to investigate interactions between both. Good examples are combinations with pest, disease and weed models (Rabbinge et al., 1989). Strong interactions between crop growth and disease or pest development make this combination potentially interesting for interactive crop management, e.g., a combination model could be used to determine the timing for spraying fungicides, to avoid unnecessary sprayings.

Simulation models can probably also be used to derive simple decision rules for farmers and extension services. For example, a wheat crop in a certain area of India is known to become water-stressed when rains fail for five consecutive days (ICAR, 1977) and irrigation is needed. This knowledge has been acquired from long experience and many field trials. Simulation models, supplied with the appropriate crop, soil and weather data can help to derive quickly such rules of thumb for new situations. Moreover, the decision rules can be made more specific and for smaller areas than is possible by relying only on field trials.

1.3.5 Basic data

The practical value of simulation results depends on the quality of the data characterizing the crop, the weather and the soil.

Collecting crop, soil and weather data suitable for simulation is not easy. The amount of data needed is always large and too much to collect first-hand. Hence, existing data sets and literature must be consulted. Much has been published in national and international journals about crops and soils; weather data are reported in bulletins. However, it is often disappointing to discover that published data are difficult to use. For example, experimental conditions are usually insufficiently described; measurements may be have been taken at the wrong time for the anticipated simulation; physical units may be difficult to convert to standard ones; and scientific terms may be used in different ways. When using published data, one must judge the measuring techniques used and on the relevance of the environmental conditions in which the observations were made. Thorough comparative analyses of species or varieties in field conditions are rare (e.g., Cook & Evans, 1983) and full sets of key characteristics for a single species obtained in one trial have not yet been published.

Sample data for several crops, soils and climates are given in the next chapters to allow the reader to get an impression of these values and to be able to use them in exercises with the models. This may be helpful to readers without access to good libraries. The crop data presented are from papers identified from an extensive literature search for well-documented experiments and measurements in realistic conditions. The physiological and morphological data per crop were all tested in the three climates presented in Section 6.1 and the resulting simulations yielded acceptable growth and yield curves. This implies that the crop data form consistent sets, not that they are good enough to give a firm prediction of crop performance under other conditions. The data presented may provide a fair starting point for research about the relative effect of changes in crop or soil characteristics (e.g., percentage of yield increase). However, care should be taken to verify and improve on the crop and soil data presented, particularly if simulation of crop performance in absolute terms is attempted (e.g., yield in kg ha^{-1}, or water use efficiency in mm water kg^{-1} dry matter). A certain amount of common sense remains indispensable for judging simulation results.

1.4 Modelling techniques

1.4.1 Simulation techniques

The simulation techniques used and the biological, soil physical, soil chemical and microclimatological systems considered in this book are simple and straightforward. This is possible because:

– the models used here dynamically simulate crop behaviour using only a general explanatory level of physiological and soil physical knowledge, with far less detail than in comprehensive models. The time period for integration is usually one day, sometimes less.

– several complex biological and physical processes are programmed as subroutines (see Appendix B). These are well defined and quantified and can be used without going into their scientific details.

– a simulation language is used which permits the modeller to focus on scientific problems, rather than on programming.

The real world as observed and quantified is simulated as closely as possible. The book contains all details needed to simulate the basic growth and water balance processes for a variety of crops and soils at Production Levels 1 and 2, as a function of weather, cultivation practices, cultivar characteristics and soil types. For those persons interested in applying the models in different situations and for different goals, alternative and more extensive formulations are given where appropriate.

The book explains little about programming techniques. However, a few considerations are given in this section and some details about programs are provided in Sections 3.4 and 5.4. All models are written in CSMP, with FORTRAN used in subroutines. A basic knowledge in simulation techniques and CSMP can be obtained from Basstanie & van Laar (1982), Goudriaan (1982a) and Leffelaar & Ferrari (1989). Extensive textbooks on dynamic simulation in CSMP are: de Wit & Goudriaan (1978), Brockington (1979), Penning de Vries & van Laar (1982) and Rabbinge et al. (1989). The manual for CSMPIII (IBM, 1975) contains full technical background and specifications. CSMP was developed for mainframe computers, but is nowadays also used on minicomputers and personal computers.

Exercises and answers are found at the end of Chapters 2-5 to allow readers hands-on experience. Emphasis in these exercises is on three crops: rice (because of its importance as a cereal crop); potato (as a common and productive tuber crop in cool regions); and soya bean (an important leguminous crop in tropical and subtropical areas).

1.4.2 Duration of integration periods

Continuous simulation implies simulation according to the state variable approach using relatively short time periods, so that the value of state variables changes only a little in each period (Subsection 1.1.2). How short is relatively short? As a rule of thumb, the time period of integration is about 0.1 times the time coefficient (Leffelaar & Ferrari, 1989). The time coefficient of a system with an exponentially changing state variable (i.e., the rate of change is proportional to the value of the state variable) is equal to the time in which that state variable increases or decreases *e*-fold (*e*, the base of the natural logarithm, equals 2.73). A young crop growing exponentially doubles its weight about once per week, so that the corresponding time coefficient is 10 days. Time periods of one day are thus sufficient for accurate simulation. The time period can be larger in later growth phases when changes in the crop are relatively slow.

Simulation with time periods as long as 10 days has been performed successfully with small (van Keulen, 1976) and large models (Jackson et al., 1983; van Keulen & Wolf, 1986). In such models it is implicitly assumed, either that the environment does not change considerably within the time period of 2-10 days, or that the effect of the average condition equals the averaged effect of changing conditions. Often neither assumption is correct for field crops. The soil water content, an important variable, can change greatly within a few days (Figure 10) and diseases can develop very quickly. Moreover, several processes respond to environmental conditions in a non-linear manner. For example, one or two days in a decade may be so dry and hot that the crop is water-stressed and photosynthesis is reduced. In such cases averaging temperature over a longer period gives a higher total photosynthesis than with a day by day calculation.

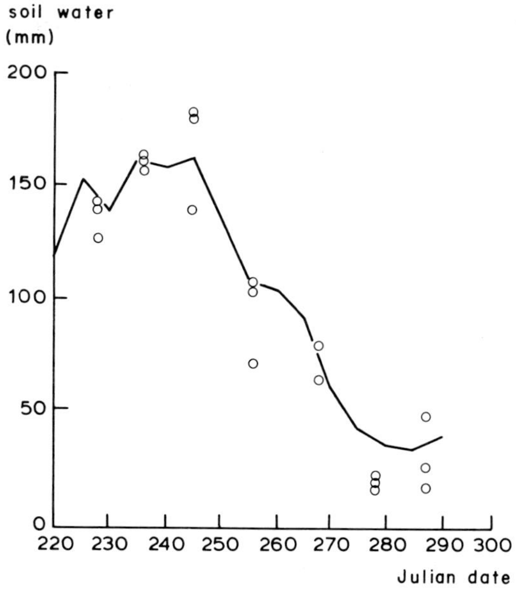

Figure 10. Simulated changes of the soil water content under a millet crop in a semi-arid zone in Mali during the rainy season (line) and observations (points) (Source: Jansen & Gosseye, 1986).

A one-day time period is, therefore, a logical choice for many crop growth models. Moreover, weather data are usually supplied on a daily basis. There is usually little interest in changes of state variables for crops over periods shorter than one day.

It can be argued that simulation with one-hour time periods would provide even more accurate results. This may be correct once comprehensive models are developed much further and computer capacity is virtually unlimited. As yet, using one-day time periods appears to be adequate. Temperature fluctuations during the growing season are usually well within the extremes of response curves for agricultural crops. The response is almost linear in this range and averaging over 24 hours is quite acceptable. But a shorter time period should be used if the temperature often exceeds the threshold beyond which important processes stop or respond differently (Figure 11).

Simulation using shorter time periods is especially necessary for crops which accumulate starch during the day, for this may lead to declining photosynthesis. For example, the canopy photosynthesis light response curve of the potato crop measured in the morning and afternoon can show hysteresis for this reason (Subsection 2.1.4). An objective of the model presented here is to be able to simulate crop growth in moderately unfavourable environmental conditions. Therefore, a version of the crop growth model was developed to handle quarter-day time periods at Production Level 1.

The rate of crop transpiration is in the order of 5 mm d^{-1}. A crop of 5000 kg dry matter ha^{-1} contains water roughly equivalent to a water layer of 5 mm. A

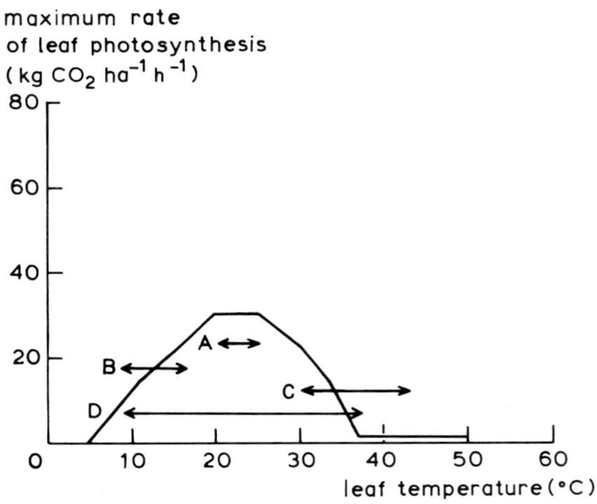

Figure 11. The response curve of maximum leaf photosynthesis to temperature. If daily temperature fluctuates within ranges A or B, one-day time periods are appropriate; if temperature often fluctuates over ranges C or D, shorter time periods should be taken.

20% decrease in available water is generally sufficient to cause severe drought stress. In other words, the daily flux of water through a crop is many times larger than the amount a crop can lose in a day without suffering. To simulate the dynamics of a crop water balance, the model should proceed with time periods of a few hours or less. Several scientists have used this method (e.g. de Wit et al., 1978). It is not necessary to follow this procedure here, because the ratio of water use to photosynthesis does not depend on the rate of water use (Section 4.1). Therefore, fluctuations in the transpiration rate during the day (due to radiation, air humidity, windspeed and precipitation) are unimportant and the total or average transpiration rate is sufficient. The simulation of the water balance for partially saturated soils requires time periods in the order of hours or less, because water fluxes between soil layers can be relatively fast. A model with short time periods is attached to the crop model with one-day time periods (Section 5.3).

1.4.3 Continuous and discrete simulation

Using a time period of one day (1 d) or a fraction of a day (e.g., 0.25 d), provides a continuous simulation of growth, despite the fact that inputs change in a discontinuous way (i.e., they have differing values from one day to the next). Such a change is a first order discontinuity (Goudriaan, 1982a). The CSMP integration method RECTangular or the Euler integration, is completely adequate in these cases. Other methods should not be used with the models described in this book.

Zero order discontinuities are sudden changes of quantities in the model. Simulation of discrete processes requires techniques other than continuous simulation, hence, CSMP is not suitable (the language SIMULA is appropriate for discrete models: van Elderen, 1987). However, elements of discrete simulation, enter into crop growth simulation models. Harvesting is a clear example for all biomass is removed in a very short time. In water balance simulation, withdrawal of water from thick soil layers is relatively slow and its simulation is continuous, but the fast withdrawal of water from top soil layers is simulated more or less discretely. Water infiltrates into soil layers quickly and this can be simulated in a discrete fashion by adding a certain amount of water to soil layers (Section 5.1). These cases are made explicit by dividing by the integration period, DELT. (This is done even when DELT equals 1.000 to maintain the correct use of dimensions in the program.) If properly carried out, there is no problem in mixing discrete and continuous simulation such as this. However, if modellers prefer to avoid mixing continuous and discrete simulation, much shorter integration periods can be used. De Wit et al. (1978) used this method with the model PHOTON, derived from BACROS. However, the same problem recurs at a more detailed physiological level, such as for stomatal opening and closing. Further elaboration of the model may overshoot the original problem for which it was built. Another method which avoids mixing discrete and continuous simulation is to construct model sections that run with different time periods (bypass method, Goudriaan, 1977, Section 5.3).

The running average concept is used a number of times in the programs presented in this book. A running average is an average of a variable in which recent values count more than older values. A running average (*AV*) of the variable (*V*) can be obtained in CSMP by the statement:

AV = INTGRL (0., (V − AV)/TC)

TC is the time coefficient of the adjustment of AV to new values of V. Because of this constant adjustment, previous values of V become less and less important. A running average can be calculated for state, intermediate, external and even for rate variables.

1.4.4 Model and data management

The models discussed in Sections 3.4 and 5.4 are presented as complete modules. Depending on the system simulated, one or more modules are to be used. Modules need to be combined with crop, soil and weather data, and with a specific set of subroutines and functions. A module with all subroutines and functions used in this book is given in Appendix B. The user can select the functions and subroutines needed, and add these between the CSMP statements STOP and ENDJOB.

Size related problems may occur on personal computer with only 512 K RAM with programs that exceed the size of the largest programs presented in this book (Section 5.4). Putting all functions and subroutines in a subroutine library reduces the program size sufficiently to avoid these problems and speeds compilation significantly. This procedure is described in the CSMP (IBM, 1975) and FORTRAN (e.g., IBM, 1984) manuals. Because of these features, using a subroutine library is recommended.

When working frequently with different sets of crop, soil and weather data, it is useful to keep the program modules, and the modules with crop, soil, and weather data in different files on the computer, and to only join sets together when a specific combination must be run. Figure 12 provides a diagram of possible choices for crop growth simulation at Production Level 1. For Production Level 2, extra crop and soil sections should be inserted before the END statement (Section 5.4). This procedure is not part of CSMP, but can be written in the operating system language of most computers. The procedure saves space and, more importantly, makes any correction or addition to a program or a data set effective immediately and wherever used.

Easy changing of structure or data in models is both an asset and a liability. It is a liability because it is tempting to make changes without fully investigating whether the change is an improvement. To ensure that making a change in the model is an asset, always document all changes in a notebook; execute dimen-

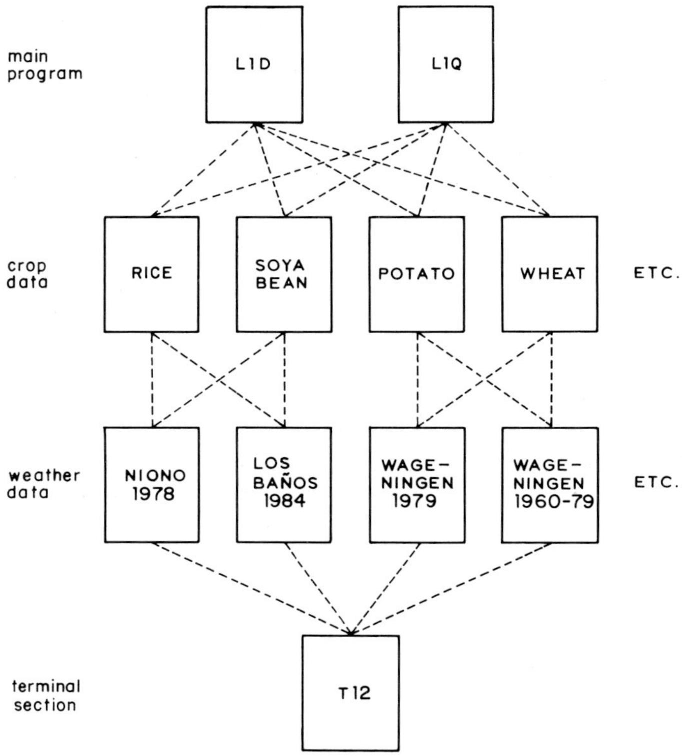

Figure 12. Alternative modules (L1D and L1Q) and alternative crop and weather data sets can be combined with a terminal section (T12) to complete a program for crop growth at Production Level 1.

sion analysis of new equations rigidly; make checks of balance calculations by comparing the accumulated net flux with the amount retained; and compare new data with values already known and evaluated.

1.4.5 Evaluation

Often, the first thorough test of a model is the comparison of its behaviour with that observed of the real world in a similar situation. This behaviour includes, for example, the general shape of the time course of variables, the presence of discontinuities and the qualitative sensitivity of output to parameter values. However, be aware that aspects of model behaviour that seem counterintuitive at first, sometimes turn out to be realistic. If the behaviour of the model qualitatively matches that of a system in the observed world, a quantitative comparison and evaluation of the predictive success of the model should be made. At this stage, statistical tools can be useful. But even when sufficient and accurate data are available, a model cannot be proven to be correct. Model behaviour can sometimes be falsified and one or more model components must then be in error. Calibrating a model is the adjustment of some parameters so that the model matches one set of observed or measured data. Simulation of this data set with the calibrated model is a very restricted form of evaluation. Calibrating several parameters simultaneously degrades simulation into curve fitting. Sensitivity analysis is a procedure in which the value of a parameter is increased or decreased by a certain percentage and the effect on the behaviour of the model recorded. Sensitivity analysis can be part of the evaluation of a model, but it is particularly useful when determining the accuracy with which parameters have to be established experimentally. Behavioural analysis is a useful form of sensitivity analysis, particularly if it is possible to critically discuss results with crop specialists.

Further information about sensitivity analysis, evaluation, validation and verification can be found in Baker & Curry (1976), Penning de Vries (1977) and Steinhorst et al. (1978).

2 Assimilation and dissimilation of carbon

This chapter discusses and quantifies the assimilatory and dissimilatory processes of crop growth. Figure 13 provides an example of the relative importance of these processes for two crops. It indicates the partitioning of the total carbon captured by harvest-ripe crops during assimilation and dissimilation processes and the corresponding amount of biomass produced. The relative importance of these processes can vary a lot between crops and growth conditions.

Photosynthesis and remobilization are carbohydrate sources; their simulation is discussed in Sections 2.1 and 2.2. Simulation of respiration processes related to maintenance and growth are discussed in Sections 2.3 and 2.4. Section 2.5 contains exercises.

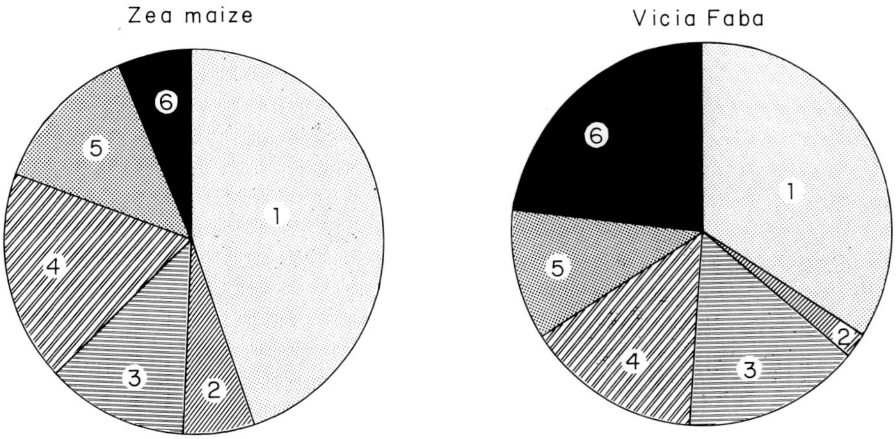

Figure 13. An example of the relative importance of assimilatory and dissimilatory processes for two crops, showing the partitioning of carbon (C) captured over these processes and the corresponding amount of biomass formed.
(1) the amount of C in biomass formed directly from assimilates; (2) the amount of C in biomass formed from remobilized carbohydrates; (3) and (4) the amount of C lost in growth and maintenance processes, respectively; (5) the excess energy involved in maintenance and NO_3^- reduction and (6) the energy involved in photorespiration are both expressed as C-equivalent. Gross photosynthesis is 1 + 2 + 3 + 4, net photosynthesis is 1 + 2, respiration is 3 + 4.

Morphological development and biomass partitioning is discussed in Chapter 3, and two complete modules for simulating crop growth at Production Level 1 can be constructed from Chapters 2 and 3. They are presented in Listing 3, a basic crop growth module with one-day time periods (L1D) and Listing 4, a crop growth module with quarter-day time periods (L1Q).

2.1 Photosynthesis

2.1.1 Introduction

Photosynthesis is the driving force behind growth at Production Level 1. In the programs presented here, daily photosynthesis is computed for whole canopies by using a summary model. Daily photosynthesis is the basis for computing the rate of crop growth. The three most important factors in photosynthesis are: the photosynthesis light response curve of leaves, the radiation intercepted by leaf canopies and the distribution of light within canopies. These are presented below, together with demonstrations of how they can be integrated and used to yield totals for each day of canopy photosynthesis under various conditions. Environmental variables considered are radiation, temperature and the ambient CO_2 concentration.

Photosynthesis was discovered in the later part of the eighteenth century by Priestley and Ingen Housz and since then has been extensively studied. The new Encyclopedia of Plant Physiology devoted two volumes (5 and 6, 1979) to this process alone. Photosynthesis comprises very complex processes by which plants reduce CO_2 and form organic molecules using absorbed radiation energy. These molecules are conveniently represented by the molecule glucose ($C_6H_{12}O_6$). Glucose molecules either serve as the building blocks for virtually all organic constituents in plants, or are respired to provide energy for metabolic processes.

Three groups of plants are often distinguished on the basis of their biochemical mechanism of photosynthesis: C_3, C_4 and CAM plants (Bidwell, 1983). Among the major agricultural crops, pineapple is the only CAM plant, so this group is ignored here. The nature of the difference between C_3 and C_4 crops is of little significance for the purpose of this book, but it is important to recognize that C_4 crops generally perform much better than C_3 crops in warm climates, but less so in temperate regions. Morphological and biochemical details can be found in the Encyclopedia of Plant Physiology, Volume 6 (1979).

'Photosynthesis' is used here for gross photosynthesis and includes photorespiration, an intensive process in C_3 plants that is intimately coupled with photosynthesis itself. Photorespiration does not lead to any product and is almost completely suppressed in C_4 plants. Suppression of dark respiration processes during photosynthesis is discussed in Subsection 2.3.3.

The rate of leaf photosynthesis is conveniently expressed per unit of leaf area (counting only upper sides). Canopy photosynthesis is the sum of the contributions of all leaves, stems and sometimes reproductive organs. Only photosynthesis of leaves, by far the most important, is considered extensively here.

2.1.2 Leaf photosynthesis

The photosynthesis light response curve

The response of leaf photosynthesis to absorbed light can be described as a curve that relates the rate of gross photosynthesis (PL, kg CO_2 ha^{-1} h^{-1}) to the intensity of absorbed radiation (PAR, J m^{-2} s^{-1}) exponentially (Figure 14). The exponential form corresponds best with most observations (Goudriaan, 1982b):

$$PL = PLMX * (1.0 - EXP(- PLEA * PAR/PLMX))$$

This type of curve is characterized by two parameters: the slope at the origin (PLEA, kg CO_2 ha^{-1} h^{-1} $(J\ m^{-2}\ s^{-1})^{-1}$) and the rate at saturated light intensity (PLMX, CO_2 kg ha^{-1} h^{-1}). The initial efficiency of the use of absorbed light characterizes, in particular, the biophysical processes and has a fairly constant value. The maximum rate depends strongly on plant properties and environmental conditions and particularly reflects biochemical processes and physiological conditions.

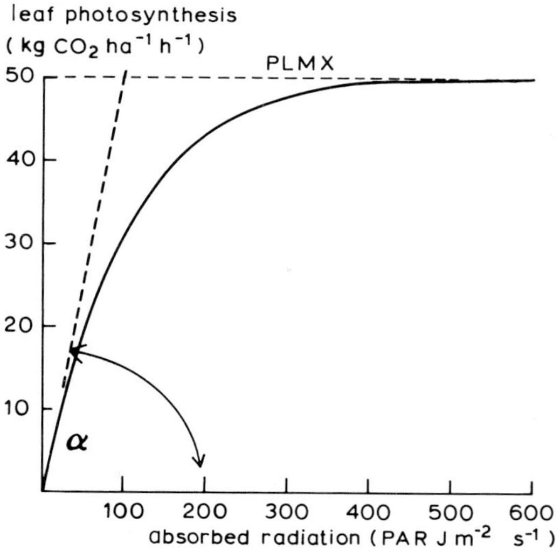

Figure 14. The response curve of gross photosynthesis of a single leaf versus the intensity of absorbed radiation (PAR). PLEA, the tangent of angle alpha, is the initial efficiency of light use, PLMX the maximum level to which the exponential curve rises.

Light absorption by leaves

Only radiation from a part of the light spectrum (400-700 nm) is effective for photosynthesis. This photosynthetically active radiation (PAR) is about 50% of solar radiation (Subsection 6.1.2).

A fraction of the radiation reaching leaves is reflected or transmitted. The values of reflectivity and transmissivity, complements of absorptivity for PAR, are remarkably similar among leaves of healthy crops (Sinclair et al., 1971). Moreover, reflectivity and transmissivity are numerically often about equal and have a value of 0.1 each (Goudriaan, 1977). When leaves are obviously yellow or are extremely thin, there is not enough chlorophyll (less than 30 microgram cm^{-2}) to absorb light and transmission doubles or quadruples (Alberda, 1969; Lin & Ehleringer, 1982). Only when thick leaves carry reflecting hairs (Ehleringer, 1976) can reflection double or quadruple. The fraction of PAR absorbed by leaves is assumed here to always be 0.80 (the complement of the constant SCV in the function FUPHOT and in the subroutine SUPHOL, Appendix B).

Initial light use efficiency

The theoretical minimum energy requirement for reduction of a CO_2 molecule is about 9.5 quanta (PAR) for light of wavelengths 540-670 nm, but the lowest values observed in C_3 plants in sunlight in the absence of photorespiration are about 13 quanta per molecule (Farquhar & von Caemmerer, 1982). The difference is caused by light absorption by non-photosynthetic pigments and by the lower (on average 10%) light use efficiency at other wavelengths (McCree, 1982). Photorespiration, induced in C_3 plants by O_2, increases the energy requirement to at least 15 quanta per CO_2 molecule. Light use efficiency (PLEI) such as this can be converted into the more practical measure of a rate of 0.48 kg CO_2 per hectare of leaf surface and per hour per J m^{-2} s^{-1} of PAR (de Wit et al., 1978). This is typical for all C_3 species at relatively low temperatures (around 10 °C). For C_4 plants the initial light use efficiency is about 0.40. It is lower than that of C_3 plants at low temperatures because suppression of photorespiration is a costly process.

The relative importance of photorespiration increases as temperature increases and the initial efficiency (PLEA) goes down as a result. The value for a C_3 crop is reduced to 0.3 at relatively high temperatures (around 30 °C) and to 0.0 at even higher temperatures. Its value remains constant at temperatures up to 45 °C in C_4 crops and at higher temperatures drops quickly to 0.0 (Ehleringer & Bjorkmann, 1976; Berry & Downton, 1982). These data, summarized in Table 3, can be of help when values for a specific case are unavailable. Some data of the initial light use efficiency for different species are included in Table 4.

Environmental factors other than temperature have little effect on how efficiently plants use radiation. For simulation of the effect of photorespiration on light use efficiency, consult Goudriaan (1982b).

Table 3. Typical values of the initial efficiency of use of absorbed light for photosynthesis by individual leaves at different temperatures for the main crop types (kg CO_2 ha^{-1} h^{-1} per J m^{-2} s^{-1} of absorbed PAR).

Crop type	Temperature (°C)					
	0	10	20	30	40	50
C_3	0.5	0.5	0.45	0.3	0.1	0.01
C_4	0.4	0.4	0.4	0.4	0.4	0.01

Maximum rate of leaf photosynthesis

The value of the maximum rate of leaf photosynthesis (PLMX) at high light intensities and normal CO_2 concentration is of major importance for crop growth simulation. Its value is usually 25-80 kg CO_2 ha^{-1} h^{-1}, but can exceed this range. An example is given in Subsection 4.1.5 for a very high maximum photosynthesis rate of a maize crop.

To quantify PLMX, it can be considered that during photosynthesis a physical and a biochemical process run parallel. CO_2 diffuses from the ambient air to the carboxylation sites in the cells, and the total diffusion resistance and the concentration gradient set an upper limit to PLMX. Simultaneously CO_2 is converted into glucose in a biochemical chain reaction. RuBPCase is the most prominent and presumably the most rate-limiting enzyme (it may form 50% or more of leaf protein in C_3 plants). Its concentration and maximum activity also determine the maximum value of PLMX. The capacities of the physical and biochemical processes tend to adjust to each other so that, in theory, either could be used to quantify PLMX. Yet, predicting the value of PLMX for a particular case from physiological or physical parameters is still inaccurate. Therefore, accurate measurements of PLMX are indispensable. Table 4 provides some data on PLMX determined under standard conditions.

The maximum rate of leaf photosynthesis per unit of leaf area is strongly related to leaf thickness and temperature. Quantifying these relations experimentally requires sophisticated equipment and much work. If PLMX measurements are lacking, the influence of leaf thickness and leaf temperature may be quantified by considering PLMX at standard temperature and thickness to be a value characteristic to a species or cultivar. Leaf thickness (more properly, specific leaf weight in kg dry matter per ha of leaf surface) plays a role because thicker leaves usually have more RuBPCase per unit surface than thin leaves. Differences in leaf thickness are the major cause of differences between the maximum rates of photosynthesis of plant cultivars (Charles-Ed-

Table 4. The initial efficiency of use of absorbed light (PLEI, measured at a low temperature) and the maximum rate of leaf photosynthesis (PLMX) of individual leaves of different species at an optimum temperature at 340 vppm CO_2, and for a characteristic specific leaf weight (given in Table 19). Values refer to field grown crops.

Species	C_3/C_4	PLEI $kg ha^{-1} h^{-1} *(J m^{-2} s^{-1})^{-1}$	PLMX $kg ha^{-1} h^{-1}$	Temp °C	Reference
Barley	C_3	0.40	35	25	Dantuma, 1973
Cassava	C_3	0.50	35	25	estimate
Cotton	C_3	0.40	45	35	Muramoto et al., 1965
Cowpea	C_3	0.50	64	25	estimate
Faba bean	C_3	0.48	35	25	van Laar CABO pers.com.
Groundnut	C_3	0.50	50	30	Bhagsari & Brown, 1976; Bhagsari et al., 1976; Pallas & Samish, 1984
Maize	C_4	0.40	60	25	Sibma CABO pers.com.
Millet	C_4	0.40	70	25	Jansen & Gosseye, 1986
Potato	C_3	0.50	30	20	Teubner, 1985
Rice	C_3	0.40	47	25	Yoshida, 1981
Sorghum	C_4	0.45	70-10	30-35	Eastin, 1983
Soya bean	C_3	0.48	40	30	Beuerlein&Pendleton, 1971; Dornhoff & Shibles, 1970
Sugar-beet	C_3	0.56	38	20	Sibma CABO pers.com.
Sugar-cane	C_4	0.40	70	25	estimate
Sunflower	C_3	0.45	60	28	Rawson & Constable, 1980
Sweet potato	C_3	0.50	30-35	25	Hahn & Hozyo, 1983
Tulip	C_3	0.50	40	15	Benschop, 1986
Wheat	C_3	0.50	40	10-25	van Keulen&Seligman, 1987 Joliffe & Tregunna, 1968

wards, 1981) and between plants grown in the field and in phytotrons. The simplest form of the PLMX – thickness relation is adopted here by using a strict proportionality for a normal range of leaf thicknesses of 200-600 kg ha^{-1} (Listing 3 Line 53; this listing contains a complete crop module (L1D, see Chapter 7)). This implies that leaf thickness is a crucial variable in the simulation programs in this book (Section 3.3). Evidence to support this view for many crop species with an adequate nutrient supply is provided by Khan & Tsunoda

(1970), Figure 15, Gulmon & Chu (1981), Bjorkman (1981) and Gifford & Evans (1981).

Examples to the contrary also exist, for example in sugar-beet PLMX is negatively correlated with leaf thickness, unless the latter is corrected for the increase in organic acid content associated with age (Sibma, CABO, personal communication). The proportionality factor for C_3 crops at an optimal temperature for photosynthesis is about 0.1 kg CO_2 ha^{-1} h^{-1} per kg leaf ha^{-1}, while that of C_4 crops is 0.15-0.2. The thickness – photosynthesis ratio depends strongly on the nitrogen content of the leaf, but this effect is not relevant here because this ratio is applied at Production Levels 1 and 2 where nutrients are in ample supply and because the rate is measured at a fixed stage of crop development.

Attention should always be given to the influence of temperature on photosynthesis. The curve relating PLMX to temperature shows an optimum (Figure 16). C_3 plants generally perform better than C_4 plants at low temperatures (less than 15 °C), and vice versa at high temperatures (more than 25 °C). Table 5 presents data on this relationship for several crops. However, these relationships should not be copied rigidly as many exceptions have been recorded (e.g., Berry & Downton, 1982). Breeding may have modified the PLMX temperature response curve. Miedema (1982) indicates that photosynthesis at 10 °C and at 20 °C in cold-tolerant maize is similar to that in subtropical maize

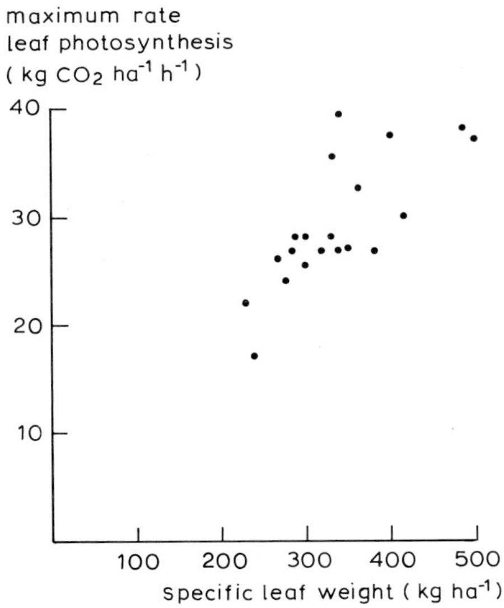

Figure 15. The maximum rate of leaf photosynthesis as a function of specific leaf weight in wheat (Source: Khan & Tsunoda, 1970).

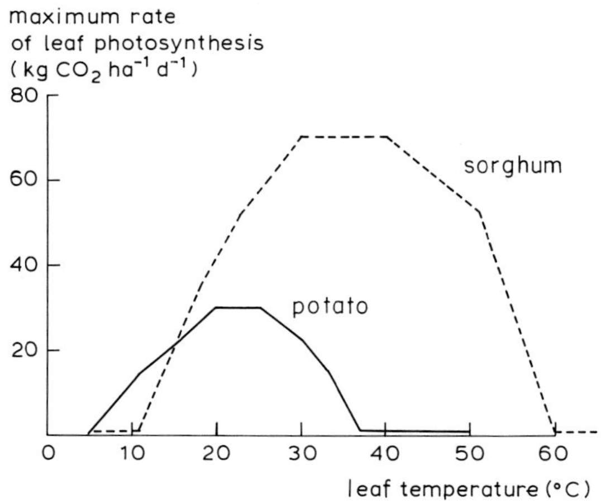

Figure 16. The maximum rate of leaf photosynthesis as a function of leaf temperature for a potato (C_3) and a sorghum (C_4) crop. Data from Tables 4 and 5.

at temperatures 5 °C higher (Table 5). A shift of this magnitude has contributed greatly to the large increase in maize production in the Netherlands since 1970. A shift of 3-4 °C in the temperature – PLMX response curve was found by Kwon (1984) between an Indica-Japonica and a more cold-tolerant Japonica rice variety. Natural selection has led some C_3 species to have a response curve similar to that of typical C_4 species (Werk et al., 1983).

Leaf temperature can be several degrees above or below air temperature, depending on environmental conditions and the moisture status of the soil (Subsection 4.3.2). However, for simplicity it is assumed here that average leaf temperature is equal to average air temperature.

Adaptation of individual plants to other temperature regimes can lead to modification of the actual temperature response curve (Berry & Downton, 1982). To allow for adaptation such as this, the response introduced in a model may have a temperature optimum that is 5 °C broader than that determined in a short-term experiment.

2.1.3 *Canopy photosynthesis*

The principle of canopy photosynthesis

If the photosynthesis rate was proportional to the light intensity and if all leaves had identical properties, then canopy photosynthesis would simply be equal to the multiple of the quantity of light absorbed and the light use efficiency. However, leaves become saturated at high light intensities and they are all exposed differently to radiation. Hence, the relation of canopy photosyn-

Table 5. The effect of leaf temperature on the maximum rate of leaf photosynthesis of several crop species. (See original publication for the cultivar used.)

Barley (Joliffe & Tregunna, 1968)
FUNCTION PLMTT = -20.,0.001, 0.,0.01, 5.,0.4, 10.,0.7, 15.,0.9,...
20.,1., 25.,1., 30.,0.9, 35.,0.8, 40.,0.5

Cotton (El-Sharkawy & Hesketh, 1964a; Ludwig et al., 1965)
FUNCTION PLMTT = 0.,0.001, 10.,0.3, 20.,0.6, 25.,1., 30.,1.,...
35.,0.8, 40.,0.5, 50.,0.001

Phaseolus vulgaris (Jones, 1971)
FUNCTION PLMTT = –10.,0.01, 0.,0.01, 10.,0.59, 15.,0.76,...
20.,0.93, 25.,1., 30.,0.92, 35.,0.84, 40.,0.75

Groundnut (Bhagsari & Brown, 1976; Pallas & Samish, 1974)
FUNCTION PLMTT = 0.,0.001, 10.,0.3, 20.,0.6, 25.,1.,...
30.,1., 35.,0.8, 40.,0.5, 50.,0.001

Maize (Hofstra & Hesketh, 1969)
FUNCTION PLMTT = 0.,0.01, 5.,0.01, 10.,0.1, 15.,0.5, 20.,0.8,...
25.,1., 35.,1., 40.,0.9, 45.,0.75, 50.,0.07

Millet (Jansen & Gosseye, 1986)
FUNCTION PLMTT = 0.,0., 10.,0., 20.,1., 40.,1., 50.,0.

Potato (van Heemst, CABO, personal communication)
FUNCTION PLMTT = –20.,0., –5.,0.01, 5.,0.02, 15.,0.8, 20.,1.,...
25.,1., 30.,0.8, 37.,0.0

Rice (van Keulen, 1976)
FUNCTION PLMTT = 0.,0.01, 10.,0.01, 20.,1., 35.,1., 42.,0.01

Sorghum (El-Sharkawy & Hesketh, 1964a)
FUNCTION PLMTT = 0.,0.001, 10.,0.01, 15.,0.3, 20.,0.6, 25.,0.9,...
30.,1., 40.,1., 45.,0.9, 50.,0.8, 55.,0.4, 60.,0.001

Soya bean (Hofstra & Hesketh, 1969)
FUNCTION PLMTT = 0.,0.001, 10.,0.3, 20.,0.6, 25.,0.8, 30.,1.,...
35.,1., 40.,0.8, 45.,0.4, 50.,0.001

Sugar-beet (Hofstra & Hesketh, 1969; Hall & Loomis, 1972)
FUNCTION PLMTT = 0.,0.01, 5.,0.01, 10.,0.75, 20.,1., 35.,1.,...
40.,0.9, 45.,0.01

Sunflower (El-Sharkawy & Hesketh, 1964a)
FUNCTION PLMTT = 5.,0.1, 10.,0.5, 15.,0.7, 20.,0.9, 25.,0.95,...
30.,1., 35.,1., 40.,0.7, 45.,0.3, 50.,0.01

Sweet potato (van Heemst, CABO, personal communication.)
FUNCTION PLMTT = –5.,0.01, 5.,0.02, 15.,0.8, 20.,0.9,...
25.,1., 30.,1., 35.,0.9, 40.,0.5, 45.,0.001

Wheat (van Keulen & Seligman, 1987)
FUNCTION PLMTT = 0.,0.0001, 10.,1., 25.,1., 35.,0.01, 50.,0.01

---

thesis to light intensity is curved and varies greatly for different situations (Figure 17).

Canopy photosynthesis is the sum of the rates of photosynthesis of all leaves. It is expressed in kg CO_2 per hectare of ground surface and per day. To compute its value, canopies are thought to be divided into relatively thin 'layers' of leaves containing 0.1-1 m^2 of leaf surface per square meter of ground surface. Light intensity in the top layers is highest and decreases towards the base of the canopy. Light must be distinguished as a fraction of direct light coming from a point source (the sun) and a fraction of diffuse light coming from all directions (Subsection 6.1.2). Extinction of light (PAR) occurs with different coefficients: 0.50 for direct and 0.72 for diffuse radiation when the leaf angle distribution is spherical (i.e., leaf surfaces distributed like the surface on a globe) (Figure 18). Its average extinction coefficient is about 0.6 for a canopy with erect leaves and 0.8 for one with horizontal leaves (Goudriaan, 1977).

Leaves grow at varying angles from the horizontal. This can be represented by a cumulative leaf angle distribution curve (Figure 19). The intensity of radi-

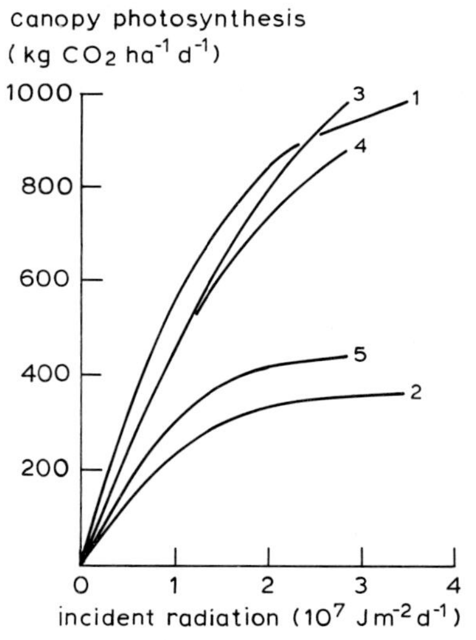

Figure 17. Photosynthesis light response curve of canopies with different characteristics and at different latitudes at June 15:

	PLEA	PLMX	ALV	LAT
Curve 1	0.5	40	5	50
Curve 2	0.3	20	2	50
Curve 3	0.4	100	5	0
Curve 4	0.4	70	5	0
Curve 5	0.4	35	2	0

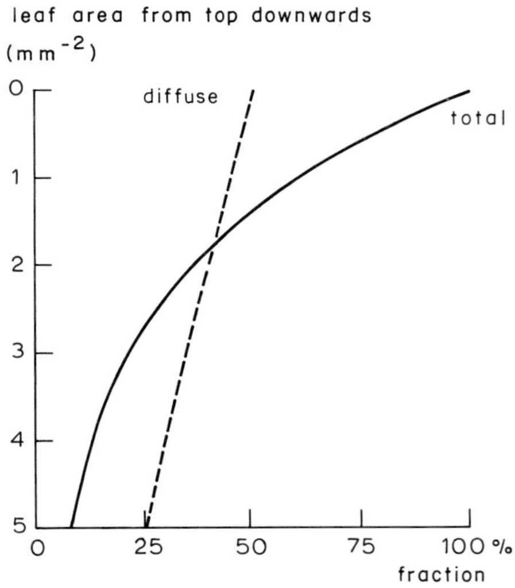

Figure 18. The approximate reduction of the intensity of the total global radiation from the top of the canopy downwards (full line), and the diffuse radiation fraction of the total global radiation (dashed lined) at each level of the canopy. The direct and diffuse component are equal above the canopy. The curves are computed with an extinction coefficient of 0.5 for total radiation and 0.7155 for diffuse radiation.

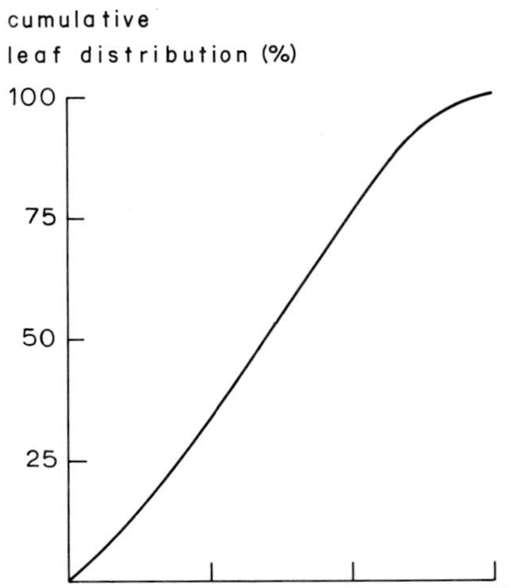

Figure 19. Cumulative leaf angle distribution of a maize crop (Source: de Wit, 1965).

ation on individual leaves is anywhere between the intensity of diffuse light only (shaded leaves) and that of diffuse plus direct light (sunlit areas perpendicular to solar rays). The actual leaf angle distribution within a normal range has little effect on canopy photosynthesis (de Wit, 1965). Clustering of leaves, relatively important for young plants and row crops, reduces canopy photosynthesis less than may be expected.

The contribution of photosynthetic area other than that of leaves to canopy photosynthesis is discussed in Section 3.3.

Modelling canopy photosynthesis

Many simulation models have been developed for canopy photosynthesis (Hesketh & Jones, 1980). A pioneer model by de Wit (1965) has been improved and expanded (Goudriaan, 1977, 1986; de Wit et al., 1978; Spitters, 1986; Spitters et al., 1986) and because of its versatility and documentation its approach is followed in this book.

An excellent way to calculate the daily gross photosynthesis of a canopy with a summary model was presented recently by Goudriaan (1986, 1988). It uses a specific way of integrating the instantaneous rate of leaf photosynthesis in time (three points between noon and sunset, times two, assuming the morning to be equal to the afternoon) and in space (three depths in the canopy). The rate of leaf photosynthesis is estimated from the actual leaf photosynthesis light response curve and the amount of direct and diffuse light at that time of day and depth in the canopy. The path of radiation intensity during the day is assumed to be sinusoidal. Results obtained using this summary model were extensively compared with those of the comprehensive models and found to agree very closely. The FUPHOT function calculates canopy photosynthesis in this way (Listing 3 Line 52, Listing 4 Line 72). Both crop characteristics PLMX and PLEA are adjusted for temperature and other conditions. (A decrease of PLMX and PLEI with depth in the canopy is discussed in Subsection 2.1.4). The extinction of PAR in the canopy is characterized by the extinction coefficient; its value is a constant in the standard photosynthesis function (KDIF in FUPHOT).

Canopy photosynthesis calculations have been computed year-round for different geographical latitudes and for leaf areas up to 10 $m^2\ m^{-2}$ (Figures 20 and 21). They show how much canopy photosynthesis varies as a result of variations in light and that the variations are not proportional with leaf area. The relation between canopy photosynthesis and the maximum rate of leaf photosynthesis (PLMX) also shows a less than proportional relationship (Figure 22). Canopy photosynthesis does not increase significantly above a leaf area index of 4 or 5, with a spherical leaf distribution at any value of PLMX. A three – to – four-fold increase in PLMX, from 20 kg CO_2 $ha^{-1}\ d^{-1}$, would only double the rate of canopy photosynthesis on a clear day. The effect on a cloudy day is even smaller.

On days that are not fully overcast or clear, radiation can be distributed

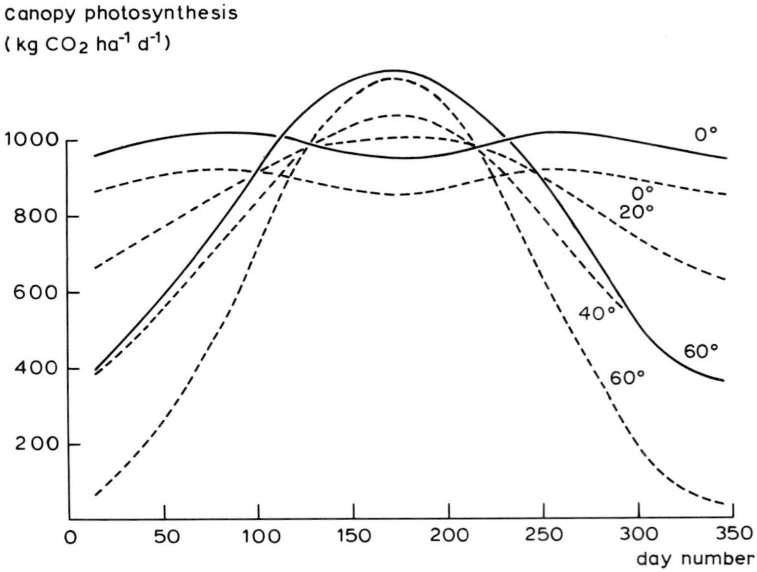

Figure 20. The simulated course of daily canopy photosynthesis throughout the year at 0°, 20°, 40° and 60° northern latitude for a C_3 crop (PLEA = 0.5, PLMX = 40., dashed lines), and at 0° and 60° for a C_4 crop (PLEI = 0.4, PLMX = 90., full lines), all at a leaf area of 5 m^2 m^{-2}.

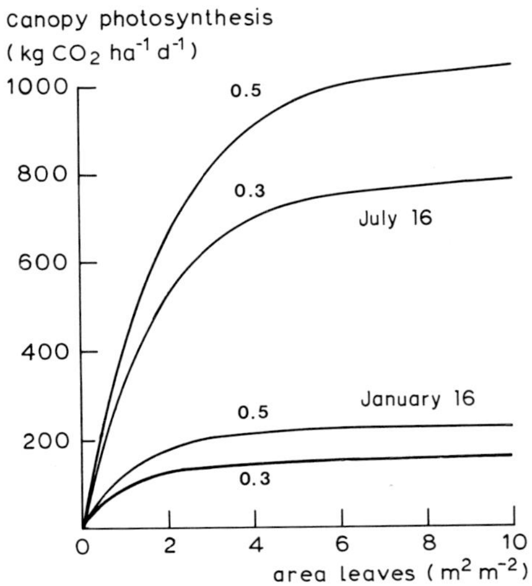

Figure 21. The relation of canopy photosynthesis with leaf area at two dates and for two values of PLEA (0.3 and 0.5), and both with LAT = 50., PLMX = 40.

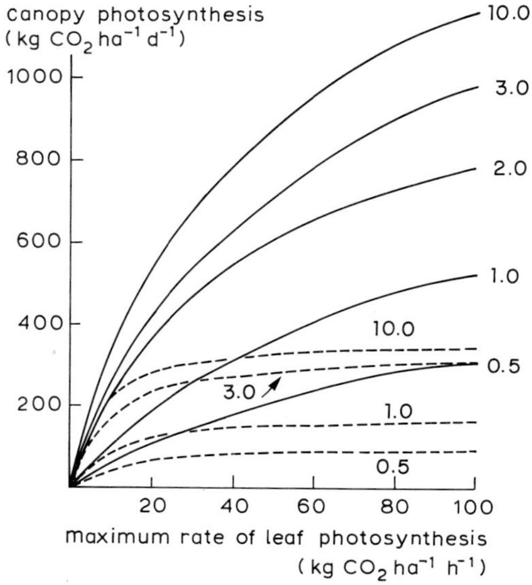

Figure 22. The simulated rate of canopy photosynthesis on a fully clear 21st June in Wageningen for different values of the maximum rate of leaf photosynthesis and at several values of the area of the leaves (in m^2 m^{-2}). Dashed lines are the response curves under overcast conditions.

quite unevenly over the day. Using a comprehensive model it was established that daily canopy photosynthesis is some 10% lower when all radiation is concentrated at noon on a partly cloudy day, as compared to a constantly cloudy day (Figure 23). This has only a moderate effect on canopy photosynthesis and is disregarded here.

Certain parameters are almost invariable in the comprehensive model for canopy photosynthesis. In other parameters the variability normally encountered has little effect, and such features are either not mentioned or kept constant. Table 6 provides a list of the major assumptions in the summary model.

The radiation intercepted on any day gives rise to a certain amount of photosynthesis (PCGW). This photosynthesis rate is used as is to compute the rate of crop growth (Listing 3 Line 39), or, in the module with quarter-day time periods (Listing 4 Line 71) it is divided by daylength to express it as a rate per 24 hours of constant radiation.

2.1.4 *Special cases for photosynthesis*

Carbohydrate accumulation

Starch and glucose sometimes accumulate in leaves. Net photosynthesis is

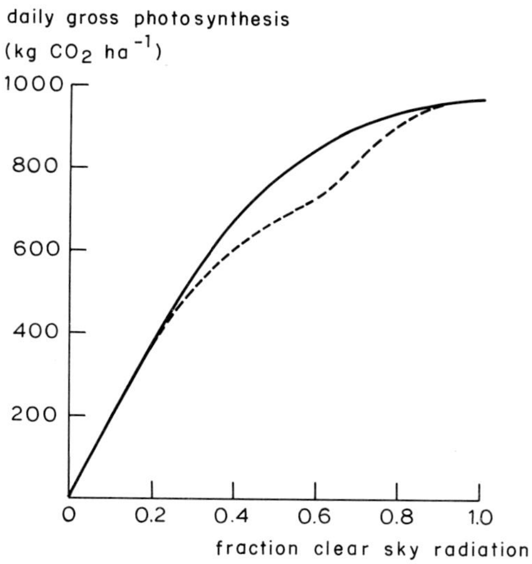

Figure 23. The simulated rate of canopy photosynthesis as a function of the fraction of clear sky radiation when the level of cloudiness is constant all day (full line) or when the same amount of radiation is concentrated around noon (dashed line).

Table 6. Assumptions concerning computation of daily canopy photosynthesis in a summary model.

- Leaf photosynthesis responds instantaneously and fully to changes in light intensity, i.e., leaf movements in the wind or clouds are without after-effects.
- PLMX and PLEI have a constant value from top to bottom of the canopy (FUPHOT) unless specified per layer (SUPHOL).
- The leaf angle distribution is spherical (FUPHOT) unless specified differently (SUPHOL).
- Leaf thickness matters a great deal; leaf form and size are unimportant.
- Leaves do not cluster much. Crops are not grown in rows.
- The CO_2 concentration inside the canopy equals that above it.
- Leaf temperature is on average equal to air temperature.

impaired when the glucose level is too high; starch probably has less effect. This reduced photosynthesis is partially caused by the glucose concentration stimulating respiration (Azcon-Bieto et al., 1983) and partially by a lowering of PLMX. To simplify simulation of this phenomenon, only the effect on PLMX is quantified (an effect of sugar concentration on maintenance respiration is discussed in Subsection 2.3.2). The carbohydrate level fluctuates during the day and PLMX may fluctuate consequently. Van Keulen & Seligman (1987) use 30% of leaf weight as the carbohydrate level above which PLMX is reduced to almost zero and this approach is followed in module L1Q (Listing 4, Line 75). The cut off level (30%) and the extent to which photosynthesis is reduced (30%) are estimates only and are used for all species for lack of information. Moreover, no distinction is made between starch and glucose which is probably necessary. If carbohydrate accumulation is important, use the L1Q quarter-day time period module (Section 3.4).

Carbohydrate accumulation in leaves can occur:

- in crops where the export of carbohydrates can be slower than its production (as in potato crops where leaf photosynthesis, even at constant external conditions, decreases during the day and the canopy photosynthesis light response curve shows hysteresis, Bodlaender (1986));
- in situations where daytime photosynthesis is less restricted than growth during any 24-hour period (as in spring in temperate climates when night temperatures are still low, but leaf temperatures during the day permit high photosynthesis);
- as a result of water or nutrient stress;
- as a result of small sink size of storage organs, such as immediately after their initiation (elaborated in Subsection 3.2.5); and
- as a result of phloem transport being blocked by insect pests or diseases.

High CO_2 concentrations

At the site of carboxylation CO_2 is bound to an enzyme in competition with O_2. Carboxylation leads to photosynthesis, oxidation leads to photorespiration. The higher the CO_2/O_2 ratio, the lower the photorespiration and the higher the maximum rate of photosynthesis and the initial light use efficiency. The maximum rate of leaf photosynthesis is about proportional to the CO_2 concentration below the normal level of 340 cm^3 CO_2 m^{-3} air (340 vppm). The proportionality holds up to CO_2 levels of about 700 vppm in many C_3 species. On the other hand, in C_4 plants concentrations beyond 340 vppm increase photosynthesis little or not at all (Figure 24). These effects have been well investigated and reviewed (Encyclopedia of Plant Physiology, Vol 6, 1979). The CO_2 concentration in greenhouses can be much higher or lower than 340 vppm. The ambient CO_2 concentration rises annually by 1-2 vppm (Goudriaan, 1987), which, in many cases, causes a slow increase in photosynthesis. This topic has been discussed extensively in the literature (e.g., Berry & Downton, 1982; Gates et al., 1983; Subsection 6.1.7).

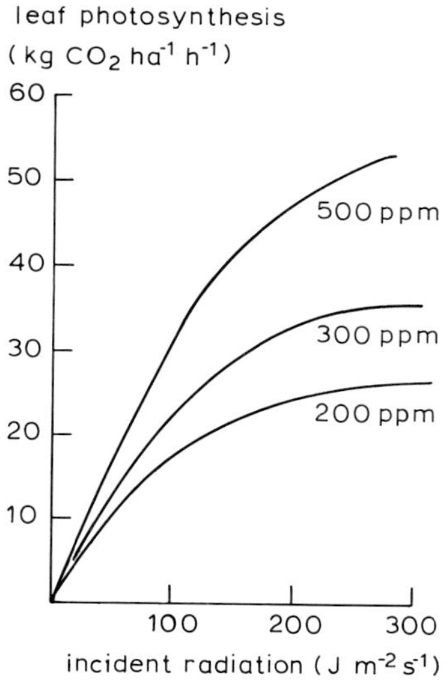

Figure 24. The photosynthesis light response curve of sunflower leaves at three ambient CO_2 concentrations (Source: Goudriaan & van Laar, 1978b).

The consequences of high or low CO_2 concentrations on crops can be simulated by adjusting PLMX in relation to the ratio of the new CO_2 concentration and 340 vppm. In this way Goudriaan et al. (1984) established that the overall effect of an increased CO_2 concentration (C) on net assimilation (A) of a canopy can be described by

$$A_x = A_o \, (1 + \beta \ln (C_x / C_o)) \qquad \text{Equation} \quad 1$$

where o refers to 340 vppm and x to the new situation; β is about 0.4 for C_4 crops and 0.8 for C_3 crops.

A layered canopy

Maximum leaf photosynthesis in a senescing crop declines in time. The oldest leaves in the base of the canopy are affected first. Many diseases also affect the crop in this way. In some cereal varieties the top leaves are clearly more erect than leaves in the lower layers. To deal with layers with different characteristics, Goudriaan (1986, 1988) extended the FUPHOT function to the SUPHOL subroutine. Computations are per leaf layer. The radiation at the bottom of one layer is the input to the next lower layer. Up to five layers are distinguished, each with its own values for maximum leaf photosynthesis and initial light use efficiency. The same variable names as for FUPHOT are retained, but these are now names of arrays with as many elements as there are layers. The areas of active and dead leaves are also specified per layer. Dead leaves in any layer are supposed to provide shade to the leaves in that layer. The CSMP modules Listing 3 (L1D) and Listing 4 (L1Q), presented in Chapter 7, can be expanded to simulate a crop with two different leaf layers as follows:

```
STORAGE PCGCL(5),PLMX(5),PLEA(5),ALVL(5),ALVDL(5)
PCGC,PCGCL = SUPHOL (2, PLMX,PLEA,ALVL,ALVDL,F1,F2,...
                RDTM,DATE,LAT)
PROCEDURE ALVL, ALVDL = ALVPRO (ALV)
ALVL(1) = AMIN1 (2.5, ALV)
ALVL(2) = ALV - ALVL(1)
ALVDL(1)= 0.0
ALVDL(2)= 0.2 * ALVL(2)
ENDPROCEDURE
```

The number of layers considered (two in this example) is the first number of the SUPHOL inputs. The STORAGE declaration must precede the new arrays. The statements for leaf area are in a 'procedure', meaning that they are sorted as a group with the variable inputs and outputs defined in the first line. CSMP cannot sort indexed variables and using a procedure is a correct alternative method.

Leaf photosynthesis characteristics can also be computed per layer:

```
PROCEDURE PLMX,PLEA = PLMPRO (PLMXP,PLEI,SLA,TPAD)
  PLMX(1) = PLMXP * (SLA / SLC) * AFGEN(PLMTT,TPAD + 1.0)
  PLMX(2) = PLMXP * (SLA / SLC) * AFGEN(PLMTT,TPAD - 1.0)
  PLEA(1) = PLEI * AFGEN(PLETT,TPAD + 1.0)
  PLEA(2) = PLEI * AFGEN(PLETT,TPAD - 1.0)
ENDPROCEDURE
```

A difference between the layers is created here by calculating different temperatures. The original statements to calculate PLMX and PLEA are eliminated at this point.

The distribution of leaf angles per layer must be specified. The common situation (and implicit in FUPHOT) is a spherical distribution, that is 13.4% of leaf area with angles from the horizontal to 30 degrees (first angle class), 36.6% with angles between 30 and 60 degrees (second class) and 50% with almost erect leaves (third class). A crop with erect leaves, such as the rice variety IR8, has an angle distribution of 8%, 17% and 75% respectively, averaged over all layers. A crop with an erect leaf angle distribution in the top layer and a spherical distribution in the bottom layer can be specified as:

```
STORAGE F1(5), F2(5)
TABLE F1(1-2) = 0.08,0.134, F2(1-2) = 0.17,0.366
```

This defines the fraction of first angle class of the upper and of the lower layers (F1) and that of the second angle class (F2) in the upper and the lower layer. The value for the third angle class is, by definition, the complement of the first plus the second, and is not specified.

The SUPHOL subroutine is included in Appendix B and is an alternative to FUPHOT. The rate of photosynthesis per leaf layer (PCGCL) is also an output of the subroutine (because there is more than one output this calculation is in the form of a subroutine rather than a function). The CSMP output statements can handle indexed variables, so that the photosynthesis per layer (PCGCL(1) and PCGCL(2)) can be printed.

Altitude

Reduced leaf photosynthesis with increasing altitude is rarely considered. Yet, at higher elevations the absolute CO_2 concentration is lowered (Subsection 6.1.7). The effect on photosynthesis is not identical to correspondingly lowering the CO_2 concentration at sea level, because the ratio of the partial pressures of CO_2 and O_2 remains the same and photorespiration is not stimulated. As a result, photosynthesis in C_3 crops decreases with increasing elevation only at about half the rate as when CO_2 is lowered at sea level (Figure 25).

Air humidity and pollution

A direct effect of low air humidity on stomata, and hence on photosynthesis, has been reported (e.g., El Sharkaway et al., 1984). It may occur in sensitive

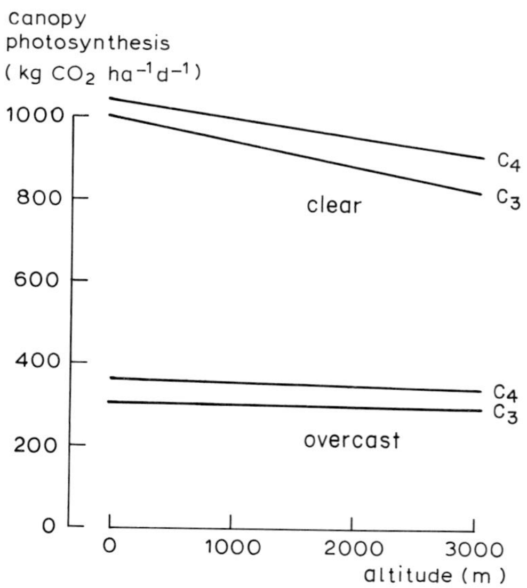

Figure 25. The simulated rate of canopy photosynthesis under clear and overcast conditions for a C_4 and a C_3 crop as a function of elevation.

crops in very dry weather. It is further discussed and modelled in Subsection 4.1.7.

Air pollution, in particular SO_2, has been reported to lower leaf photosynthesis in several species (Berry & Downton, 1982; Kropff, 1987). However, it is still difficult to quantify this effect for crop models.

2.2 Remobilization

2.2.1 Introduction

The first source of carbohydrates for maintenance and growth is photosynthesis (Section 2.1); the second source is internal mobilization or redistribution. The relative importance of remobilization, synonymous with redistribution was shown in Figure 13. Redistribution permits glucose formed before flowering and stored as polysaccharides, such as starch, to enter storage organs (seeds, fruits, tubers) and can allow a high growth rate to continue for a period, in spite of low radiation levels. In simulation, this glucose can be treated in the same way as glucose from photosynthesis. Towards the end of the growing season and in conditions of acute energy shortage, remobilization may also involve protein breakdown and degradation. Protein breakdown is an active and complex process that yields amino acids as well as glucose. In calculating energy balance, however, the error made by assuming that all remobilization is starch hydrolysis is small.

Efficiency and the rate of remobilization in established plants and during germination or sprouting are discussed here briefly; pre-flowering remobilization of starch is not considered. Postflowering redistribution of protein (nitrogen) can be an important determinant of the duration of the reproductive period (see Subsection 3.2.6).

Accumulation of carbohydrate reserves is discussed in Subsection 3.2.4.

2.2.2 Temporary storage

A common feature among crops is that some glucose is deposited in stems or roots as starch and some of this is mobilized weeks later. At flowering, 20% or more of the weight of vegetative organs may consist of mobilizable starch, particularly in cereals. There is little data published on stem reserve contents around flowering, but Table 7 shows some indicative values. The magnitude of the fraction in any particular case is probably also dependent on weather and crop husbandry.

To be mobilized starch must be hydrolyzed into glucose. This is a 'passive' process (i.e., it does not require additional energy). Mobilization of glucose requires only a small amount of energy (Subsection 2.4.3). The amount of glucose produced in remobilization is included in Listing 3, Line 39 and Listing 4, Line 30.

Table 7. The fraction of stem weight at flowering consisting of remobilizable carbohydrates (starch, sucrose plus glucose). Data are unpublished results provided by scientists at the Centre for Agrobiological Research, (CABO), Wageningen, unless indicated otherwise.

Species	Fraction	Source
Barley	0.3	
Cotton	0.1	estimate
Faba bean	0.45	
Maize	0.35	
Millet	0.1	estimate
Potato	0.2-0.4	
Rice	0.25	
Sorghum	0.2	Hodges et al., 1979
Soya bean	0.18	Hanway & Weber, 1971
Sugar-cane	0.5	
Sunflower	0.1	
Sweet potato	0.35	Hahn & Hozyo, 1983
Tulip	0.1	Benschop, 1986
Wheat	0.4	

Most stored carbohydrates are redistributed to the storage organs, so that regulating this process is not crucial to simulating yield. But regulation does affect growth dynamics and is therefore briefly considered. Redistribution is probably induced when, on consecutive days, the total demand for sugars exceeds the supply. A simple view is that redistribution starts once stems stop growing, and then continues at a rate of 0.1 d^{-1} of the redistributable starch (Listing 3 Line 35).

An alternative hypothesis is that starch is remobilized when the growth rate of the developing storage organ drops below a certain level (in kg ha^{-1} d^{-1}, not a relative rate). This level can be set to the highest two-day running average (Subsection 1.4.3) of the growth rate that the storage organs previously attained. After induction, remobilization proceeds at a rate of 0.2 d^{-1}, or at 0.1 d^{-1} in crops with a relatively long reproductive period. This level and rate are chosen without an experimental basis, but in many cases yield a reasonable pattern of stem weight loss. This hypothesis is programmed in Listing 4 Lines 42 and 45-47 of L1Q, the quarter-day time period module. (The provision that the average storage organ growth rate must be at least 10 kg ha^{-1} d^{-1} more than the maximum that it previously attained, avoids triggering remobilization too early). The level of triggering and the remobilization rate could be made dependent on crop type or external conditions (such as temperature), however, this is not attempted because of the lack of basic data.

Leaves lose weight during senescence. A sizeable fraction of the biomass is broken down and used for respiration or remobilization (functionally of similar value) before individual leaves die. This fraction is estimated at 0.5 (Listing 4 Lines 24, 30). It can be assumed that the same regulation holds for remobilization from leaves during senescence, as for that of starch from stems, but it probably becomes effective at a later stage. This is achieved by triggering the process when the storage organ reaches 80% of the maximum average growth level previously attained. The effective remobilization rate from leaves is estimated at 0.15 d^{-1} (Listing 4 Lines 40, 43). The amount of glucose that results from the breakdown of structural leaf material is affected by its carbon content (Listing 4 Line 30). It is assumed that dying roots do not contribute carbohydrate to growing points.

Remobilization from leaves during senescence is ignored in the simpler program of Listing 3 (module L1D).

2.2.3 Germination

Reserves in seeded or planted material are reconverted and mobilized as glucose and amino acids in the very early stages of plant growth. It is also a key process in regrowth after cutting or ratooning.

The rate of germination and early development in field conditions is difficult to simulate adequately. This is partly because environmental conditions for very young plants are difficult to assess. Crop growth modellers generally avoid this problem by initializing the simulation run at the time that 10-100 kg of dry matter has already been formed (see also Subsection 3.1.3).

The thermodynamic efficiency of germination is high (Penning de Vries & van Laar, 1976), but the efficiency attained in the field is lower because the fraction remobilized from the seed is often incomplete and because organic components leak into the soil. The germination process of highly viable (more than 80%) seed is estimated to yield about 0.25 g dry weight of seedling per g dry seed in cereals, 0.35 g g^{-1} in seeds of leguminous species and 0.45 g g^{-1} in seeds rich in lipids. In this approximation, it is assumed that seedlings do not fully exhaust their seeds, because photosynthesis takes over the carbohydrate supply before that stage. It is expected that the efficiency of the sprouting process of tuber and root crops corresponds with that of cereals. Ng & Loomis (1984) and Ingram & McCloud (1984) simulated the sprouting process of potato.

Bulb crops provide an extreme example. Most of the vegetative biomass is formed from carbohydrates in the motherbulb, and even in darkness a beautiful plant can grow from it. After flowering, photosynthesis provides the carbohydrates to fill the new bulbs. Benschop (1986) provides an example of simulating the growth of a tulip bulb crop.

2.3 Maintenance

2.3.1 Introduction

Respiration, like photosynthesis, has been studied for almost 200 years (Steward & Bidwell, 1983). However, the regulatory mechanisms of the processes at the whole plant level are still fairly new territory. Traditionally, but inappropriately, respiration has been regarded as complex, but basically a single process. However, maintenance respiration and growth respiration are processes that occur at different rates and with their own regulation, but they have CO_2 production in common (Figure 26). Quantifying the intensity of maintenance processes suffers considerably from a lack of understanding its basis, yet this process alone consumes 15-30% of the assimilates of a whole

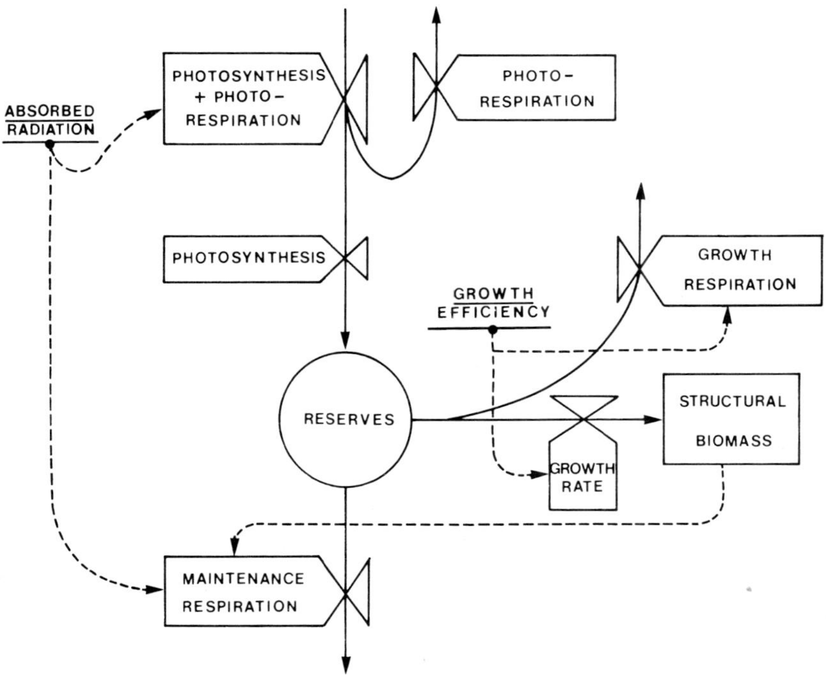

Figure 26. A relational diagram of respiration processes in crop growth. Valve symbols indicate fluxes, the circle an intermediate variable, the rectangle a state variable and the underlined variables are input variables or constants.

growing season (Figure 13). Therefore, the processes and assumptions underlying simulation of maintenance respiration are discussed here more extensively than those of other carbon balance processes.

Living organisms continuously use energy to maintain their current biochemical and physiological states. Respiration provides this energy. Though CO_2 is only a byproduct, measuring the CO_2 production rate is, nevertheless the best way to quantify maintenance respiration.

Maintenance can be considered at different levels of biological organization. Maintenance is only considered at the cellular level in crop growth modelling. Maintaining the biomass of leaves and root system is regarded as a balance of separate growth and loss processes.

2.3.2 Biochemistry and regulation of maintenance

Three components of maintenance at the cellular level are distinguished: maintenance of concentration differences across membranes, maintenance of proteins and a component related to the intensity of metabolism (Penning de Vries, 1975).

Concentration differences

This maintenance process is made up of activities which maintain the concentration differences of organic and inorganic ions and of neutral molecules across cell membranes. These processes keep up electrical and pH gradients and counteract spontaneous leakage. The membranes involved are those of cell organells (mitochondria, chloroplasts) and of tonoplast and plasmalemma (enveloping the vacuole and cytoplasma, respectively). This process requires considerable energy, because though membranes only measure 8-25 nm across, the difference between both sides of the membranes can be appreciable (90 mV, 1 pH unit) and because the total membrane surface is large (typically 2-20 m^2 per g dry matter). Active transport of one molecule through one membrane probably requires, on average, the energy of one ATP (irrespective of the gradient), while two molecules follow passively.

Protein turnover

Decomposition of some proteins and synthesis of others is continuous. Respiratory processes provide energy for this turnover. The rate of enzyme turnover varies: from almost none for the bulk of the proteins, to several times per day for some enzymes in key metabolic positions. Their overall average turnover rate has been estimated to be about 0.10 d^{-1} in active leaf, stem and root tissue, but is probably much smaller (deducted from data by Huffaker & Miller, 1978; Wittenbach et al., 1982; Bidwell, personal communication). Protein degradation uses an insignificant amount of energy; the cost of protein resynthesis resembles that of synthesis.

Metabolic activity

The biochemical basis of metabolic activity, the third component of maintenance respiration, is poorly understood. The component is thought to be related directly to the overall metabolic activity of the crop (Penning de Vries, 1975; McCree & Kresovich, 1978; Amthor, 1984). The few data available indicate that its value is low in phytotron plants grown at low light. However, in field crops with high growth rates it is roughly equal to the sum of the other components and is therefore significant. Gross photosynthesis may be used as a measure of the overall metabolic activity in a crop. The third maintenance respiration component is then equal to 10-20% of daily photosynthesis (an estimate, based, as yet, on few observations). The course of CO_2 production in a crop suggests that this maintenance component follows gross photosynthesis with a time coefficient of 1-2 days. The process takes place mainly in the leaves.

An alternative explanation of the maintenance respiration component is that a high photosynthetic activity leads to a high glucose level in leaves, which in turn can partially uncouple ATP production from glucose oxidation, i.e., lower the efficiency of energy production (Azcon-Bieto et al., 1983). Up to three times as much glucose can then be combusted to provide the same amount of energy. CO_2 production increases correspondingly. This promising hypothesis awaits direct experimental support.

Rates of maintenance respiration

The unit energy costs of the first and second maintenance components (concentration differences and protein turnover) are reasonably well quantified. Thermodynamic efficiency of the process is high (about 40%) if uncoupling does not occur. But the rates and regulation of those processes are insufficiently known to calculate the intensity of maintenance processes. Direct measurements of the energy required for maintenance are also unavailable. Even measuring the CO_2 production rate is difficult, because this rate is low and often confounded with CO_2 evolution from other processes. The best available measurements are those in which other sources of respiration are avoided as much as possible (cf. Forward, 1983; Amthor, 1984). Values of the rate of maintenance respiration of field grown leaves measured in this way are 0.03-0.08 g CO_2 per g dry matter per day at 20 °C (for temperate crops) or at 30 °C (for tropical crops) and less for plants grown in phytotrons. In general, the more active the tissue and the higher the nitrogen concentration, the higher the rate of maintenance respiration.

Table 8 provides respiration rates for several field crops. These rates are the result of maintaining concentration differences and of protein turnover in leaves. The metabolic component of maintenance respiration is assumed to be proportional to the photosynthesis rate. The values of Table 8 may not be applicable to any specific simulation and observed rates should be obtained if possible.

Table 8. Rates of maintenance respiration of leaves of field crops at a reference temperature.

Species	Rate of maintenance respiration ($g\ CO_2 g^{-1} d^{-1}$)	Temperature (°C)	Reference
Barley	0.03	23/18	Ryle et al., 1973
Cotton	0.038	30	Amthor, 1984
Faba bean	0.017	25	Penning de Vries, 1975 (phytotron plants)
Field bean	0.018	25	Amthor, 1984
	0.027	25	Penning de Vries, 1975 (phytotron plants)
Maize (temperate)	0.032	24/18	Amthor, 1984
	0.026	25	Penning de Vries, 1975 (phytotron plants)
Millet	0.03	25	Jansen & Gosseye, 1986
Rice cv IR58	0.02	25	estimate based on Yoshida, 1981
Sorghum	0.01	30	McCree, 1974 (phytotron plants)
	0.029, 0.073	30	Amthor, 1984
Sunflower	0.025	20	Amthor, 1984(shoot)
	0.060	25	Penning de Vries, 1975 (phytotron plants)
Wheat	0.016	20	van Keulen & Seligman, 1987

In spite of such uncertainties there are indications of differences in the rates of maintenance respiration between lines and cultivars of several crop species (e.g., Wilson, 1982; Gifford & Jenkins, 1982; Spitters, Stichting voor Plantenveredeling (SVP) personal communication). In some cases the differences are as large as 30-50% and this is of interest to plant breeders.

Non-leaf tissues

The literature provides insufficient observations on the rate of maintenance respiration of non-leaf tissue to provide a list of species-specific data; values range from 0.005-0.09 g $CO_2\ g^{-1}\ d^{-1}$. For the roots of annual plants use 0.015 g $CO_2\ g^{-1}\ d^{-1}$ (at 20 or 30 °C for temperate and tropical crops, respectively) and

0.010 or less for stem tissue. (The rate in young stem tissue is 1.5-2 times higher; this is ignored here for the consequences are small.) The composition of storage organs up to 1000 kg ha^{-1} is similar to that of young stems and has the same relative maintenance cost. All biomass above this threshold is assumed to be biochemically stable and maintenance free. Amthor (1984) lists data for maintenance respiration of roots that are generally higher than those of leaves. However, these rates include the cost of ion uptake by roots, which are included here in the cost of growth (Subsection 2.4.3).

Temperature

Temperature has a direct effect on the rate of maintenance respiration. It corresponds to a doubling of the rate for each 10 °C rise in temperature (McCree, 1974) up to temperatures that will kill plants (45-60 °C). This dependence between temperature and rate of maintenance respiration corresponds with the biological concept of a Q_{10} with the value of 2.0 (Listing 3 Line 67, Listing 4 Line 92). Lower and higher Q_{10} values are sometimes reported (Amthor, 1984), but the value of 2.0 appears to be a reasonable average.

It is assumed that the third component of maintenance respiration (metabolic activity) is not directly dependent on temperature.

Air pollution might increase maintenance respiration (Berry & Downton, 1982), but the extent is not yet quantified.

2.3.3 Reduction of maintenance respiration during photosynthesis

At high light intensities the photosynthesis light response curve deviates considerably from the straight line set by the initial efficiency. Leaves then absorb more energy than can be channeled into CO_2 reduction. The excess energy equals the difference between normal canopy photosynthesis and that of a canopy with a very high maximum photosynthesis rate (Figure 27). Part of the excess is used for maintenance (cf. Graham & Chapman, 1979, vol 6 p. 154; Bidwell, 1983; Amthor, 1984). However, this side benefit of photosynthesis is limited to upper leaves during the bright hours of the day, for this energy cannot be stored or exchanged between cells or organs. It is estimated that this excess energy covers half the daytime cost of the basic maintenance processes in leaves and half the cost of metabolic activity. For a whole season this side benefit of photosynthesis amounts to as much as one third of the total cost of maintaining the crop, or a yield gain of 1000-3000 kg ha^{-1}. (It is expected that energy for nitrate reduction also comes from this source (Subsection 2.4.3). Photosynthesis appears to be more than only CO_2 reduction! Figure 13 sector 5 reflects this extra benefit.)

This view of a direct interaction between photosynthesis and respiration has no implications for interpreting photosynthesis light response curves or for measuring the maximum value of the curves.

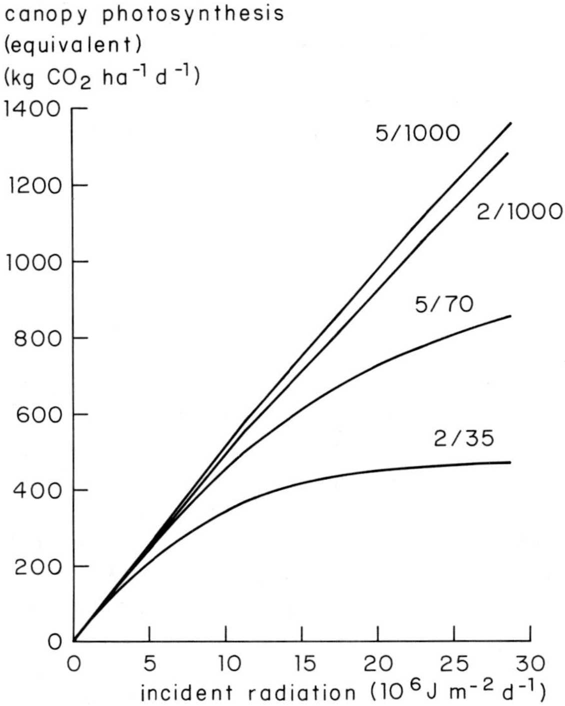

Figure 27. Photosynthesis computed with a normal maximum leaf photosynthesis rate (35 and 70 kg CO_2 ha^{-1} h^{-1} respectively) and with a very high value (1000), for a leaf area of 2 and 5 m^2 m^{-2}, respectively. The vertical distance between corresponding lines represents the energy available for other energy-consuming processes in green cells.

2.3.4 Programming maintenance processes

Maintenance has absolute priority over growth and related energy-demanding processes. All carbohydrates required for crop maintenance are subtracted from daily photosynthesis and the remainder are left for growth (Listing 3 Line 39). In the module L1Q with a daily cycle in stored sugars, priority for maintenance processes is achieved by stopping growth processes when carbohydrate reserves drop below a certain threshold (5% of leaf dry weight), while maintenance continues (Listing 4 Lines 29, 58).

The basic processes of maintenance respiration are simulated in a straightforward manner (Listing 3 Lines 61-66, Listing 4 Lines 84-91). Table 8 presents species-specific data for leaves. The metabolic component is assumed to be equal to 20% of the daily gross photosynthesis (Listing 3 Line 69) or to 20% of the value of the running average of gross photosynthesis with a time coefficient of one day (Listing 4 Lines 94-95). To simulate the course of individual processes more realistically, this fraction of maintenance respiration is not subtracted directly from photosynthesis, though numerically it would be the same. Daytime maintenance respiration of leaves is estimated to be reduced by 50% due to excess energy. This is achieved in the model by multiplying with 0.75 when computing on a 24-hour basis (Listing 3 Line 63) or with 0.5 for daytime respiration only (Listing 4 Line 88). In both cases, the cost of the metabolic component is reduced by 50%.

The effect of air temperature on maintenance is expressed as an exponential function without limits. The reference temperature at which the relative effect of temperature is unity is chosen to be a fixed value (see Table 8, Subsection 2.3.2) and not subjected to adaptation processes by the plant.

If insufficient photosynthetic products are available to meet maintenance respiration demands, which may occur in heavy crops on very cloudy days, structural material is sacrificed to provide energy to maintain some processes at a lowered rate. Whole chloroplasts can be consumed in the process (Wittenbach et al., 1982). After a few days of energy shortage, irreparable damage occurs and many cells die. A full dynamic simulation of these processes is not yet possible. Therefore simulation is stopped by a FINISH condition when leaf maintenance respiration requires more energy than photosynthesis supplies for three consecutive days (Listing 3 Lines 46,47,114; Listing 4 Lines 65-67, 145). Negative net photosynthesis causes some structural material, particularly proteins, to breakdown. To allow this situation to occur for three consecutive days reflects that limited damage is recoverable; the length of this period is chosen arbitrarily.

2.3.5 *Special cases*

Wasteful respiration has been demonstrated to occur in several species. This involves a normal mitochondrial respiration, but electron pairs are led along a pathway that yields only one, instead of three, ATP. This is called uncoupled respiration. It has been suggested that this mechanism allows elimination of excess carbohydrates, such as those reaching roots, as a result of inaccurate translocation regulation (Lambers, 1979). Wasteful respiration is expected to be relatively unimportant in field crops at Production Level 1 when sink size limitation is not common (Chapter 3.1). This is because, as a rule, simulated potential yields do not exceed significantly experimental potential yields. Wasteful respiration is therefore not explicitly included in the models described here. Its eventual effect can be simulated by increasing the rate of maintenance respiration up to threefold and by slightly decreasing the conversion efficiency (Penning de Vries et al., 1974). When carbohydrates accumulate to high levels in leaves, wasteful maintenance respiration can also be expected to occur. In these cases, the models here reduce the photosynthesis rate (Subsection 2.1.4). However, stimulating maintenance respiration would have been an alternative solution. There is no data on the effect on maintenance respiration of high carbohydrate levels in the storage cells of sugar-beet or sugar-cane, but it is assumed not to induce wasteful respiration.

2.4 Crop growth

2.4.1 Introduction

'Growth' is defined here as the biochemical conversion of reserve substances into 'structural dry matter'. Structural dry matter consists of the organic components that remain at the end of the plant's life, that is, are not normally broken down. In contrast, 'reserves' are components that only exist temporarily and are used for maintenance or growth within hours or days (available reserves) or after a few weeks (shielded reserves).

Two aspects of crop growth need particular consideration here: the rate of growth of the entire crop and the efficiency of the growth process. The rate of growth respiration is related to both (Figure 26). (See Subsection 3.2.2 for calculation of the growth rate of separate organs).

The crop growth rate is the multiple of the assimilates used for growth and the efficiency of the process. Efficiency is quantified in Subsection 2.4.3, and can be characterized by separate parameters for leaves, stems, roots and storage organs.

2.4.2 Rate of crop growth

Simulation models using a one-day time period of integration, usually assume that any glucose produced during a day and remaining after the day's maintenance processes, will be used for growth. This is adequate for practical purposes. The carbohydrates available for growth processes are then related directly to daily photosynthesis (Listing 3 Line 39). Temperature does not affect this.

Glucose production and consumption do not occur at the same rate over quarter-day time periods. Simulating this process requires introducing a buffering pool of reserves. All products from photosynthesis (and from remobilization, if any) are treated as if they are assembled in this pool from which all consuming processes draw. This pool of reserves is physically dispersed in the crop. Most are found in the active leaves, where reserves may account for up to 30% of the dry weight. (At these values net photosynthesis is reduced to avoid a further build up, see Subsection 2.1.4). In many cases, a starch or glucose pool is also formed in stems and roots. This pool is not available short term, but over many days or weeks. These are shielded reserves (see also Subsection 2.2.2 Temporary storage and Subsection 3.2.4 Assimilate partitioning).

With time periods in the order of hours, the growth rate can be directly related to the level of available reserves (e.g. de Wit et al., 1978; Penning de

Vries et al., 1979; Ng & Loomis, 1984). Neither of these approaches to growth rate control is quite appropriate for the module with a quarter-day time periods. However, as simulation of the dynamics of the reserve pool is not the objective, its programming can be a compromise. Hence, the rate of use of available reserves in the pool is set at 0.85 d^{-1} above the lower threshold of 5% (Listing 4 Line 58; $0.85 = 1.0 - (1.0 - 1.5 \cdot \text{DELT})^4$). Both numbers are somewhat arbitrarily chosen, but yield acceptable patterns of daily fluctuations in the leaf glucose level (Figure 28), and crop yields are insensitive to these numbers. (By changing the factor 1.5 d^{-1} to 0.5, a crop with slow carbohydrate export is simulated; and changing it to 4.0 (= 1. / DELT) avoids any accumulation. The effect of temperature on the magnitude of this fraction may be considered). The reserve level fluctuations are somewhat exaggerated because there is no feedback between growth rate and level during the quarter-day time period while, in reality, there is. The difference in structure of a module with 24-hour and quater-day time periods is schematized in Figure 29.

Implicit in this formulation is that the crop's growth rate at Production Level 1 is almost fully source-dependent, that is, a higher rate of photosynthesis results in a higher growth rate. However, there are two important exceptions: low temperature may reduce the growth rate more than photosynthesis, and

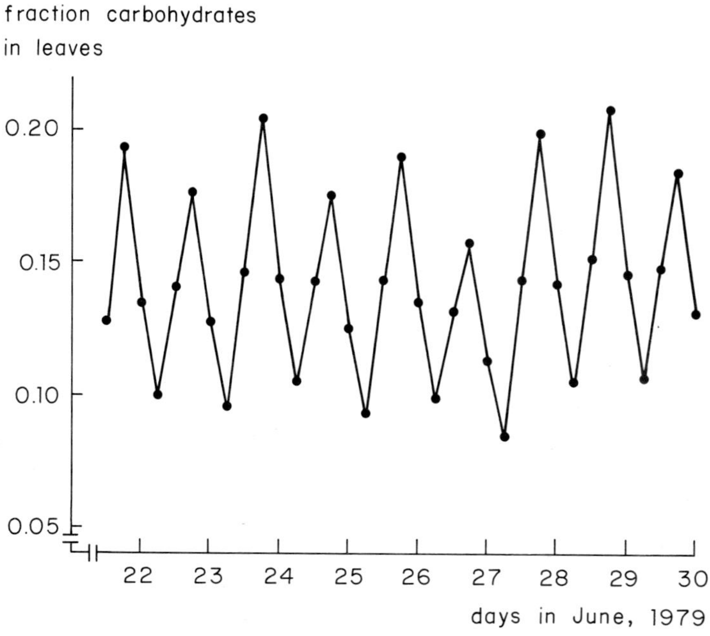

Figure 28. The simulated course of the amount of reserve carbohydrates relative to leaf weight in wheat, in Wageningen, using quarter-day time periods.

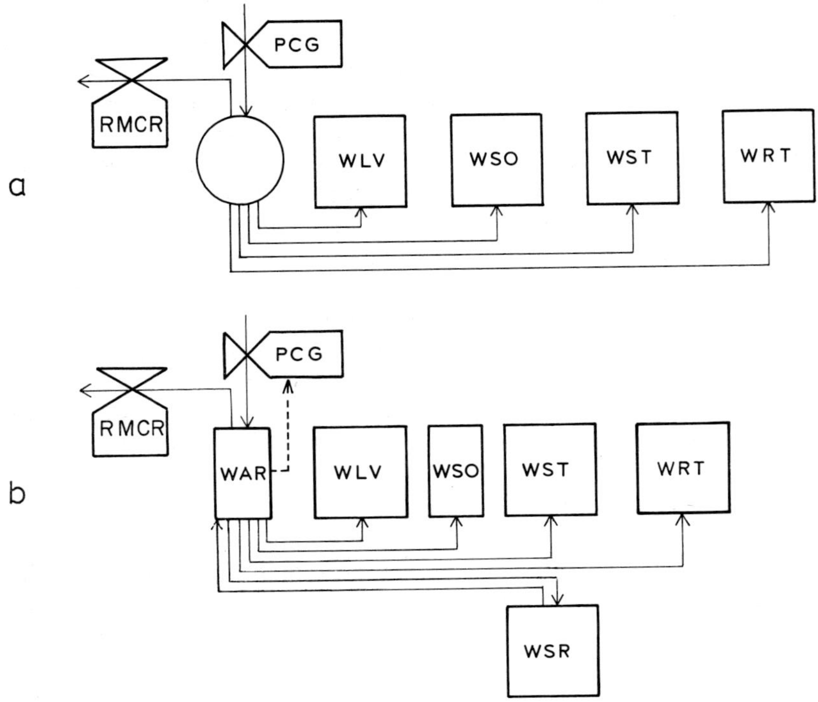

Figure 29. A relational diagram of major growth processes when simulating with one-day time periods (a) and with quarter-day time periods (b). The empty circle represents the daily and complete partitioning of gross photosynthesis (PCG) over maintenance respiration and growth processes. For full names of variables see Listing 12.

there may not be enough growing points to accept the carbohydrates. In both cases, the real growth rate is below that which carbon balance permits. A pool of shielded reserves provides a buffer for carbohydrates not immediately used for growth. The quarter-day time period module distinguishes these as potential rates of carbohydrate use for growth (Listing 4 Lines 58-63) and realized growth rates (Lines 51-56). Where the growth rate limits photosynthesis rather than vice versa, care must be taken to properly quantify this constraint.

A dynamic approach to simulating the sink size for carbohydrate in cereals is given in Subsection 3.2.5. Though, in general, sink size has no effect on growth, there is insufficient knowledge of non-cereal crops to simulate sink size properly. In these cases, for example, when organs are too small or too few in number to accept all carbohydrates available to them, a rough approximation must be made. This type of situation may be handled by calculating a maximum growth rate for the storage organ (Listing 4 Lines 35-36). Though individual seeds of cereals often grow at a constant rate, the number of growing seeds changes. Hence, the simplest approximation is similar for all species: the maximum relative growth rate, times the actual weight of the storage organ, plus a small basic amount (to permit very small storage organs to grow relatively faster). The value of 0.35 g g^{-1} d^{-1} may be used for this maximum relative rate for temperate crops and 0.50 g g^{-1} d^{-1} for tropical crops unless better data are available. A similar reasoning and programming of sink-size limitation could be applied to other organs after pruning young leaves and growing points damaged by insects. Ng & Loomis (1984) provide an example for potato.

Temperature usually has no direct effect on daily growth. However, low night temperatures may provide an exception, for reserves can then accumulate to levels high enough to reduce photosynthesis or stimulate wasteful respiration (Subsections 2.1.4, 2.3.5). This feature is an important addition for modelling spring and winter crops (cf., van Keulen & Seligman, 1987).

2.4.3 *Growth efficiency and growth respiration*

Biochemical research has led to a relatively accurate procedure to derive the growth efficiency and the concomitant CO_2 evolution (Penning de Vries et al., 1974, 1983; McDermitt & Loomis, 1981; Forward, 1983). The approach, applicable to all species of higher plants, is summarized in Listing 1.

'Growth' consists of biosynthetic processes *per se*, that is, conversion of glucose into other organic components, plus translocation of the glucose from the source to the growth site, plus (in the case of legumes) the cost of nitrogen reduction. The weight ratio of product to substrate ranges from 0.35 to 1.0 g g^{-1}; the lowest values apply to compounds with a high combustion heat. The efficiency of carbon use during the growth of these crops is 68-86%, while 78-86% of the combustion heat of the substrate is retained; leguminous crops score about 10% lower in both respects (these numbers are produced with Listing 1).

Growth respiration is defined as the CO_2 evolution resulting from growth processes. Growth efficiency and growth respiration are two aspects of the same process (see Figure 26, Subsection 2.3.1).

Biosynthesis

Innumerable different organic components of structural dry matter exist. For considerations of biosynthesis efficiency, however, only five relatively uniform groups need to be distinguished: nitrogenous compounds (particularly proteins, but also nucleic acids, nucleotides, free amino acids and peptides), carbohydrates (cellulose, hemicellulose, starch), lipids (fats, fatty acids, oils), lignin and organic acids. Table 9 indicates some characteristics of these five groups. The fact that the molecules of the five groups are assembled in superstructures, such as organelles, has no implications for the cost of synthesis.

Table 9. Some characteristics of the five major groups of plant components and minerals.

	Heat of combustion ($kJ\ g^{-1}$)	Nitrogen content ($g\ g^{-1}$)	Carbon content ($g\ g^{-1}$)	Found particularly in
Carbohydrates	17.3	0.0	0.451	cell wall,vacuole
Proteins	22.7	0.151	0.532	enzymes,membranes
Fats	37.7	0.0	0.774	membranes
Lignins	29.9	0.0	0.690	cell wall
Organic acids	13.9	0.0	0.375	vacuole
Minerals(K,Ca,P,S)	0.0	0.0	0.0	vacuole,cytoplasma

Pathways over which plants synthesize the most common compounds have been unravelled and quantified, and balance equations have been made for biosynthesis of simple and more complex organic molecules using stoicheiometry (e.g., Dagley & Nicholson, 1970). By weighing these equations according to the relative occurrence of specific molecules in its group, equations were obtained that characterize synthesis of such a group as a whole. For instance:

$$3.030\text{ g glucose} + 0.352\text{ g } O_2 \rightarrow 1.000\text{ g fat} + 1.606\text{ g } CO_2 + 0.776\text{ g } H_2O$$

This balance equation provides the amount of carbohydrates required for synthesis of fat ($3.030\ g\ g^{-1}$) and the CO_2 production factor for this process ($1.606\ g\ g^{-1}$). Although intricate for large molecules, such derivations are basically straightforward computations. Only protein synthesis is calculated in two steps: proteins are formed from amides and glucose at the growth site, and amides and glucose are formed in the photosynthesizing cells. (Amides are not explicitly considered here to avoid confusion, for the implications for growth simulation are negligible). Table 10 provides the data that characterize biosynthesis of the five groups of organic constituents (for their derivation see Penning de Vries et al. (1983), but with two corrections: ion uptake uses more energy (see below) and protein synthesis from amides and glucose was not explicit).

Sensitivity analysis has shown that the variability in the amino acid constitution of different proteins results in substrate requirements that usually differ less than 5% from the average. This is probably also correct for other organic components. For consistency within the programs used here and for checking the carbon balance (Subsection 3.4.4), very precise numbers are given in Table 10. They reflect an average situation and rarely need to be adapted. The num-

Table 10. Glucose required and CO_2 produced during formation of organic components (1); plus related transport (2), expressed in g per g of the product formed (1+2). Values for N reduction (3) applying to leguminous crops includes the cost of transport.

	Carbohy hydrates	Proteins **	Fats	Lignins	Organic acids	Minerals
Biosynthesis(1)*						
glucose	1.211	1.793	3.030	2.119	0.906	0.000
CO_2	0.123	0.679	1.606	0.576	– 0.045	0.000
Transport (2)						
glucose	0.064	0.094	0.159	0.112	0.048	0.120
CO_2	0.093	0.138	0.234	0.164	0.070	0.176
Growth (1+2)						
glucose	1.275	1.887	3.189	2.231	0.954	0.120
CO_2	0.216	0.817	1.840	0.740	0.025	0.176
Reduction (3)						
glucose		0.897				
CO_2		1.316				

* from Penning de Vries et al. (1983)

** synthesis via amides, no N reduction

bers in Table 10 are thought to be correct for all species of higher plants. Moreover, they are insensitive to the rate of growth *per se* and to temperature, and probably also to water stress and other environmental factors.

Transport and uptake

Transporting carbohydrates from sources to sinks requires at least three active steps: loading the phloem, transport within the phloem and uptake by the sink cell. The first and last processes consist of traversing cell membranes, but exactly how many is unknown. Here the lowest reasonable estimate is used. Unloading the cells is passive and loading is active across one membrane, which requires 5.3% of the energy content of transported glucose (i.e., one ATP per glucose molecule per passage). This estimate was used to compute the values of Table 10. (If the cost of transport is higher in particular cases, these numbers can be increased and new growth requirements computed). In C_4 species, some transport may be passive and the cost of loading correspondingly lower (Moorby, 1981). Transporting sucrose in the phloem is an active process, but the amount of energy involved is probably insignificant.

The uptake of inorganic constituents from the soil solution is an active process for some ions and passive for others. It has been estimated that, on average, three monovalent ions are brought across one membrane per ATP, but it is difficult to estimate the number of membranes to be passed. Assuming the minimum of three crossings (into and out of the endodermis, and into the sink cell) and an average molecular (equivalent) weight of 40, 0.12 g glucose per g minerals is needed (Table 10). In high salt plants, this figure could be twice as high (Veen, CABO, personal communication). In a few crops (e.g., rice) silica (SiO_2) makes up to 20% of the vegetative dry weight. It is assumed that silica is absorbed as silicate at the same cost as other minerals.

It is assumed that there is no metabolic energy involved in water uptake.

Nitrogen reduction

Nitrogen (N) enters plants in the form of nitrate (NO_3^-), ammonium (NH_4^+), or bi-nitrogen (N_2). The first and last form need reduction before N can be assimilated. The direct cost of reduction of nitrate-N to amino-N equals 1.27 gram glucose per gram nitrate, or 0.852 gram glucose per gram protein (Table 1 in Penning de Vries et al., 1974). This glucose is actively transported. N reduction could be an important addition to the cost of protein synthesis (Table 10). However, in crops amply supplied with nutrients, the bulk of nitrate reduction occurs in the leaves during photosynthesis. Excess $NADPH_2$ and ATP generated by chloroplasts (see also Subsection 2.3.3 and Figure 13 in the introduction to Chapter 2), is probably used for this process and N reduction comes essentially free to the crop. This assumption is implicit in the modules for simulating at Production Levels 1 and 2.

Rhizobia-bacteria in nodules on roots reduce N in leguminous crops. These bacteria live symbiotically with the host plant, reducing N_2 and providing amino-N to the host while receiving carbohydrates in exchange. The rate of N_2 reduction by the rhizobia in good conditions can be as high as the rate of absorption of NO_3^- from the soil (i.e., in the order of 10 kg ha^{-1} d^{-1}). The costs of reducing N_2 per gram N are about the same as those for reducing NO_3^- in an effective combination of rhizobium strain and host cultivar on good soils and with an appropriate water supply. The cost can be many times higher in other conditions (Minchin et al., 1981). For simulations of legume growth, the minimum reduction cost is incorporated in the computation of the carbohydrate requirement for growth. The direct cost of N_2 reduction to a leguminous crop is 1000-2000 kg glucose per hectare per season. In the presence of NO_3^- in the soil, many legumes obtain most of their N in that form and reduce it in the leaves at no energy cost. The mechanism may not be quite clear, but from an energy balance point of view NO_3^- reduction in leaves is advantageous (de Visser, 1984).

Simpler biochemical analyses

Biochemical analysis of the bulk of major organs appears to be sufficient to derive growth efficiency and growth respiration. However, performing such biochemical analyses routinely and accurately is laborious. An alternative method was developed to characterize the biochemical composition of biomass (Vertregt & Penning de Vries, 1987). This method only requires measurement of the carbon content of the dry matter (*C*, in g kg^{-1}) and its ash content (*A*, oxidation at 550 °C, g kg^{-1}); the nitrogen content (*N*, g kg^{-1}) is determined only if the crop is expected to reduce all its N in the roots. These measurements can be taken relatively simply and quickly. The C, A and N contents, the glucose required (*CRG*, g g^{-1} dry matter) and the CO_2 produced during growth (*CPG*, g g^{-1} dry matter) can be calculated directly and with ample accuracy by:

$$CRG = ((5.39 * C + 0.80 * A + 5.64 * N - 1191) * 1.053) / 1000$$
$$CPG = (4.24 * C + 1.17 * A + 8.28 * N - 1744 + CRG * 77.7) / 1000$$

N is to be set to 0.0 if reduction occurs in the leaves at no energy cost. (See also consistency check in Listing 5 Line 14)

2.4.4 Simulating biosynthetic processes

The relative costs of forming plant components are the sum of those for synthesis, transport and N reduction (see Table 10 Subsection 2.4.3). These characteristic values can then be used in models where the five groups of organic components are explicitly distinguished (e.g., de Wit et al., 1978). From the growth efficiency viewpoint, the composition of organs often does not change significantly under optimal nutritional conditions. Typical biochemical compositions can be given for the bulk of the leaves, stems and roots of all crops, while storage organs for each crop can also be typified. (Leaf and stem tissue of rice can contain 15-20% of the dry weight in SiO_2; this is an exception large enough to be accounted for.)

Thus, the entire growth process can be characterized by two parameters: the carbohydrate requirement and the CO_2 production factor. These parameters are assumed to be similar for the leaves of all non-legume crops, the roots of all non-legume crops, and the stems of all non-legume crops. Storage organs of crops are quite different in this respect. Table 11 gives the values for these parameters for many crops. They are produced with the program Listing 1, with the biochemical composition as input plus the parameter LEG = 1. for crops paying their own N reduction costs and LEG = 0. for all other cases. Only rarely may there be a need for the detailed approach. If there are reasons to assume that all N_2 reduction occurs in roots rather than in leaves, then either conversion constants must be adapted (recalculate them with Listing 1 and set parameter LEG = 1.), or another energy-consuming process must be added to the roots. The amount of glucose required for N reduction (see Table 10 Sub-

Table 11. Carbohydrate requirement (1) (g g^{-1} dry matter) and CO_2 production (2) (g g^{-1} dry matter) for growth of important tissues. The carbon fraction in the dry matter is given in column (3). The numbers in column (4) are percentages of the dry weight in carbohydrates, proteins, fats, lignins, organic acids and minerals respectively.

	(1) Glucose	(2) CO_2	(3) Carbon fraction	(4) Biochemical composition
Vegetative organs non-leguminous and non-rice crops:				
Leaves	1.463	0.461	0.459	52;25;5;5;5;8
Stems	1.513	0.408	0.494	62;10;2;20;2;4
Roots	1.444	0.406	0.467	56;10;2;20;2;10
Vegetative organs rice crops:				
Leaves	1.326	0.408	0.419	53;20;4;4;4;15
Stems and roots	1.326	0.365	0.431	58;8;2;15;2;15
Vegetative organs leguminous crops:				
Leaves	1.687	0.790	0.459	52;25;5;5;5;8
Stems	1.603	0.540	0.494	62;10;2;20;2;4
Roots	1.534	0.537	0.467	56;10;2;20;2;10
Storage organs:				
Cassava, tuber	1.297	0.259	0.448	87;3;1;3;3;3
Cotton, boll	1.861	0.748	0.540	40;21;23;8;4;4
Cowpea, pod+seed	1.653	0.698	0.471	61;22;2;7;4;4
Faba bean, pod+seed	1.740	0.816	0.473	55;29;1;7;4;4
Field bean, pod+seed	1.668	0.717	0.472	60;23;2;7;4;4
Groundnut, pod+seed	2.518	1.433	0.616	14;27;39;14;3;3
Maize, cob+grain	1.491	0.384	0.491	75;8;4;11;1;1
Millet, ear+grain	1.477	0.391	0.484	69;9;4;12;3;3
Pigeon pea, pod+seed	1.652	0.675	0.476	60;20;2;10;4;4
Potato, tuber	1.285	0.274	0.439	78;9;0;3;5;5
Rice, inflor.+grain	1.462	0.357	0.487	76;8;2;12;1;1
Sorghum, ear+grain	1.473	0.377	0.486	72;9;3;12;2;2
Soya bean, pod+seed	2.161	1.238	0.527	29;37;18;6;5;5
Sugar-beet, beet	1.294	0.263	0.446	82;5;0;5;4;4
Sugar-cane, whole tops	1.478	0.392	0.484	57;7;2;22;6;6
Sunflower, inflor.+seed	1.862	0.719	0.549	45;14;22;13;3;3
Sweet potato, tuber	1.328	0.287	0.453	84;5;2;3;3;3
Tomato, fruit	1.424	0.412	0.457	54;17;4;9;8;8
Wheat, ear+grain	1.415	0.347	0.471	76;12;2;6;2;2
Yam, tuber	1.286	0.272	0.440	80;6;1;3;5;5

Source: Penning de Vries et al. (1983)

section 2.4.3) is enhanced when respiration is partially uncoupled (de Visser, 1984).

The parameters for carbohydrate requirements relate carbohydrate used to biomass produced (Listing 3 Lines 28-31, Listing 4 Lines 31-35). Growth respiration is calculated in Listing 3 Lines 71-75 and Listing 4 Lines 97-101. Note that the transport-related part of growth respiration evolves in the photosynthesizing leaves, while the respiration related to biosynthesis evolves in the growing cells. Respiration due to transport of remobilized carbohydrates (Listing 3 Line 76, Listing 4 Lines 102-104) is added to growth respiration.

2.5 Exercises

Use the following data for the exercises in all chapters:

– Crops: for rice (cv IR36) those of Listing 3; for soya bean (cv Hawkeye) or for potato (cv Favorita) as indicated in various Tables. Initial leaf and stem weight, development stage and rooting depth are respectively: 35., 15., 0.0 and 0.15 for soya bean and 170., 90., 0.2 and 0.15 for potato. Suppose that RMCLV is 0.03 at 20 °C for potato. Start wet season crops at DATEB = 197. and dry season crops at 349. (except rice: DATEB = 52.).

– Weather: use radiation data of Los Baños, 1984 (Listing 11) plus the approximations:

TABLE RAINT(1-365) = 40 * 0., 50., 159 * 0.,...
300., 29 * 4., 200., 29 * 0., 15 * 20., 75 * 0., 15 * 10.
TABLE TPMT(1-365) = 365 * 32.
TABLE TPLT(1-365) = 365 * 24.
TABLE WDST(1-365) = 365 * 1.
TABLE HUAAT(1-365) = 365 * 2.

Use the data for loamy soil (Listing 10) for exercises of Chapters 4 and 5. All layers have the same characteristics. For a fine sandy soil or a light clay soil, replace the water contents at saturation, field capacity, wilting point and air dry with those from Table 27. For SAWAH, use 5 layers of 0.2 m each. For sand, use soil type type 4, set EES to 30., CSA to 0.05 and CSB to 15.. For clay, the numbers are 17, 10., 0.2 and 5., respectively.

2.5.1 Photosynthesis

T1. Make a relational diagram of the processes and variables which determine canopy photosynthesis.

T2. Use the leaf photosynthesis equation of Subsection 2.1.2 for PLMX 30. and 60., and for PLEA 0.5 and 0.3. Draw the curves.

As an alternative to the exponential relation, a Blackman curve and a hyperbola are sometimes used. Their equations are, respectively:

PL = AMIN1(AL * PLEA, PLMX), and
PL = PLMX * AL / (AL + ALH)

ALH, the radiation intensity where PL = 0.5 * PLMX, equals 100 J $m^{-2}s^{-1}$. Draw these curves for the same values of PLMX and PLEA. Compare the shapes of the curves.

T3. Determine the impact of an increase in specific leaf weight of rice from 370 kg ha^{-1} by 10% on canopy photosynthesis on a clear day in Figure 22 for leaf areas of 1, 3 and 10 ha ha^{-1}.

T4. What is the maximum value that PLMX can attain in a C_3 and a C_4 crop, assuming a minimum diffusion resistance of 100 s m^{-1} in both, an ambient CO_2 concentration of 340 vppm and a CO_2 concentration in the leaf of 136 in a C_4 crop and 238 vppm in a C_3 crop? What does this imply for the capacity of the diffusion process?

T5. Why does the maximum rate of canopy photosynthesis of temperate regions exceed that in the tropics? Why are there two dates with maximum photosynthesis at the equator and only one in temperate zones?

T6. Why is the degree of the photosynthesis reduction as a result of unequal distribution of light during a day with clouds, compared to a day with equal cloud distribution (Figure 23), not unexpected?

S1. Determine the impact of an increase in specific leaf weight from 370 kg ha^{-1} by 10% on canopy photosynthesis on day 244 (Sept 1) for leaf areas of 1, 3 and 10 ha ha^{-1}. (Suggestion: replace the definition of ALV and SLA by parameters with these values.) Compare these results with Figure 22 and exercise T2.

S2. Determine the gain in yield of a potato crop in Los Baños when breeding shifts the photosynthesis-temperature response curve higher by 5 °C. Assume planting on day 349 (Dec 15). Which process benefits most from the shift? Is the yield realistic?

S3. Compare the potential rice yield of a standard IR36 crop with a spherical leaf distribution in the dry season (DATEB = 52.) with a crop in which all leaves are vertical (leaf angles 60-90°) and one in which all are horizontal (0-30°). Use the SUPHOL subroutine as in Subsection 2.1.4 without the temperature difference between layers and without dead leaves. What percentage of photosynthesis is provided by the second layer?

S4. Suppose a disease causes rice grains to increase in weight not more than 10% d^{-1} (once the 10 kg ha^{-1} threshold is exceeded). This leads to a carbohydrate buildup in the leaves and to a large reduction in photosynthesis. How much photosynthesis would be lost during a standard season?

S5. Construct a figure similar to Figure 13 for a dry season rice crop. Use the modules of Listings 3 and 5 and Appendix B. Add integrals to accumulate the specific fluxes of carbon during the season.

2.5.2 *Remobilization*

T7. Make a relational diagram of the processes and variables which determine remobilization.

T8. Estimate the amount of reserves in rice, potato and soya bean crops at flowering. With how many days of photosynthesis to the amounts correspond?

T9. Estimate the leaf weight of a rice, a potato and a soya bean crop after emergence and before photosynthesis plays a role, assuming that the seed rate (dry matter) is 200, 330 and 400 kg ha^{-1} respectively.

S6. Determine the sensitivity of rice grain yield for the fraction stem reserves (0.0 to 0.5) in the L1D and L1Q modules. What is the reason for the difference in result between both programs? (Suggestion: print rates at intervals of 2.5 d for some daytime output.) What do you notice about the final stem weight?

2.5.3 *Maintenance respiration*

T10. Make a relational diagram of the processes and variables which determine maintenance respiration.

T11. Why is maintenance respiration in leaves more intensive than in other tissues? Is the response to temperature similar in all organs?

T12. Why is wasteful respiration too small to explain the large reduction in leaf photosynthesis at high sugar levels?

S7. Breeding might yield soya bean varieties with a basic maintenance respiration rate in leaves which is only half of that in Table 8. How much more leaves, stems and beans (pod plus beans) would this produce, supposing all other factors remain equal (wet season crop)? And how much will be produced if the metabolic component of maintenance respiration and the rates in stems, roots and storage organs are also reduced by 50%? (Suggestion: replace constants for stem, etc., by a parameter).

S8. Determine the rate of maintenance respiration of a rice crop and of its organs during the four time periods of day 95 with the module L1Q. Why are the fluctuations of rates of leaves and stems not parallel?

2.5.4 *Growth rate and growth respiration*

T13. Make a relational diagram of the processes and variables which determine the growth rate and growth respiration of the crop (not for organs). What other information flows could be indicated in Figure 26?

T14. Some of the glucose produced on a bright day is carried over to the next day. The L1D module ignores this. Does this cause errors in simulation?

T15. How should growth respiration be measured in crops and whole plants?

T16. The chemical composition of rubber may be approximated by $(C_5H_8)_n$. Estimate the amount of glucose required for growing one ton of rubber and the concurrent CO_2 production. How does rubber synthesis compare with other organic compounds?

S9. What is the effect of doubling or halving the rate coefficient of reserve use on soya bean yield in Los Baños in the wet season? What do you notice about the results?

S10. Suppose there is a soya bean cultivar identical to Hawkeye except that its storage organs have a biochemical composition the same as Faba beans (Table 11). How much more would it yield in Los Baños? Explain the difference.

S11. Suppose another soya bean cultivar is identical to Hawkeye except that its N-fixation is inefficient and requires 10 times more energy than normal. How much would it yield in Los Baños?

2.6 Answers to exercises

2.6.1 Photosynthesis

T1. Carefully distinguish the types of variables and the relations involved. Use the symbols of Figure 5 to draw the diagram.

T2. The photosynthesis light response curve calculated with the exponential function and PLMX = 30., PLEA = 0.5 is found in Figure 14. The response curve for the Blackman equation is the broken line. The hyperbolic response approaches PLMX more closely at higher light intensities than the exponential curve.

T3. At a specific leaf weight of 370 kg ha^{-1}, PLMX equals 47.0 (Listing 5). Canopy photosynthesis on a clear day is 350, 690 and 850 kg CO_2 ha^{-1} for leaf areas of 1, 3 and 10, respectively. With a 10% increase in leaf thickness, PLMX becomes 51.7, and photosynthesis is 380, 720 and 900 kg ha^{-1} respectively.

T4. For a C_3 crop: (340-238) = 112 vppm = 112 cm^3 CO_2 m^{-3} = 203 mg m^{-3}. This concentration difference across a resistance of 100 s m^{-1} yields a leaf photosynthesis rate of 73 kg CO_2 $ha^{-1}h^{-1}$. For the C_4 crop, the result is 133 kg CO_2 $ha^{-1}h^{-1}$. The diffusion process can easily keep up with the biochemical processes.

T5. Maximum daily total global radiation at 50° latitude exceeds the maximum at the equator by 11%. Moreover, this amount is received during a 34% longer daytime period. Together, this leads to a higher maximum daily total canopy photosynthesis. The radiation at the equator peaks twice a year (March 21, September 21) so that maximum canopy photosynthesis reaches a maximum also twice a year. See Figure 20.

T6. The increase in canopy photosynthesis per unit increase of radiation decreases continuously (Figure 17). Splitting an amount of radiation in unequal portions over a day leads to a lower daily total.

S1. Canopy photosynthesis is 275.58, 534.60 and 651.05 kg CO_2 ha^{-1} on day 244 for an ALV of 1, 3 and 10, respectively when SLA = SLC. When SLA increases 10%, photosynthesis increases to 285.70, 550.04 and 666.74 respectively. The relative increase is smaller than that in Figure 22 and exercise T2 because day 244 is not fully clear.

S2. The breeding effect can be evaluated by substituting TPAD with TPAD-5 in the equations for PLMX and PLEA. The normal crop yields 15136 kg ha^{-1} and the adapted crop yields 22532 kg ha^{-1}, both at Julian day 8, after more than a year of growth. The effect on PLEA is larger than that on PLMX.

The example underlines the fact that the potato data set is of limited value at these high temperatures, because, among other factors, crop development is faster and because leaf and stem ageing play a more prominent role.

S3. Potential rice yield for a spherical leaf distribution is 7851.2 kg ha^{-1}, for erect leaves 8186.3 and for horizontal leaves 6404.6. The second leaf layer contributed 23.6%, 27.2% and 10.6% to the total photosynthesis, respectively.

S4. Run L1Q with PARAM GSORM = (0.5,0.1). Restricted growth reduces photosynthesis by 5044 kg CO_2 ha^{-1}. Note the very high level of reserves in the crop and the poor yield.

S5. The relative amount of carbon involved in sectors 1-6 can be obtained by:

```
FCREM= LSTR * 1.111 * 0.947 / (PCGW * 0.682)
GAC    = GLV * FCLV + GST * FCST + GSO * FCSO +
   GRT * FCRT
*fraction C in remobilized carbohydrates and
*growth of actual C in the crop, respectively
CS1    = INTGRL(0.,GAC * (1. – FCREM))
CS2    = INTGRL(0.,GAC * FCREM)
CS3    = INTGRL(0.,12. / 44. * RGCR)
CS4    = INTGRL(0.,12. / 44. * RMCR)
CS5    = INTGRL(0.,GCS5)
GCS5   = (RMMA + RMLV * 0.33) * 12. / 44.+12. / 30. * ...
   (GLV * 0.15 + GST * 0.1 + GRT * 0.1 + ...
   GSO * 0.12) * 0.852
CS6    = INTGRL(0.,12. / 44. * (PCGC4 – (PCGW + GCS5))
PCGC4  = FUPHOT(70.,PLEI,ALV, RDTM,DAT,LAT)
CSUM   = CS1 + CS2 + CS3 + CS4 + CS5 + CS6 + 1.E-10
RS1    = CS1 / CSUM
RS2    = CS2 / CSUM
RS3    = CS3 / CSUM
RS4    = CS4 / CSUM
RS5    = CS5 / CSUM
RS6    = CS6 / CSUM
```

CS1-6 are cumulative values per sector. The pie chart of Figure 13 is constructed with RS-values at maturity. For rice, the 6 RS-values are 0.396, 0.027, 0.095, 0.160, 0.112, 0.210, respectively.

2.6.2 *Remobilization*

T7. Distinguish carefully the types of variables and relations involved. Use the symbols of Figure 5 to draw the diagram.

T8. Assuming a stem biomass of 4000 kg ha^{-1} at flowering, the amount of starch is 1000, 720 and 800 kg ha^{-1} for rice, potato and soya bean, respectively. Photosynthesis is then about 600 kg CO_2 $ha^{-1}d^{-1}$, so that the reserves correspond with 2-3 days of gross photosynthesis.

T9. With efficiencies as given in the text, crop growth starts with 100, 165 and 300 kg ha^{-1} in rice, potato and soya bean respectively. About 50% of this biomass is in leaves, 50% in roots.

S6. The grain yield simulated with the L1D module is considerably enhanced by increasing the fraction of stem reserves to 0.50 (from 7691.4 to 8898.6 kg ha^{-1}), but it decreases to 6429.1 when the stem contributes none. In the L1Q module, these yields are 7159.4, 7500.4 and 6521.3 kg ha^{-1}, respectively. The response is smaller for the high reserve fraction because high glucose levels during remobilization reduce photosynthesis. The final weight of structural biomass of the stem changes in accordance to the stem reserves.

2.6.3 *Maintenance respiration*

T10. Distinguish carefully the type of variables and relations involved. Use the symbols of Figure 5 to draw the diagram.

T11. Leaf maintenance respiration is more intensive than stem, root and storage organ maintenance processes because the fraction of non-storage protein is largest and because the large metabolic component of maintenance respiration is located in the leaves. The response to temperature of the basic processes is characterized by a Q_{10} of 2.0 The overall relative response to temperature in leaves is smaller than in other organs, because the metabolic component is not directly affected by temperature.

T12. Reduction of the normal ratio of 3 ATP to 1 oxygen (O) to only 1 ATP O^{-1} stimulates maintenance only three-fold, which is insufficient to lower leaf photosynthesis as much as is observed.

S7. The potential wet season yield of soya bean is 3692.4 kg ha^{-1} (cf., Figure 63). With leaf maintenance respiration reduced, the yield rises to 3952.8 kg ha^{-1}. All maintenance respiration rates reduced by 50% would increase yield to 4179.2 kg ha^{-1}.

S8. At day 95, the rates of leaf maintenance respiration are 19.0, 11.3, 14.6 and 25.2 kg CO_2 $ha^{-1}d^{-1}$ at 0, 6, 12 and 18 hours respectively. Stem maintenance is then 8.3, 10.0, 12.9 and 11.3. The values for the roots are 2.6, 3.1, 3.9 and 3.4. The fluctuations of the rates are not parallel because maintenance respiration in stems and roots is only affected by temperature, while the large metabolic component in leaves has its own dynamics. (Hint: use the END

CONTINUE statement at day 94 and reset the PRDEL to 0.25 to suppress excessive output).

2.6.4 Growth rate and growth respiration

T13. Distinguish carefully the type of variables and relations involved. Use the symbols of Figure 5 to draw the diagram. In Figure 26, the photosynthesis rate affects the rate of maintenance respiration (in L1D directly, in L1Q via PCGDV); temperature affects maintenance; photosynthesis in L1D and the reserve level in L1Q affect the growth rate and growth respiration rate.

T14. Too much biomass is formed on the bright day simulated with L1D. If this occurs in the vegetative period, then too much leaf area is simulated to develop and the second day starts with too much leaf area. However, the error is small unless in reality crop growth was severely limited. In these cases, use the L1Q module. The error is always negligible after the period when leaves are formed. Growth rates simulated with the L1D module fluctuate more than is expected to occur in the real crop.

T15. Most growth respiration evolves in the growing points. These are usually difficult to access or include properly in equipment. The rate also varies during the day, making a rate for any particular moment difficult to measure and to interpret. It is advisable to estimate the respiration rate of the entire plant or crop over 24 hours, and to subtract measured or estimated maintenance respiration (cf., McCree, 1974). Respiration of a growing fruit or tuber is easier to measure (Schapendonk & Brouwer, 1984).

T16. Applying the equations in Subsection 2.4.3 shows that 3.754 ton glucose is required and 2.289 ton of CO_2 is released. Rubber is even more expensive to produce than fat, assuming that the biochemical pathways involved are not exceptional from the point of view of energy efficiency.

S9. The Soya bean yield goes up from 3676.6 kg ha^{-1} to 3708.0 when the rate coefficient is 0.75 and up to 3775.8 when it is 3.0. The effect on yield is small, but in the biomass at flowering is it large. A reduced rate in the carbohydrate supply leads to a lower growth rate in the vegetative mass. Photosynthesis after flowering is not greatly affected, but a larger mass requires more carbohydrates for maintenance.

S10. The Soya-Faba-bean would yield 4274.1 kg ha^{-1}, 828.2 kg ha^{-1} more than the soya bean. The lower fat and protein content of Faba bean mean that more weight can be produced from the same amount of carbohydrates.

S11. The yield would be only 987.6 kg ha^{-1} in Los Baños due to poor growth. Most of the difference in biomass with respect to the good crop develops during the pod filling period.

3 Morphological development and assimilate partitioning

This chapter discusses simulation of the morphological development of crops. Complete models for crop growth at Production Level 1 can be constructed by combining these processes with the assimilation and dissimilation processes of Chapter 2. Listing 3 (basic module L1D for crop growth) and Listing 4 (crop growth module L1Q with quarter-day time periods) provide two complete modules for these processes.

There is less known about the mechanisms of morphological processes and their regulation in crops, in comparison with assimilation and dissimilation. There are also more differences between species and cultivars from a morphological, than from a physiological point of view.

3.1 Crop development

3.1.1 Introduction

Many changes occur as a crop grows. Some changes, such as those of weight and leaf area, are easy to quantify, while others, such as plant age and phenological development, are more difficult. Nevertheless, it is essential to quantify crop phenology because the important process of partitioning of new biomass depends directly on this expression of age.

The 'development stage' of a crop quantifies its physiological age and is related to its morphological appearance. Development stage is a state variable in crop growth models. The development stage cannot be expressed simply as chronological age, because several environmental factors, such as temperature and waterstress, can speed up or reduce the rate of phenological development. Daylength is crucial in some crops to induce flowering. Contrary to what is suggested by intuition, the rate of crop growth per se has no effect on the rate of phenological development, as long as the growth rate is not very low. The concept of development stage is used to characterize the whole crop; it is not appropriate for individual organs. The development stage has the value of 0.0 at emergence, 1.0 at anthesis and 2.0 at maturation. It is dimensionless and its value increases gradually. The development rate has the dimension d^{-1}. The multiple of rate and time period yields an increment in stage. Figure 30 shows development stages of a rice crop.

The rate of phenological development can be affected by temperature differently in the vegetative stage than in the reproductive stage. Daylength has an effect only in the vegetative stage. These differences indicate that the physiological process of development is not the same before and after anthesis.

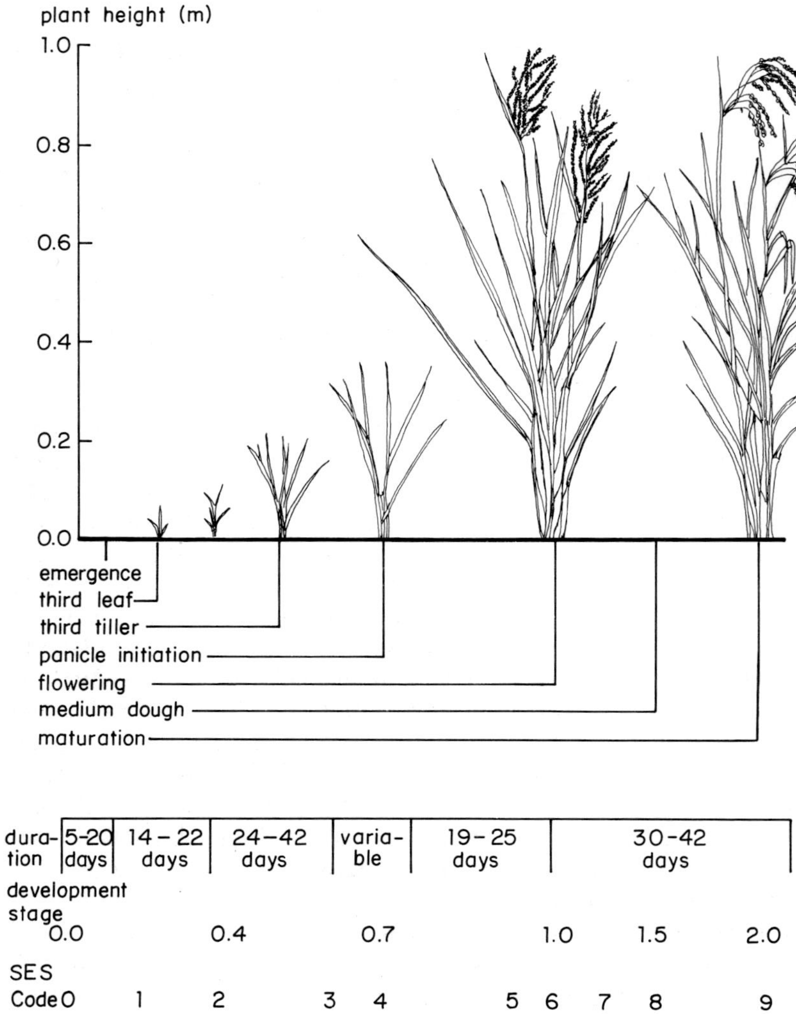

Figure 30. Development stages of a rice crop (Source: Stansel, 1975).

Physiological or biochemical methods of characterizing and measuring the development stage of the crop are yet unknown. Phenological development is still not understood enough to provide an explanatory model of this process, hence, descriptive modelling is used here.

Plants in a field do not flower simultaneously. The date of anthesis refers to the first date that 50% of the fertile tillers carry or have carried open flowers.

3.1.2 *Vegetative phase*

A first approximation of the development rate in the vegetative stage is the inverse value of the duration of the period between emergence and anthesis. This value is equal to the development rate constant, when temperature has been constant and daylength has had no effect. Table 12 gives typical values of this constant for different crops and for some important cultivars. The effects of temperature and daylength on the development rate of several species are summarized from the literature and presented in Tables 13 and 14. The extensive Handbook of Flowering (Halevy, 1985) provides a review of relevant literature and contains many data. Roberts & Summerfield (1987) analyzed the effects of temperature and daylength on the duration of flowering of several crops.

Temperature is often the dominant factor influencing plant development in temperate climates. The development stage can then be expressed as a temperature sum (degree.days, sum of average temperatures above a lower threshold, e.g. Vanderlip & Arkin (1977), Warrington & Kanemasu (1983), van Heemst (1986)). This assumes a proportionality between temperature and development rate, which is easy to grasp, but has limited validity. Here, the more flexible description of a non-linear relation of development rate with temperature is preferred (Roberts & Summerfield, 1987). Figure 31 provides an example. The relative effect of temperature can be expressed as a multiplication factor (Listing 3 Lines 97, 99, Listing 4 Lines 125, 127). It is difficult to extract data on the temperature-development rate relationship for different crops from the literature and only a few are presented in Table 13. Moreover, responses among cultivars can differ considerably. Modellers must therefore often make their own approximations. As long as daylength does not influence development, a reasonable approximation of the development rate constant and the effect of temperature on it can be obtained from field data plus the intuitive knowledge of crop experts.

The temperature that affects the phenological development process can be taken as equal to the daily average air temperature at the height of the shoot's growing point. Only when day or night temperatures regularly reach values where the response is non-linear, is another procedure of weighing temperatures required or should shorter time periods be taken. The temperature of the growing point can be higher or lower than the air temperature at two metres due to insolation, transpiration from the growing point and heat transfer from the soil. This topic is relatively unimportant and solutions may be species specific.

Daylength is important for photoperiod-sensitive crops that are common in the tropics. Subsection 6.2.2 gives a calculation of the photoperiodic daylength. Long days speed up the development rate in long-day plants, but reduce it in short-day plants. Some plants require a certain minimum or maximum night length before flowering is triggered, but in most cultivars phenolog-

Table 12. The rate of crop development in the vegetative stage at a reference temperature and reference daylength.

Species	Rate (d^{-1})	Tempe-rature (°C)	Day-length (h)	Reference
Barley	0.032	19	16	
Cassava	0.017	23	12.5	
Cotton cv BarLX1	0.016	30	12	
Cowpea cv TVu1188	0.022	23	12	
Faba bean	0.033	20	12	
cv Minica	0.030	25	–	Grashoff,CABO, pers.comm.
Groundnut				
cv Robut 33-1	0.033	25	–	Saxena et al., 1983
Maize, grain				
cv XL45	0.025	28	12	Warrington & Kanemasu, 1983
cv Pioneer	0.0265	25	–	Sibma,CABO, pers.comm.
Maize, silage,cv LG11	0.016	25	–	Sibma,CABO, pers.comm.
Millet, early	0.015	30	12	Jansen & Gosseye, 1986
late	0.010	30	12	
Potato cv Mara (late)	0.035	18	–	van Heemst, 1986
cv Favorita (early)	0.0555	18	–	van Heemst, 1986
Rice	0.014	25	10	
cv Nipponbare	0.020	28	10	Horie, 1987
Sorghum CSH6	0.020	30	13	Huda et al., 1984
Soya bean cv Hawkeye	0.025	23	?	
cv Jupiter	0.024	27	12.6	Patron, pers.comm
Sugar-beet	0.0069	20	16	(vernalization important)
Sugar-cane	0.025	25	12.5	
Sunflower cv Sobrid	0.013	22	13	
cv Relax	0.0195	18	16	unpublished results
Sweet potato	0.006	27	13	Hahn & Hozyo, 1983
Tulip cv Apeldoorn	0.010	15	–	Benschop, 1986
Wheat, winter				
cv Arminda	0.015	20	13	de Vos,CABO, pers.comm.
cv UQ189	0.0195	20	14	Angus et al., 1981
Wheat,spring				
cv Miriam	0.020	25	14	de Vos,CABO pers.comm.

Source: Halevy (1985), unless otherwise specified
– daylength not relevant
? unknown

Table 13. The ratio of the development rate in the vegetative phase at a certain temperature to that rate at reference temperature (Table 12) for different crops. Data in CSMP FUNCTION style: temperature first and the multiplication factor next.

Species	Temperature response
Barley	4.,0.01, 19.,1.0, 24.,1.0, 30.,0.01
Cassava	15.,1., 30.,1.
Cotton	12.,0.01, 30.,1.0, 40.,1.0
Cowpea cv TVu1188	19.,0.77, 23.,1., 27.,1.32 (use smallest values from Tables 13, 14 at any date).
Faba bean	-10.,0.01, 5.,0.01, 12.,0.5, 20.,1.0, 30.,1.2
cv Minica	-10.,0.01, 0.,0.01, 20.,0.9, 25.,1.0, 35.,1.2
Groundnut	6.,0.01, 20.,1., 32.,1.2, 40.,1.2
Maize cv XL45	8.,0.01, 14.,0.25, 19.,0.65, 28.,1.0, 35.,0.9
cv Pioneer,LG11	0.,0.3, 10.,0.3, 15.,0.75, 25.,1., 35.,1.2
Millet	10.,0.5, 20.,1.0, 30.,1.2, 40.,1.2 (Jansen & Gosseye, 1986)
Potato	7.,0.01, 18.,1.0, 29.,0.01 (van Heemst, 1986)
Rice	10.,0.1, 19.,0.8, 25.,1., 27.,1.1, 32.,1.2, 40.,1.0
cv Nipponbare	16.,0.25, 20.,0.6, 24.,0.85, 28.,1., 32.,1. (Horie, 1987)
Sorghum CSH6	7.,0.01, 30.,1.0 (Huda et al., 1984)
Soya bean	0.,0.01, 10.,0.5, 20.,0.9, 27.,1.0, 30.,1.1, 40.,1.2
Sugar-beet	0.,0.01, 5.,0.5, 20.,1., 25.,0.5
Sugar-cane	10.,0.01, 19.,0.01, 23.,1.0, 27.,1., 32.,0.01, 50.,0.01
Sunflower	
cv Sobrid	7.,0.01, 12.5,0.62, 22.,1.0, 35.,1.3
cv Relax	10.,0.3, 15.,0.75, 18.,1.0, 25.,1.1, 35.,1.2
Sweet potato	10.,0.1, 16.,0.5, 27.,1.0, 36.,0.8
Tulip	
cv Apeldoorn	−10.,0.01, 0.,0.2, 5.,0.4, 10.,0.7, 15.,1.0, 20.,1.4, 25.,2.0 (Benschop, 1986)
Wheat, winter	
cv's UQ189, Arminda	2.,0.0, 10.,0.68, 15.,0.89, 20.,1.0, 25.,1.03, 30.,1.05, 35.,1.06, (Angus et al, 1981)
Wheat, spring	
cv Miriam	−10.,0.01, 0.,0.01, 20.,0.9, 25.,1.0, 35.,1.2

Source: based on Halevy (1985), unless otherwise specified

Table 14. Effect of daylength on the development rate of sensitive cultivars, expressed by the acceleration relative to the reference daylength (Table 12). Data in CSMP FUNCTION style: daylength (h) first and the relative multiplication factor second. The type of sensitivity indicated in the second column: LDP = long day plant, SDP = short day plant, IDP = intermediate daylength plant, DNP = daylength neutral plant.

Species	Type	Sensitivity
Barley	LDP	0.,0.1, 11.,0.1, 16.,1., 20.,1.2, 24.,1.2
Cassava	LDP	strong effect but highly variable among cv's
Cotton	SDP,DNP	0.,1., 12.,1., 12.5,0.5, 14.,0.1, 24.,0.1
Cowpea	SDP,DNP	0.,1.4, 10.7,1.4, 11.7,1.2, 13.3,0.9,... 15.,0.6, 24.,0.6
cv TVu1188	(use smallest values from Tables 13, 14 at any date)	
Faba bean	DNP,SDP	0.,1., 12.,1., 13.,0.75, 17.,0.5, 24.,0.1
Groundnut	almost none	
Maize cv XL45	SDP,DNP	12.,1., 14.,0.95, 16.,0.9 (Warrington & Kanemasu, 1983)
Millet	SDP,DNP	0.,1.0, 12.,1.0, 12.5,0.9, 13.,0.6, 24.,0.6
Potato	almost none	
Rice	DNP,SDP	8.,1., 12.,1., 13.,0.5, 14.,0.33, 24.,0.33
cv BPI76		(Vergara & Chang, 1985; sensitive only in period DS 0.2-0.7, otherwise always 1.0)
cv Nipponbare		0.,1.0, 10.,1.0, 12.,0.9, 14.,0.7, 16.,0.01,... 24.,0.01 (Horie, 1987)
Sorghum cv CS.	SDP,DNP	0.,1.0, 13.6,1.0, 14.3,0.65, 24.,0.1 (Huda et al., 1984)
Soya bean cv Jupiter	SDP	12.6,1.0, 13.,0.85, 13.5,0.75, 13.6,0.5 (Patron, personal communication)
Sugar-beet	LDP	12.,1.0, 16.,1.0, 24.,10. (large differences among cv's)
Sugar-cane	IDP	0.,0.1, 11.5,0.1, 12.,0.5, 12.5,1., 13.5,0.5,... 13.5,0.5, 14.,0.1, 24.,0.1 (when days shorten and plenty water; otherwise 0.1)
Sunflower cv Sobrid	SDP,DNP	0.,1.0, 13.,1.0, 17.,0.9, 24.,0.9
Sweet potato	SDP	10.,1.25, 13.,1., 17.,0.1, 24.,0.1
Tulip	almost none	
Wheat cv UQ189	SDP,DNP	10.,0.29, 11.,0.55, 12.,0.75, 13.,0.89,... 14.,1.0, 15.,1.08, 16.,1.14, 17.,1.18 (Angus et al., 1981)

Source: based on Halevy (1985), unless otherwise indicated

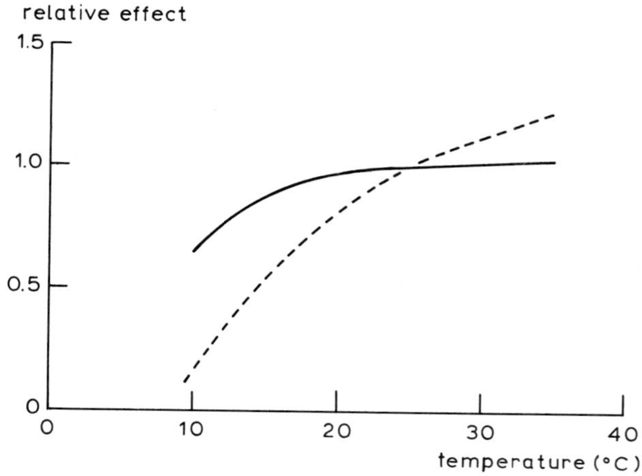

Figure 31. The response curve of the development rate to temperature for the wheat cultivar QU189. The full line represents the relationship before flowering, the dashed line the relationship after flowering. Values are relative to that at 25 °C. (derived from Angus et al., 1981).

ical development is a continuous process slowed by unfavourable daylengths. An example for a moderately sensitive wheat cultivar is shown in Figure 32. Daylength effect can also be translated into a multiplication factor of the development rate (Robertson, 1968). Ng & Loomis (1984) use this approach to simulate the photoperiodic induction of potato tubers. Roberts & Summerfield (1987) consider the effect of daylength on the development rate should be added to, rather than multiplied by, that of temperature. Given the inaccuracy of the data, it is difficult to judge which assumption is most appropriate.

Sensitivity to daylength can change during the vegetative phase. This is well documented for rice (Table 14; Vergara & Chang, 1985). When plants are only sensitive between certain development stages (such as 0.2 and 0.7 for rice) DRED in Listing 3 Line 97 or in Listing 4 Line 125 can be replaced by:

INSW((DS − 0.2) * (0.7 − DS), 1., DRED)

(note that the daylength effect on rice is exerted over a fraction of the vegetative period, so that its values should be derived correspondingly).

A list of daylength sensitivity in crop species is given in Table 14. In spite of much research, neither the mechanism of photoperiodicity, nor the reasonable data base needed for modelling the effects of daylength exists. The degree of sensitivity is characteristic for a variety, and can be modified or suppressed by breeding. Many modern crop varieties are bred to have little, or no sensitivity to daylength to allow easier manipulation in cropping systems and over larger

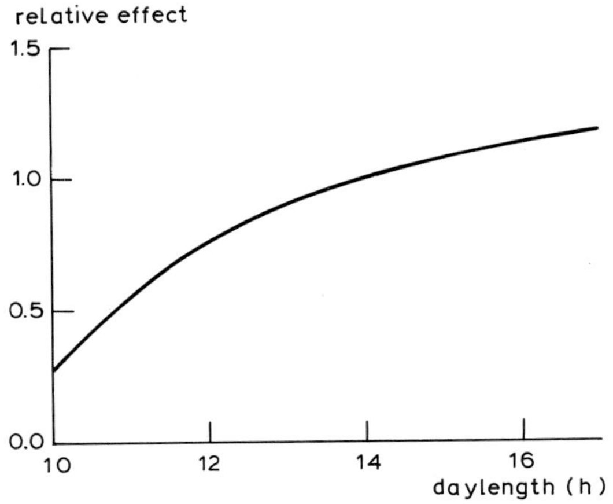

Figure 32. The response curve of crop development rate to daylength for wheat cultivar QU189. Values are relative to that at a daylength of 14 hours. (derived from Angus et al., 1981).

areas. However, sensitivity may be a desirable trait for crops in areas where the planting date can vary a lot, but the harvest date must be constant. Modellers must be aware of the different effects of daylength among individual crop varieties, so that new observations are required for new varieties.

Sugar-beet and sugar-cane are harvested before they reach the flowering stage. In these crops photoperiodicity can be strong. Although the development stage can still be expressed as a fraction of the time from emergence to anthesis, quantifying the development rate is more difficult. It may then be estimated that the development stage at harvest is close to 1.0. An alternative is to relate the development stage to the number of leaves formed (as effected by Loomis et al., (1979) for sugar-beet), or simply to the temperature sum.

3.1.3 Initialization

The development stage begins at 0.0 when simulation starts at seedling emergence. Often, however, simulation starts when young plants are already well established (Subsections 2.2.3 and 3.3.3), using observed or assumed quantities of leaves and roots as initial values. Values of 0.1-0.25 for the initial development stage of field crops are common, and as high as 0.5 for transplanted rice. The effect of transplanting in rice can be approximated by reducing its development stage by 0.2. For winter wheat in a temperate climate, simulation may start in spring at development stage 0.33. No generally valid initial values can be given, because they depend strongly on the experimental situation and on local management practices.

In tuber and bulb crops, seeding material has already undergone a certain physiological ageing before planting. A normal development stage for a potato crop at emergence is about 0.2.

An estimate of the initial development stage in a specific situation can be made with the development rate constant (Table 12), the response to temperature (Table 13) and the actual temperatures. It can also be obtained as a 'temperature sum' in cool climates (e.g., Gupta et al., 1984). Its calculation is basically straightforward, but requires good temperature data of the soil at the appropriate depth.

Vernalization is not considered here. It is implicitly assumed that crops requiring low temperatures before flowering is induced, such as winter cereals, have been sufficiently exposed to cold.

3.1.4 Reproductive phase

The reproductive period is defined here as the period after flowering until maturity. Simulating this development process proceeds in the same manner as that of the vegetative period. The development rate constant for the reproductive period and for the vegetative period are numerically different, as is the effect of temperature. Daylength has no effect. Some data are presented in Tables 15 and 16; note that such data are difficult to obtain and not very precise; they are also specific to cultivars within species, although less so than in the vegetative phase. If specific data are lacking, it can be assumed that there is no effect from temperature (i.e., duration of the reproductive phase is fixed) or that the effect from temperature is the same as in the vegetative phase.

This simulation of the development process contains no explanatory mechanism for the crop ripening. Simulation is halted by imposing an end when the development stage reaches the value of 2.0, by including the statement FINISH DS = 2.0, or FINISH DS = 0.95 for sugar crops and for sweet potato that are harvested before flowering. In the module TIL (Listing 2) for tillering and grain formation, simulation is ended when grains reach their maximum weight (cf., Subsection 3.2.5).

Crops that continue to produce branches plus leaves and flowers, while fruits or seeds are being filled are called indeterminate crops (Subsection 3.2.2). Many cultivars are fully or largely indeterminate particularly among leguminous crops and cotton. Reproductive and vegetative growth are parallel as long as conditions remain favourable. This situation can be described as a very slow progress of the development stage after flowering.

Table 15. The development of crops after flowering at a reference temperature for each crops.

Species	Rate (d^{-1})	Temperature (°C)	Reference
Barley cv Grit	0.021	16	
Cotton cv BarLX1	0.01	?	
Cowpea	0.03	?	
Faba bean cv Minica	0.0185	18	Grashoff, CABO, pers. comm
Groundnut cv Robut 33-1	0.012		Saxena et al., 1983
Maize			
cv Pioneer	0.017	25	Sibma, CABO, pers.comm.
cv LG11	0.021	25	Sibma, CABO, pers.comm.
Millet	0.024	25	Jansen & Gosseye, 1986
Potato			
cv Mara (late)	0.015	18	van Heemst, 1986
cv Favorita (early)	0.0225	18	van Heemst, 1986
Rice	0.038-0.046	28	
Sorghum cv CSH6	0.050	27	Huda et al., 1984
Soya bean			
cv Hawkeye	0.014	27	
cv Jupiter	0.027	28	Patron, pers.comm.
Sunflower cv Relax	0.028	18	
Tulip cv Apeldoorn	0.010	15	Benschop, 1986
Wheat, winter			
cv UQ189	0.0275	20	Angus et al., 1981
cv Arminda	0.0255	20	de Vos, CABO, pers.comm.
Spring cv Miriam	0.020	25	van Keulen, CABO, pers.com

– = not relevant
? = unknown

3.2 Assimilate partitioning

3.2.1 Introduction

Assimilate partitioning is the process by which assimilates available for growth are allocated to leaves, stems, roots and storage organs. Though only part of the total biomass is harvested, all components are important for allocation of new dry matter even before the economic products are formed. New biomass invested in leaves gives a quick high return from photosynthesized

Table 16. The ratio of the development rate after flowering at a certain temperature to that rate at reference temperature (Table 15) for different crops.

Species	Temperature response
Barley	16.,1.0
Chickpea	21.,1.0, 37.,2.0
Faba bean	0.,0.0, 18.,1.0, 25.,1.23, 35.,1.5
Groundnut	0.,0.0, 10.,0.5, 21.,1.0, 30.,1.2, 40.,1.2
Maize	as in Table 13 for Pioneer & LG11
Millet	25.,1., 35.,1.
Potato	7.,0.0, 18.,1. 29.,0.0
Rice	10.,0.45, 19.,0.75, 25.,0.9, 28.,1.0,... 30.,1.1, 40.,1.1
Sorghum	7.,0.0, 27.,1.0, 47.,0.0 (Huda et al., 1984)
Soya bean	27.,1.0, 35.,1.0
Sunflower	0.,0.3, 10.,0.3, 15.,0.75, 25.,1., 35.,1.2
Sweet potato	0.,0., 20.,0.9, 23.,1., 32.,1.2
Tulip	as in Table 13
Wheat, winter,	cv's UQ189 (Angus et al., 1981), Arminda 10.,0.14, 15.,0.66, 20.,1.0, 25.,1.23,... 30.,1.4, 35.,1.5
Wheat, spring,	cv Miriam as in Table 13

products, whereas biomass invested in roots and stems gives a slower, more indirect return (when there is no water or nutrient shortage).

Crop growth and development, discussed in Sections 2.4 and 3.1, should not be confused with distribution of new biomass over plant organs for these are different processes. The distribution pattern is a function of physiological age. The form and number of leaves, stems, roots and storage organs are not generally considered at this level of crop growth modelling. The 'storage organ' includes the economically valuable product and its hull or supporting tissue; the relative weights of these parts differ greatly between crops (Table 17).

There is little quantitative information on the internal control of carbohydrate distribution in crops, but there are interesting explanatory approaches to this problem (see e.g., Horie et al., 1979; Cock et al., 1979; Dayan et al., 1981; Vos et al., 1982; Ng & Loomis, 1984; van Keulen & Seligman, 1987). There are now some submodels but these are still too specific to implement them into the summary models here. The models here, therefore, rely on quantifications that are, at best, halfway between descriptive and explanatory.

A starting point here is the assumption that the growth rate is basically

Table 17. Major economic components of storage organs of different crops. Farm yields refer (in most cases) to products with a non-zero moisture content.

Crop	Storage organ	Major component of dry weight
Cassava	tuber	tuber 100%
Cotton	bolls	lint 35%, seed 65%
Cowpea	pods	seed 75-85%
Faba bean	pods	seed 75-85%
Field bean	pods	seed 75-85%
Groundnut	pods	seed 60-75%
Maize	cobs	grain 70%
Millet	ears	grain 60%
Pigeon pea	pods	seed 75-85%
Potato	tuber	tuber 100%
Rice	panicle	rough rice 100%, paddy 80%, polished rice 72%
Sorghum	panicle	seed 70-75%
Soya bean	pods	seed 60-80%
Sugar-beet	beet	beet 100%
Sugar-cane	millable cane	sugar 9-13%
Sunflower	heads	seed 44%
Sweet potato	tuber	tuber 100%
Wheat	ears	grain 85%

Source: Penning de Vries et al. (1983)

source-dependent: the more carbohydrates supplied, the faster is growth. Temporary storage of starch around flowering occurs in many crops and is considered in Subsection 3.2.4.

Two ways of simulating assimilate partitioning are presented. The first method (Subsection 3.2.2) is appropriate for determinate crops. With modification, this method of biomass distribution also approximates partitioning in indeterminate crops. The second method, specifically referring to cassava, uses a principle probably applicable to most crops, though to a smaller extent (Subsections 3.2.3 and 3.2.4).

Simulating the formation of sink size in rice is discussed in Subsection 3.2.5.

Loss of biomass during senescence and by root exudation are briefly considered in Subsections 3.2.6 and 3.2.7.

3.2.2 *Allocation of new biomass in source-limited crops.*

When simulating the allocation of new biomass in determinate crops a distribution key is used. Carbohydrates available for growth in any time period are partitioned over organs according to this key, independent of the available amount of carbohydrate. The key changes slowly with the development stage of the crop. Generally, the pattern is such that the largest share is initially attributed to leaves and roots, then to stems, and ultimately to the storage organ. Figure 33 shows a typical distribution pattern for rice. Because the sum of the fractions of the shoot is 1.0 by definition, a cumulative presentation is easier to read (Figure 34). Assimilate distribution is shown for soya bean as an example of a leguminous crop (Figure 35) and potato as an example of a tuber crop (Figure 36). The potato example shows why late-planted seed potatoes, in which the phenological development is already well advanced, immediately produce tubers and have small leaves and stems.

The idea of a fixed dry matter distribution pattern has been used by different authors (cf., Vanderlip & Arkin, 1977; van Keulen et al., 1982; van Heemst, 1986). In this model, glucose rather than biomass is partitioned to organs, as this is closer to reality. Note that a certain pattern of carbohydrate allocation does not translate into a predetermined distribution of dry matter at harvest time and into a fixed harvest index. Data on this distribution pattern are crucial

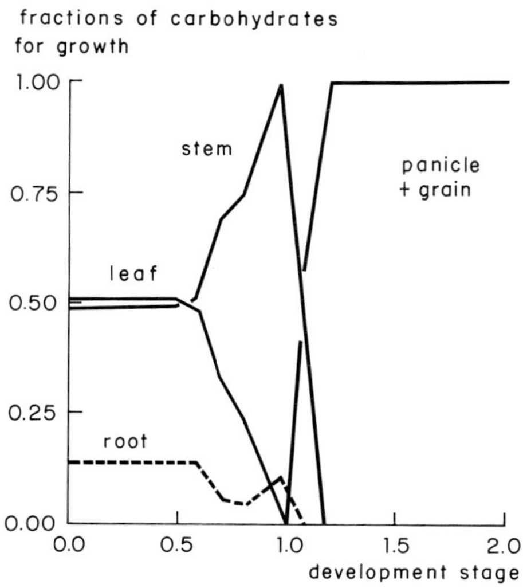

Figure 33. The pattern of carbohydrate partitioning to the organs in rice as a function of development stage. The fraction root is the fraction from the total crop growth; the sum of the fractions for growth of leaf, stem and panicle plus grain always equals 1.00.

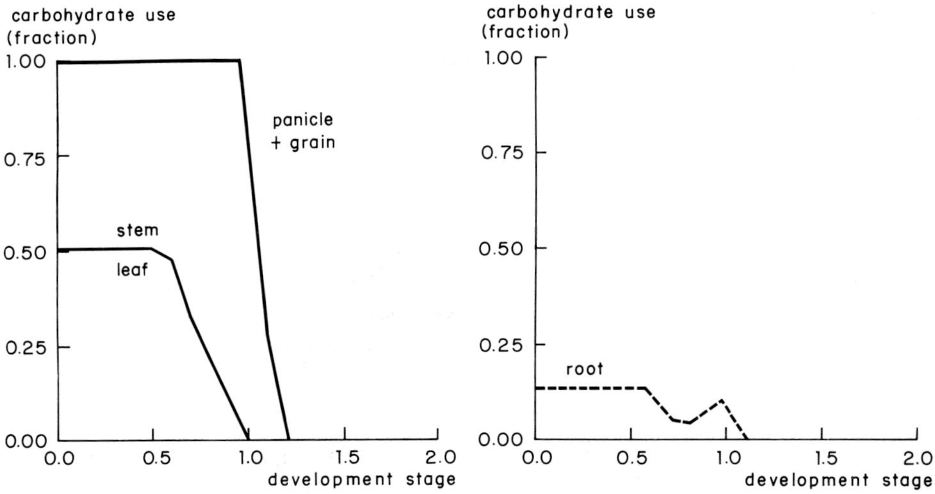

Figure 34. The pattern of carbohydrate partitioning to the organs in rice as a function of development stage. Data from Figure 33; shoot fractions are presented cumulatively.

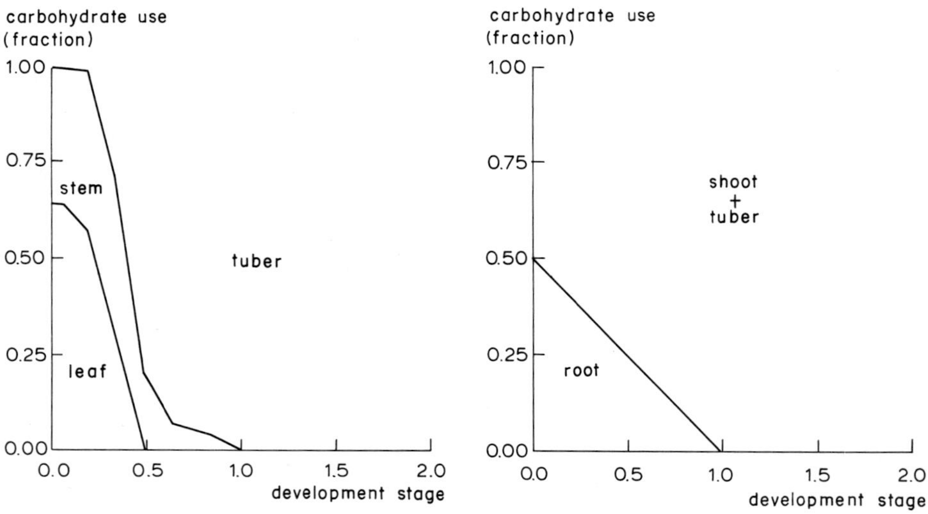

Figure 35. The cumulative pattern of carbohydrate partitioning in soya bean as a function of development stage.

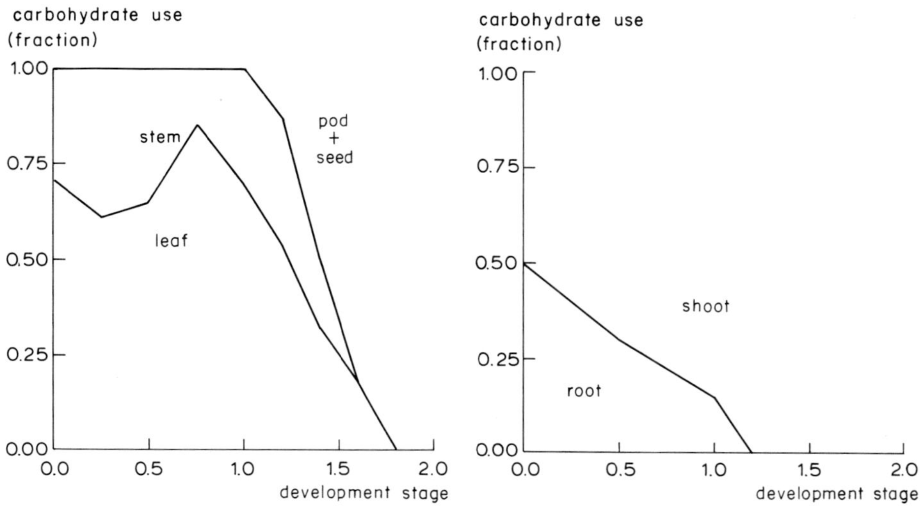

Figure 36. The cumulative pattern of carbohydrate partitioning in potato as a function of development stage.

inputs to simulation models and deserve proper attention. It is practical to first distinguish the distribution between shoot and root, and then to distinguish leaves, stems and storage organs as shoot subfractions. (The advantage of this two-step procedure becomes more evident when water stress is simulated: Subsection 4.3.3). 'Stems' are defined in a functional rather than a morphological manner and include stem proper, leaf sheaths and stem-like petioles. Below-ground storage organs, such as tubers and beets, are treated as part of the shoot. Programming of this method of carbohydrate partitioning is straightforward (Listing 3 Lines 40-44, Listing 4 Lines 59-63).

Simulating the allocation of assimilates applies equally well to crops with a fixed development pattern, such as cereal crops, many legume crops and also to bulb crops, such as tulips. A growth phase of only leaves, stems and roots is followed by a period in which these organs and the reproductive or storage organ grow together. In the final phase only the storage or reproductive organ increases in weight. The middle phase is short in cereals and can be long in legumes. The time at which the reproductive or storage organs begin to grow often coincides with flowering, but is directly related only in cereals. In other crops the relation is indirect; potato tubers are induced before flowering, legume pods start to fill weeks after the first flowers appear, and in sugar-beet the flowering stage should not be reached before harvest.

Data on carbohydrate allocation patterns, such as those of Figures 34-36, are obtained from field crops with a series of harvests and crop growth analyses. Data on dead and removed material should be added to the biomass data of the standing crop. Table 18 presents a number of such patterns, derived from the patterns of increase in biomass (i.e., the derivative to time of the weight of live and dead biomass), accounting for the efficiency with which glucose is converted into structural dry matter (Table 11). Note that this analysis applies to the growth process of organs; death or biomass removal are governed by other factors and must be treated separately (Subsection 3.2.6).

Significant differences in allocation patterns can exist between cultivars as Table 18 shows for maize, potato, rice and wheat and it is important to establish these patterns for the simulated cultivar whenever possible. Van Heemst (1986) found that the pattern of the fraction of assimilates going to leaves occurred about 0.1 unit of development stage later in a late potato cultivar, than in an early one and that the fraction for stems was 0.2 later. The overall pattern remained the same (but there is a large difference in their development rates, Subsection 3.1.2, Table 12). Van Heemst also showed that the lines can shift a little in response to management practices, but that chemical growth retardants had little or no effect on the allocation pattern.

The direct effects of environmental conditions on the allocation pattern of new biomass are probably small (van Heemst, 1986) and are not considered here. The effect of water stress is discussed in Subsection 4.3.3. A lowered shoot-root ratio is often reported for growth at unfavourably low temperatures, and the indirect effect (through the development rate) of actions or circumstances that increase or decrease soil and canopy temperatures on partitioning can be considerable. Direct effects of moderate nutrient shortage on carbohydrate allocation patterns are probably small.

The descriptive basis of the carbohydrate allocation patterns implies that this approach may be inappropriate in conditions in which a deviating morphological development of the crop can be expected, for example, after severe pruning.

In indeterminate crops, growth continues as long as environmental conditions remain favourable. Physiological aging slows down and vegetative and reproductive organs grow simultaneously. Existing parts age, but new branches or tillers are formed and rejuvenate the crop. The pattern of assimilate allocation then remains stable. This may be approximated by reducing the rate of phenological development to a low value (Subsection 3.1.4).

3.2.3 Allocation of new biomass in cassava

Another method of regulating dry matter allocation has been described and modelled for cassava by Cock et al. (1979). Cassava is a tropical crop that grows in areas with periodic droughts and takes 8-20 months to mature. Stems and leaves are produced all the time, and storage roots, the economic product, are formed once the plants are a few months old. Starch can make up 80% or more of the storage roots. The protein-rich leaves are also sometimes eaten.

Table 18. Distribution keys for glucose allocation to leaves (CALVT), to stems (CASTT) and to the entire shoot, including storage organs (CASST), as a function of the development stage in healthy crops of different species.
Data are in CSMP style: in each pair, development stage is the first number and the fraction of assimilates the second.

Barley
FUNCTION CALVT = 0.,0.82, 0.25,0.70, 0.51,0.55, 0.6,0.50,...
0.72,0.23, 0.83,0.01, 0.95,0.0, 2.1,0.0
FUNCTION CASTT = 0.,0.18, 0.25,0.30, 0.51,0.45, 0.6,0.50,...
0.72,0.77, 0.83,0.99, 0.95,1.0, 1.21,0.0, 2.1,0.0
FUNCTION CASST = 0.,0.35, 0.51,0.45, 0.72,0.85, 0.95,1.0,...
2.1,1.0

Cotton
FUNCTION CALVT = 0.,0.59, 0.32,0.63, 0.55,0.56, 0.77,0.45,...
1.,0.37, 1.12,0.09, 1.24,0.28, 1.47,0.02, 1.6,0.0, 2.1,0.0
FUNCTION CASTT = 0.,0.41, 0.32,0.37, 0.55,0.44, 0.77,0.55, 1.,0.61,...
1.12,0.48, 1.24,0.58, 1.47,0.31, 1.6,0.43, 2.1,0.36
FUNCTION CASST = 0.,0.5, 0.77,0.75, 1.,1.0, 2.1,1.0

Cowpea
FUNCTION CALVT = 0.,0.81, 0.16,0.80, 0.39,0.85, 0.61,0.83,...
0.86,0.64, 1.06,0.50, 1.45,0.30, 1.73,0.26, 2.0,0.07, 2.1,0.
FUNCTION CASTT = 0.,0.19, 0.16,0.20, 0.39,0.15, 0.61,0.17,...
0.86,0.36, 1.06,0.49, 1.45,0.57, 1.73,0.48, 2.0,0.22, 2.1,0.
FUNCTION CASST = 0.,0.50, 0.16,0.63, 0.39,0.76, 0.61,0.80,...
1.06,0.85, 2.1,0.85

Faba bean
FUNCTION CALVT = 0.,0.5, 0.54,0.60, 1.,0.25, 1.2,0.01, 2.1,0.
FUNCTION CASTT = 0.,0.5, 0.54,0.40, 1.,0.50, 1.2,0.25,...
1.42,0.14, 1.51,0.01, 1.71,0., 2.1,0.
FUNCTION CASST = 0.,0.5, 0.54,0.7, 1.,0.8, 1.2,1., 2.1,1.

Groundnut
FUNCTION CALVT = 0.,0.63, 0.47,0.56, 1.02,0.58, 1.25,0.52,...
1.48,0.30, 1.7,0.07, 2.1,0.0
FUNCTION CASTT = 0.,0.37, 0.47,0.44, 1.02,0.42, 1.25,0.4,...
1.48,0.22, 1.7,0.07, 2.1,0.07
FUNCTION CASST = 0.,0.4, 0.47,0.70, 1.25,0.95, 1.5,0.99, 2.1,1.0

Maize (silage maize, cv LG11)
FUNCTION CALVT = 0.,0.66, 0.47,0.69, 0.56,0.66, 0.65,0.58,...
0.82,0.29, 1.,0.0, 2.1,0.0
FUNCTION CASTT = 0.,0.34, 0.47,0.31, 0.56,0.34, 0.65,0.42,...
0.82,0.71, 1.,0.70, 1.16,0.0, 2.1,0.0

FUNCTION CASST = 0.,0.5, 0.51,0.75, 1.24,1.0, 2.1,1.0

Maize (grain maize, cv Pioneer)

FUNCTION CALVT = 0.,0.49, 0.35,0.59, 0.67,0.20, 1.0,0.12,...
1.18,0.09, 1.37,0.0, 2.1,0.0

FUNCTION CASTT = 0.,0.51, 0.35,0.41, 0.67,0.66, 1.0,0.64,...
1.18,0.31, 1.37,0.0, 2.1,0.0

FUNCTION CASST = 0.,0.5, 0.67,0.75, 1.37,1.0, 2.1,1.0

Millet

FUNCTION CALVT = 0.,0.7, 0.26,0.7, 0.40,0.67, 0.58,0.64,...
0.7,0.60, 0.84,0.45, 1.,0.28, 1.24,0.05, 1.5,0.0, 2.1,0.0

FUNCTION CASTT = 0.,0.3, 0.26,0.3, 0.40,0.33, 0.58,0.36,...
0.7,0.40, 0.84,0.50, 1.,0.61, 1.24,0.50, 1.5,0.0, 2.1,0.0

FUNCTION CASST = 0.,0.5, 0.25,0.5, 0.50,0.75, 1.,1.0, 2.1,1.0

Potato, cv Favorita (early)

FUNCTION CALVT = 0.,0.77, 0.37,0.67, 0.83,0.39, 1.14,0.10,...
1.35,0.0, 2.1,0.0

FUNCTION CASTT = 0.,0.16, 0.37,0.15, 0.83,0.16, 1.14,0.15,...
1.35,0.02, 1.57,0.0, 2.1,0.0

FUNCTION CASST = 0.,0.5, 1.,1.0, 2.1,1.0

Potato, cv Mara (late)

FUNCTION CALVT = 0.,0.80, 0.15,0.80, 0.46,0.65, 0.76,0.38,...
1.03,0.05, 1.17,0.0, 2.1,0.

FUNCTION CASTT = 0.,0.20, 0.15,0.20, 0.46,0.23, 0.76,0.22,...
1.03,0.22, 1.17,0.0, 2.1,0.

FUNCTION CASST = 0.,0.5, 1.,1.0, 2.1,1.0

Rice, IR36. See Listing 5, Chapter 7.

Rice, IR64

FUNCTION CALVT = 0.,0.577, 0.515,0.577, 0.625,0.520,...
0.71,0.409, 0.82,0.278, 0.995,0.0, 2.5,0.0

FUNCTION CASTT = 0.,0.423, 0.515,0.423, 0.625,0.480,...
0.71,0.591, 0.82,0.722, 0.995,1.0, 1.0,0.39, 1.25,0.09,...
1.4,0.0, 2.5,0.0

FUNCTION CASST = 0.,0.862, 0.515,0.862, 0.625,0.844,...
0.71,0.844, 0.82,0.940, 1.25,1.0, 2.5,1.0

Sorghum

FUNCTION CALVT = 0.,0.55, 0.4,0.55, 0.6,0.56, 0.74,0.56,...
0.86,0.06, 1.,0.07, 1.2,0.05, 1.4,0.01, 1.6,0., 2.1,0.

FUNCTION CASTT = 0.,0.45, 0.4,0.45, 0.6,0.44, 0.74,0.42,...
0.86,0.78, 1.,0.56, 1.2,0.24, 1.4,0.04, 1.8,0.0, 2.1,0.0

FUNCTION CASST = 0.,0.5, 0.4,0.7, 0.6,0.8, 1.3,1.0, 2.1,1.0

Soya bean

FUNCTION CALVT = 0.,0.71, 0.25,0.61, 0.5,0.65, 0.75,0.85,...
1.,0.70, 1.2,0.54, 1.4,0.32, 1.6,0.18, 1.8,0., 2.1,0.

FUNCTION CASTT = 0.,0.29, 0.25,0.39, 0.5,0.35, 0.75,0.15,...
1.,0.3, 1.2,0.33, 1.4,0.18, 1.6,0., 2.1,0.

FUNCTION CASST = 0.,0.5, 0.5,0.7, 1.,0.8, 1.2,1., 2.1,1.

Sugar-beet

FUNCTION CALVT = 0.0,0.7, 0.35,0.36, 0.52,0.08, 0.7,0.07,...
0.86,0.05, 1.1,0.0

FUNCTION CASTT = 0.0,0.3, 0.35,0.32, 0.52,0.23, 0.7,0.20,...
0.86,0.05, 1.1,0.0

FUNCTION CASST = 0.,0.6, 0.35,0.9, 0.52,1.0, 1.1,1.0

Sunflower cv Relax

FUNCTION CALVT = 0.,0.74, 0.42,0.68, 0.51,0.50, 0.57,0.40,...
0.66,0.35, 0.73,0.30, 0.82,0.23, 0.90,0.20, 1.,0.14,...
1.28,0.05, 1.54,0.0, 2.1,0.0

FUNCTION CASTT = 0.,0.26, 0.42,0.32, 0.51,0.50, 0.57,0.60,...
0.66,0.65, 0.73,0.67, 0.82,0.69, 0.90,0.66, 1.,0.48,...
1.28,0.10, 1.54,0.10, 2.1,0.0

FUNCTION CASST = 0.,0.5, 0.73,0.75, 1.,1.0, 2.1,1.0

Sweet potato

FUNCTION CALVT = 0.,0.66, 0.22,0.63, 0.49,0.59, 0.69,0.31,...
1.,0.07, 1.22,0.03, 1.35,0.01, 2.1,0.0

FUNCTION CASTT = 0.,0.34, 0.22,0.37, 0.49,0.40, 0.69,0.38,...
1.,0.24, 1.22,0.42, 1.35,0.11, 2.1,0.0

FUNCTION CASST = 0.,0.5, 1.,1.0, 2.1,1.0

Wheat, winter

FUNCTION CALVT = 0.,0.90, 0.33,0.85, 0.43,0.83, 0.53,0.75,...
0.62,0.56, 0.77,0.20, 0.95,0.09, 1.14,0.05, 1.38,0., 2.1,0.

FUNCTION CASTT = 0.,0.10, 0.33,0.15, 0.43,0.17, 0.53,0.25,...
0.62,0.44, 0.77,0.80, 0.95,0.64, 1.14,0.62, 1.38,0., 2.1,0.

FUNCTION CASST = 0.,0.5, 0.33,0.5, 0.53,0.75, 1.,1., 2.1,1.

Wheat, spring

FUNCTION CALVT = 0.,0.90, 0.19,0.83, 0.26,0.85, 0.45,0.82,...
0.6,0.32, 0.86,0.15, 1.,0.26, 1.26,0.0, 2.1,0.0

FUNCTION CASTT = 0.,0.10, 0.19,0.17, 0.26,0.15, 0.45,0.18,...
0.6,0.68, 0.86,0.85, 1.,0.34, 1.26,0.27, 1.5,0.0, 2.1,0.

FUNCTION CASST = 0.,0.4, 0.32,0.5, 0.6,0.75, 1.,1.0, 2.1,1.0

Cock et al. suggest that cassava leaves and stems usually grow at a rate below that permitted by the carbohydrate supply. The rate of leaf and stem growth is not source-limited, but sink-limited. The growth rate of leaves plus branches is proportional to leaf area (Figure 37). 'Excess' carbohydrates are stored as starch in storage roots as a reserve for times of energy shortage ('shielded reserves'). The accessibility of these reserves is much lower than that of glucose in the leaves which are called 'available reserves' (Subsection 2.4.2). (The relations of the processes in which reserves are involved are illustrated in Figure 29 part b Subsection 2.4.2). Cock's formulation implies that the growth rates of leaves and stems are constant during much of the growing season, while that of shielded reserves fluctuates considerably with weather and crop conditions. These shielded reserves provide a buffer which the crop uses to regrow leaves after severe insect attacks.

To simulate this type of assimilate distribution, the rate of leaf and stem growth must be computed independently of the carbohydrate supply. Cock's model computes the rate of leaf production and growth per leaf, relates stem growth to this, and derives the amount of carbohydrate involved. The excess

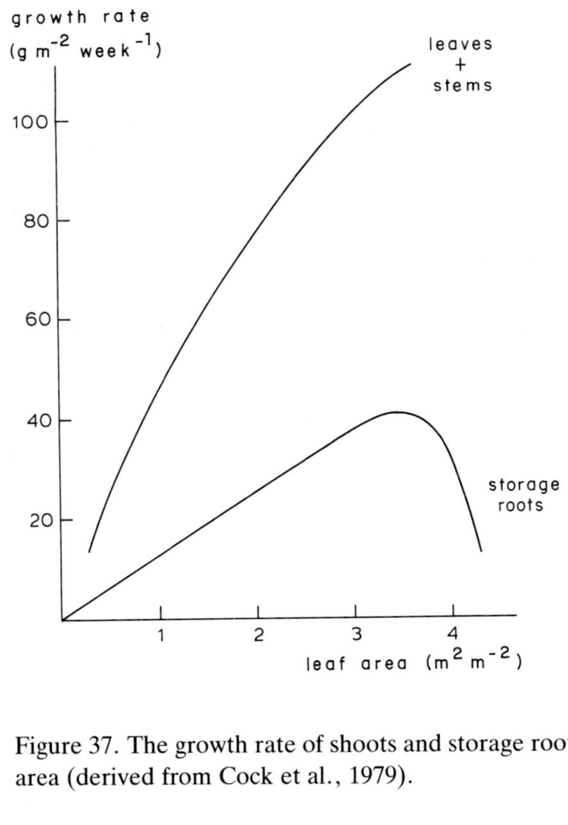

Figure 37. The growth rate of shoots and storage roots in cassava as a function of leaf area (derived from Cock et al., 1979).

photosynthate goes to the storage roots. The model realistically simulates dry matter distribution.

3.2.4 *Formation of shielded reserves*

The storage behaviour of cassava is not unique, for storing starch or sucrose occurs to a certain extent in many species. However, it is usually noticed less because it is a temporary phenomenon that is over at harvest time (except for sugar-cane and sugar-beet). Carbohydrates are commonly stored in stems or roots for some weeks, particularly when young storage organs are insufficiently developed to handle the total flow of assimilates. A sizeable amount of shielded and available reserves can then be built up in vegetative tissues. Reserve buildup may be to such an extent that it may temporarily or even permanently hamper canopy photosynthesis (Subsection 2.1.4). Van Heemst (1986) reported the growth rate of a late potato cultivar which appeared to be somewhat lower than that of an early cultivar, possibly due to the lack of a sufficiently large sink size when tubers were initiated. A permanent reduction of photosynthesis of this nature is not included in the model; it could be approximated by making PLMX a state variable, the value of PLMX decreasing when the reserve level is too high.

A simple way to deal with the formation of shielded reserves is to assume that a certain fraction of the increase in stem weight will be available for redistribution after flowering (Listing 3 Lines 17, 35). This fraction is assumed to consist only of starch. Some data on the magnitude of the remobilizable fraction are given Table 7 (see Subsection 2.2.2).

Formation of shielded reserves can be simply simulated by adding those carbohydrates that growing organs cannot absorb to the fraction of the stem weight that consists of starch. This may continue until the maximum level (estimate: Table 17 + 10%) is reached. This assumption is programmed in Listing 4 Lines 32-33, 37-38; where the complex calculation of the growth of stem weight accounts for the different amounts of glucose required to produce one kilogram of starch or stem. The sink size of leaves, stems and roots in a vegetative crop is usually large enough to accept all carbohydrates provided. But young storage organs may not have the capacity to grow; though they may be large in number, they have too small a sink size.

3.2.5 *Modelling morphological development in rice*

Carbohydrate production in cereals can be limited by the capacity of the grains to use them (Cock & Yoshida, 1973; Evans & Wardlaw, 1976). In the case of rice this is easy to understand, as the size of grains is physically restricted by the size of the hull. The maximum size of the grain is a variety-specific characteristic. To simulate this effect, the 'sink size' of the grains must be quantified (i.e., their capacity to absorb available carbohydrates). It is then essential to keep track of the number of grains. Grain setting is the end result of a series of events, so that the processes of tillering and floret formation must be considered. Tillers dying is only approximated as a dynamic simulation of age groups with different light interceptions is not justified. A module for morphological development in rice is presented to simulate the phenomenon that is often referred to as the sink-source relationship (Listing 2, module TIL to simulate development of tillers, florets and grains in rice). This approach, developed by van Keulen & Seligman (1987) for wheat also facilitates modelling damage by pests and diseases.

In the module TIL, the formation rate of plant parts, such as tillers, florets and grains, is assumed to depend on the net carbohydrate supply to the crop. The larger this supply, the higher the organ formation rates, so long as their numbers do not exceed certain limits. Formation rates of organs, in numbers per hectare per day, are equal to the difference between potential number and current number, divided by an appropriate time constant (Lines 7, 16, 23). The time coefficient is 15 days for tiller formation, 7 days for floret formation, and 3 days for grain initiation. The time coefficient for tillers dying is set at 14 days (Line 8), assuming that tillers only die slowly and when carbohydrates are lacking. Each type of organ forms during a restricted developmental period. The potential numbers of these plant parts at any moment, equals the carbohydrate supply of that day, divided by the daily requirement of carbohydrates for forming and maintaining one tiller, one floret and one grain. The carbohydrate required for florets and grains is a constant. The older the plant and the larger most of the early tillers, the more carbohydrates are required to initiate new tillers. This effect is mimicked by making the carbohydrate required for initiating new tillers a function of the development stage of the crop (Line 12). The number of tillers that will be formed in a rice crop in a specific simulation depends on environmental conditions. The maximum number of grains equals the number of florets (Line 25).

The sink size of the storage organ equals the number of grains multiplied by the maximum growth rate of one grain (Line 32). The maximum growth rate per individual grain is estimated to be the weight of mature kernels of that variety, divided by half the grain filling period. Hence, if sufficient grains have been formed sink size will rarely limit growth.

The module itself is straightforward. Acronyms are explained in Listing 12, Chapter 7. The data for this module are derived from field experiments with the rice varieties IR36 and IR64; data may be different in varieties with other tillering characteristics or panicle structures. Simulation stops when the grain weight reaches its maximum value (Line 42). The TIL module (Listing 2) interacts with the main module L1Q (Listing 4) by providing the maximum growth rate for storage organs, while using as inputs the carbohydrate supply for growth and the morphological development rate. The initial weight of leaves and planting density are related. When adding the module TIL to the crop growth model L1Q, Line 36 (Listing 4) has to be deleted. It is not appropriate to combine this module with the simpler model of Listing 3 (see Table 1 in the Reader's guide).

3.2.6 *Senescence and death*

Senescence refers to the loss of capacity to carry out essential physiological processes and to the loss of biomass. It is particularly important in the case of leaves, for even at Production Level 1 senescence occurs towards maturation. The fundamental processes involve physiological ageing and protein (enzyme) breakdown. These processes are difficult to quantify. Hormones are important as messengers, but it is not known how (de Wit & Penning de Vries, 1983). In addition, nutrient remobilization, in particular nitrogen, often plays a crucial role.

If senescence is not very important for the crop or for the research objectives, the simple descriptive approach in which the relative rate of loss is a function of the development stage can be used (Listing 3 Lines 33-34). Listing 5 contains a numerical example that corresponds with observations on a rice crop in the Philippines (functions LLVT and LRTT). Description such as this usually results in a loss of 40-60% of leaf area at harvest time. Loss of absorbing roots is handled in the same descriptive manner. These numbers can be used as default values, but should be calibrated to mimick specific situations. High temperatures accelerate senescence. This environmental effect can be easily added to the program provided that data are available to quantify it. The contribution of senescing leaves and roots to the pool of reserves is disregarded here.

A more mechanistic approach to senescence is by setting the death rate of leaves and roots to a certain fraction per day once the conditions for growth deteriorate; except for their reserves, stems do not lose weight. Deteriorating conditions are defined as a drop in the growth rate of the storage organ below a previously attained level (Listing 4 Line 43) (see also Subsection 2.4.2). Other definitions of deteriorating conditions, or of values for relative death rates are also possible.

Leaves that drop from the plant have generally lost some weight, even without diseases. This indicates that some biomass was used before the leaves died either for respiration or remobilization (proteins). Both processes increase the amount of carbohydrates available for growth. It is assumed that the leaves that dropped contributed half their original weight to the carbohydrates pool (Listing 4 Lines 24, 30).

When the biomass from live leaves becomes smaller, the specific leaf weight, the maximum leaf photosynthesis and canopy photosynthesis consequently decrease. This favours remobilization, so that senescence is characterized by positive feedback. The crop dies when photosynthesis becomes lower than leaf maintenance on several consecutive days (Subsection 2.3.4). Sinclair & de Wit (1976) argued that the process of nitrogen and carbohydrate redistribution from leaves is the major cause of senescence in leguminous crops.

Cassava leaves have a fixed life span of about 80 days. The so-called boxcar train method (Rabbinge et al., 1989) can be used to mimick this type of leaf ageing.

3.2.7 *Absorbing roots and excretion*

The absorbing roots of annual crops grow almost exclusively in the vegetative stage and stop when the storage organ starts to gain weight. (Storage roots are treated as part of the shoot, Subsection 3.2.2.). At initialization, root weight is often taken to be equal to shoot weight. The root mass has a dry weight of 500-2000 kg ha^{-1} around flowering. For lack of good field data, partitioning of assimilates between root and shoot is carried out so that the above mentioned pattern and values are obtained. Though this may not be quite accurate, root mass is usually only one tenth or less of the final above-ground biomass at harvest, so errors do not cause many problems. Proper data to simulate the growth of root systems are difficult to obtain. The effect of water stress on root growth is discussed in Subsection 4.3.3.

Excretion of organic substances into the soil by the roots occurs through exudating and sloughing off of root tips. Estimates for excretion run from 0-20% of the assimilated carbon (Woldendorp, 1978). This process is not included in the programs here, but could be incorporated as a fixed fraction of root growth or of gross photosynthesis (just as for maintenance related to metabolic activity, Subsection 2.3.2.). The crop carbon balance check should then be adjusted accordingly (Subsection 3.4.4).

3.3 Leaf area

3.3.1 *Introduction*

The leaf area of a crop is usually expressed as the total surface of live leaves, one-sided, per unit of soil surface. It is given in ha ha^{-1} or m^2 m^{-2} and often called leaf area index. The leaf area of an established crop has a value of 3-6 ha ha^{-1}, or even up to 10 ha ha^{-1} in very dense canopies. More than 80% of light is intercepted when the leaf area reaches 3 ha ha^{-1}, and 5 ha ha^{-1} in a canopy with erect leaves, which is called a closed canopy situation (see Figure 21 Subsection 2.1.3).

The amount of leaves and the rate at which leaves are formed at the start of the growing season are of considerable importance for the final yield. This makes partitioning of new biomass between leaves and other organs important. Growth of leaf weight is computed in earlier Subsections (2.4.2, 3.2.2). It must now be determined with how much leaf area it corresponds. 'Specific leaf weight' is used for this purpose. This is defined as the dry weight of leaves (no reserves, only structural dry matter) with a total one-sided leaf area of one hectare. Petioles and leaf sheaths around the stems are not included. The specific leaf weight for individual leaves ranges from 200 to 800 kg ha^{-1}, though the average for entire canopies is rarely more than 600 kg ha^{-1}.

The growth of leaf area is related here to growth in leaf weight. The specific leaf weight of new leaves may change with crop age. Growth of leaf area can also be simulated independently of leaf weight (e.g. Johnson & Thornley, 1983). Both approaches yield a pattern of leaf area development that is approximately realistic provided that parameter values are properly chosen. A general and explanatory simulation of the development of the leaf area of different crops does not yet exist; crop-specific models have been constructed by Cock et al. (1979), Horie et al., (1980), and Ng & Loomis (1984). A descriptive rather than an explanatory approach cannot be avoided.

All leaf area is usually treated as being equally effective in photosynthesis and transpiration if exposed to the same conditions. When using the SUPHOL subroutine, however, leaf layers can have different characteristics (Subsection 2.1.4).

3.3.2 Thin and thick leaves

When leaves are formed in a young crop, a certain area of relatively thin leaves seem to be more effective for quick growth than only half the area with leaves twice as thick. However, if the maximum rate of leaf photosynthesis increases proportionally with leaf thickness (as discussed in Subsection 2.1.2), this compensates for the lack of a large leaf area. Consider a leaf biomass of 2000 kg ha^{-1} with a leaf surface of 3 ha of thick leaves, or 5 ha of normal leaves, or 8 ha of thin leaves. Each thickness corresponds with a maximum rate of leaf photosynthesis. The canopy photosynthesis rates that correspond with these combinations lie on a hyperbola when plotted as a function of leaf photosynthesis and leaf area. Computation of canopy photosynthesis on fully clear days for other amounts of biomass in which leaf thickness is varied, shows similar curves. Compensation of area by thickness appears to be almost complete, because the canopy photosynthesis isolines follow almost rectangular hyperbolas (Figure 38). In full sunlight there is no difference in whether the crop invests biomass in thin or thick leaves; only very thick leaves are disadvantageous. However, on overcast days greater leaf area is always more effective than thicker leaves.

Factors other than radiation level also affect the importance of thin versus thick leaves, such as shading competitors (making thin leaves more advantageous) and water use efficiency (making a minimal area advantageous). This calls for a general pattern of thin leaves early in the life of the crop and thick leaves later. Crops in humid climates often show this phenomenon distinctly, but crops in semi-arid regions less so, or not at all. Observations indicate that there is a tendency for leaves formed early to be thinner than later ones, but the extent varies a great deal between crops. For example, Sibma (1987) re-

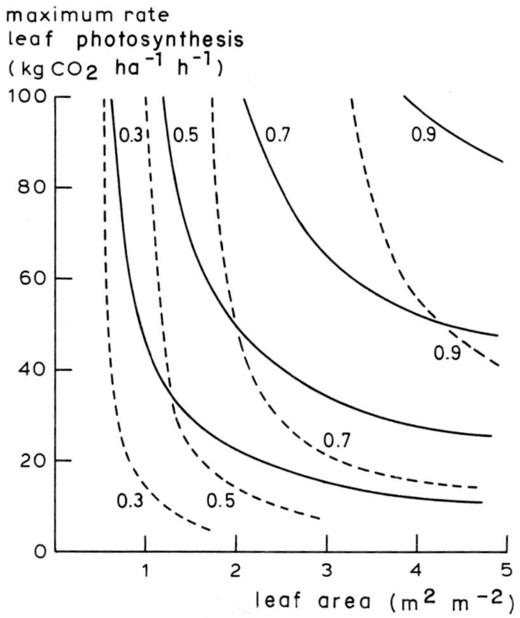

Figure 38. Combinations of leaf area and maximum rate of leaf photosynthesis that lead to a certain fraction of the maximum daily total of canopy photosynthesis (indicated next to the lines). Isolines are shown for fully clear skies (full lines) and overcast skies (dashed lines). The amount of photosynthesis is expressed relative to the maximum of 1244 kg CO_2 ha^{-1} d^{-1} in clear, and to 349 in overcast situations. The values are computed for a canopy with ALV = 10., PLMX = 100., PLEA = 0.4 and at LAT = 50. and DATE = 166.

ported that the specific leaf weight of the first leaves of maize was 200 kg ha^{-1} or less and of the last leaves almost four times as much (Figure 39). On the other hand, cassava in a dry climate produces leaves with a fairly constant specific leaf weight.

Leaves of plants grown indoors usually have thinner leaves than field plants, because they are grown at a much lower light intensity.

3.3.3 Modelling growth of leaf area

Several approximations can be used to simulate growth of leaf area. The simplest is to assume that the specific leaf weight is a crop characteristic and that it is constant in time and throughout the canopy. This value is called the specific leaf weight constant. Leaf area is determined by dividing the weight of live leaves by the specific leaf weight (e.g., van Keulen et al., 1982). In many

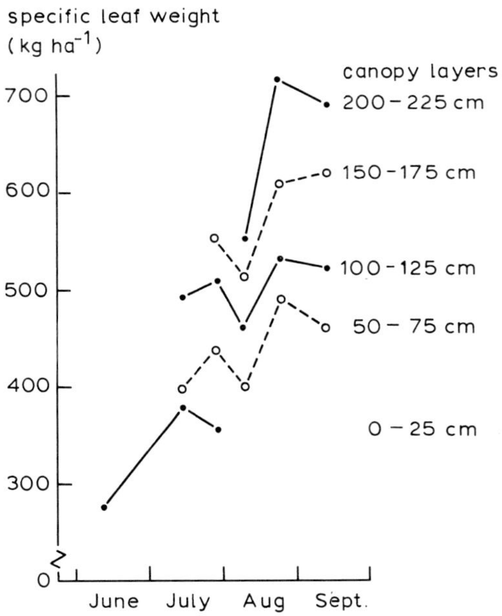

Figure 39. The specific leaf weight at different levels in a maize canopy during a growing season in Wageningen (Source: Sibma, 1987).

cases, this is a fair approximation provided that its value is measured at a proper time, such as at the end of the phase when most assimilates go to the leaves (e.g., development stage 0.5 for rice, Figure 34). Table 19 presents values of specific leaf weight constants for different crops at about this stage.

A more realistic approximation of development of leaf area takes into account that new leaves formed early in the life of the plant are thinner than leaves formed later. The specific leaf weight of new leaves is then found by multiplying the specific leaf weight constant with a factor that depends on the development stage of the crop (Listing 3 Line 91, Listing 4 Line 119). Figure 40 gives an example of the relation between specific leaf weight and development stage. Insufficient data were found in the literature to derive more than a few of these crop specific relations (Table 20). Descriptive functions such as these should be used carefully and checked whenever possible. A low specific leaf weight at the beginning of leaf growth speeds up growth of leaf area. Hence this method of simulating leaf area development enhances the simulated date of canopy closure by several days (or even as much as two weeks), in comparison to using a constant value of the specific leaf weight (Figure 41). However, the effect on leaf biomass and on final yield tends to be small.

The average specific leaf area of the crop is used to derive the standard

Table 19. The specific leaf weight constants for different crops are average values for the entire canopy. Leaf refers to leaf blades (one surface only) excluding petioles and leaf sheaths. The specific stem weight is an average for the growing season and includes stem proper, petioles, branches and leaf sheaths.

Species	Specific leaf weight ($kg\ ha^{-1}$)	Specific stem weight ($kg\ ha^{-1}$)	Reference
Barley	325	625	
Cassava	450		
Cotton	490	2000	Hearn, 1969
Cowpea	450	2000	
Faba bean	315	2114	Grashoff et al., 1987
Groundnut	600	2000	Bhagsari & Brown, 1976
Maize	450	2775	Sibma, 1987
Millet	440	1200	Jansen & Gosseye, 1986
Potato	300	2000	
Rice (IR36)	440	1000	
Sorghum	400	2500	Sivakumar et al., 1979; McCree, 1983
Soya bean	400	2100	Dornhoff & Shibbles, 1970 Hanway & Weber, 1971
Sugar-beet	500	1900	
Sugar-cane	700		
Sunflower	540	7825	Rawson & Constable, 1980
Tulip	710	2900	Benschop, 1986
Wheat, winter	425	1080	
spring	500	1080	

Values without reference were obtained from colleagues at CABO, Wageningen.

maximum rate of leaf photosynthesis that serves as an input in computing canopy photosynthesis. The fact that thicker leaves are usually at the top of the canopy and thinner leaves at the bottom has only small implications for canopy photosynthesis and is disregarded here.

Different growing conditions, such as those caused by different plant densities or fertilization level in maize (Sibma, 1987), and by irrigation in potato (Ng & Loomis, 1984), have little effect on specific leaf weight. The influence of these environmental factors may also be small in other crops; they are disregarded here. The change in the specific weight of new leaves with development stage may be a reflection of the change in carbohydrate supply to

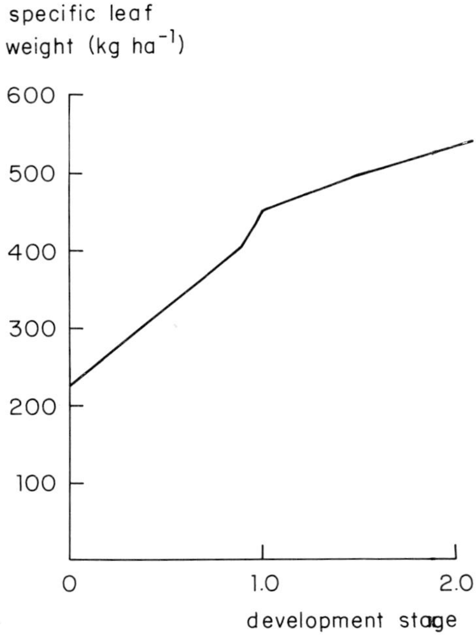

Figure 40. The specific weight of new leaf area as a function of crop development stage in maize (Source: Sibma, 1987).

growing tissue (cf de Wit et al., 1970).

The rate of leaf area loss is computed in direct relation to the rate of leaf weight loss, assuming that the average value of the specific leaf weight applies (Listing 3 Line 89, Listing 4 Line 117). Van Keulen & Seligman (1987) calculated the rate of leaf area loss in wheat independently of leaf weight loss. They put it at 5% d^{-1} once the leaf area exceeds the value of 6 m^2 m^{-2} to account for mutual shading. The constant life span of cassava leaves (Subsection 3.2.6) is also reported to be due to young leaves that are produced at a constant rate shading old leaves (Cock et al., 1979).

Simulation of leaf area for a specific case sometimes appears to be unrealistic and disturbs progress in a study. It is then advisable to cut the positive feed back loop (leaf weight – leaf area – photosynthesis – growth – leaf weight) and introduce the observed (or a chosen) leaf area development as a forcing function.

Table 20. The specific leaf weight relative to the values of Table 19, as a function of development stage. Of each pair of values, the first is the development stage, the second the fraction.

```
Barley
  FUNCTION SLT = 0.,1., 0.51,1.05, 0.60,0.71, 0.72,1.41, 2.1,1.
Cotton
  FUNCTION SLT = 0.,0.85, 0.32,0.85, 0.55,1.04, 0.77,0.95,...
                 1.,1.27, 1.12,1.25, 1.24,1.22, 2.1,1.22
Faba bean
  FUNCTION SLT = 0.,1.1, 0.54,1., 2.1,1.
Groundnut
  FUNCTION SLT = 0.,1., 0.47,1.05, 1.02,0.85, 1.25,1.29,...
                 1.48,0.95, 1.70,0.93, 2.1,0.93
Maize (cv.LG11)
  FUNCTION SLT = 0.,0.6, 0.5,1., 2.1,1.2
Maize (cv.Pioneer)
  FUNCTION SLT = 0.,0.5, 0.5,0.9, 1.,1., 1.5,1.1, 2.1,1.2
Millet
  FUNCTION SLT = 0.,0.64, 1.,1., 2.1,1.
Potato
  FUNCTION SLT = 0.,1.2, 0.07,1.2, 0.9,0.83, 1.0,1.0, 2.1,1.0
Rice
  FUNCTION SLT = 0.,0.5, 0.5,1., 1.,1.2, 2.1,1.2
Sorghum
  FUNCTION SLT = 0.,0.6, 0.4,0.8, 0.6,1.3, 1.,1.1, 2.1,1.
Soya bean
  FUNCTION SLT = 0.,0.8, 1.,0.8, 1.2,1., 1.5,1.1, 1.8,1.25,...
                 2.1,1.25
Sugar-beet
  FUNCTION SLT = 0.,1.0, 0.35,1.1, 0.52,1.2, 0.7,1.2,...
                 0.86,1.3, 1.1,1.3
Sunflower
  FUNCTION SLT = 0.,1., 0.42,1., 0.51,1.48, 0.57,0.79, 0.66,0.73,...
                 0.73,0.9, 0.82,1.39, 0.9,1.32, 1.,1.53, 1.16,1., 2.1,1.
Wheat, winter
  FUNCTION SLT = 0.,1., 0.33,1.1, 0.36,1.06, 0.43,1.5,...
                 0.53,1.05, 0.62,1., 0.77,0.85, 0.95,1.07, 1.14,1., 2.1,1.
Wheat, spring
  FUNCTION SLT = 0.,0.67, 0.55,0.67, 0.6,1., 2.1,1.
```

Source: see Table 19

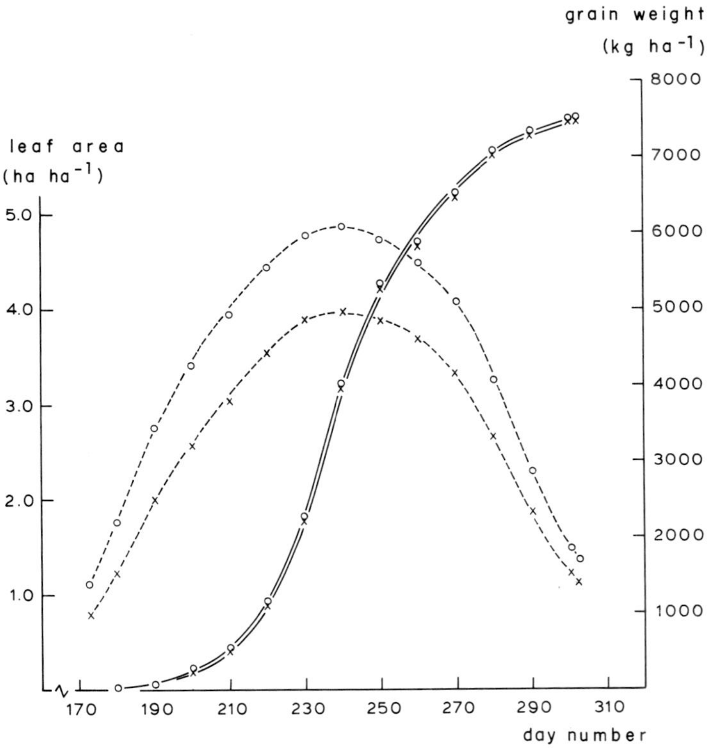

Figure 41. Development of leaf area (dashed lines) and grain weight (full lines) of a maize crop simulated with a specific leaf weight increasing as in Figure 40 (o) and simulated with a fixed value of 450 kg ha^{-1} (x).

3.3.4 Green surfaces other than leaves

The overall contribution of organs other than leaves to gross photosynthesis of the canopy can be positive or negative. It is positive when additional photosynthesis outweighs additional shading, as is the case in very open crops and when most leaf area has already died. The net contribution of organs other than leaves is negative when the light absorbed by the non-leaf surface is used less efficiently than it would have been by leaves.

Non-leaf photosynthesis is often only a minor addition to gross photosynthesis and may then be disregarded. However, the contribution of stems to the total green area can be significant for small grains and is therefore included in both crop modules (Listing 3 Lines 87,90, Listing 4 Lines 115, 118). The effective surface area of stems is determined by dividing their weight by a 'specific stem weight'. This constant has a value varying from 625 to 8000 kg ha^{-1} depending on species (see Table 19 Subsection 3.3.3). The factor 0.5 is added to the area calculation because, unlike for leaves, only the upper surface is active. The specific stem weight is not given as a function of crop development, but this relation may be introduced if the need arises. It is assumed that stem area has the same photosynthesis properties as leaf area.

Fruits usually intercept less than a few percent of radiation. However, sunflower provides an interesting exception. Its large heads all point in the same direction and intercept 10-20% of radiation, but barely photosynthesize. It can be simulated that a modified sunflower crop with flowers low in the canopy would yield about 10% more than the existing crop.

The contribution of green non-leaf area to photosynthesis can be evaluated with the programs presented. If stems are in the same positions with respect to radiation as leaves, then their contribution is proportional to their share in the green area. If the distributions of the angles of leaf and non-leaf areas are quite different, and, if these areas are at different positions in the canopy, the contributions of non-leaf area to photosynthesis can be evaluated with the SUPHOL subroutine (Subsection 2.1.4).

There is no reason to treat products resulting from photosynthesis by stems or fruits any differently from those of leaves, for the assimilates are transported in the same basic forms and are not formed in their destined cells. Hence all growth costs are the same.

3.4 Two modules to simulate potential crop growth

3.4.1 Introduction

The background of different processes of crop growth and suggestions for their programming have been given in Chapters 2 and 3. The major components are combined in two modules (L1D, Listing 3 and L1Q, Listing 4) which can be used to simulate the growth of annual crops. The first module is basically the sum of the simplest approaches. The time period for this module is one day (24 hours) and since it simulates at Production Level 1, it is called L1D. The second module (L1Q) includes some of the more detailed approaches and its time period is a quarter of a day.

L1D and L1Q must be supplemented with data sets for a crop and weather. Data to typify crops are given in many of the preceding Tables. Listing 5 uses rice as an example of how crop data can be added; Listing 11 (Chapter 7) shows this for weather data. Not all crop characteristics in this data set are required for all modules, but including more data than is required causes no problems. The simulation program is to be completed with module T12 (Appendix B), which contains several special functions. Figure 12 in Subsection 1.4.4 illustrates how sections and data of Chapters 2 and 3 may be combined into a full program. Table 1 in the Reader's guide shows how sections discussed throughout the book can be combined.

The simulation language CSMP is used throughout and references are provided in Subsection 1.4.1. FORTRAN is used in the functions and subroutines. Though, in the main, the programs presented are straightforward or explained in the text, a few peculiarities are discussed in this Section.

Line numbers in listings are used only for identification and are not part of the CSMP program. Lines starting with an asterisk (*) contain comment. All names of variables are explained in Listing 12, Chapter 7). Biological and physical constants and precisely known biological data are given in the listings with three or more significant digits. Approximate values are given with one or two digits. The modules contain a number of constants. They are not combined as each represents a single process: 1.467 stands for g CO_2 produced per g of glucose, 0.682 is its inverse; 0.053 refers to the fraction of glucose sacrificed during intercellular transport to provide energy for this process, 0.947 is its complement; 1.111 represents the yield in glucose from starch hydrolysis, 0.900 is its inverse; 0.2727, 0.400, and 0.444 are the carbon fractions in CO_2, glucose and starch respectively. The small value 1.E-10 is added in some cases to avoid division by 0.0, which would halt the simulation. The function AINT is used to truncate values; the output of the function AMOD is equal to its first input except when it exceeds the second input; the second input is then subtracted a number of times until a value between zero and the second input remains.

Data in CSMP AFGEN functions should cover a range of values that is wider than the range in which inputs are expected. This ensures that extrapolation outside the data, causing unexpected results or irrelevant warnings, does not occur. This is important when using data from Tables 13 and 14 (Subsection 3.1.2), Table 16 (Subsection 3.1.4) and others.

3.4.2 *Basic crop growth module with one-day time periods (L1D)*

Module L1D (Listing 3) can be used when growth limited by sink size does not need to be considered and when environmental conditions for crop growth are favourable. Hence, it will often be appropriate. L1D should not be used when environmental conditions are unfavourable (e.g., when day or night temperatures considerably exceed the range where the temperature response curves for photosynthesis, respiration and phenological development are more or less linear). It is comparable with the SUCROS models by van Keulen et al. (1982) and by Spitters et al. (1989). Careful reading and practice are required to become familar with this module. Explanations and background were given in Chapters 2 and 3.

The module contains an initial and a dynamic part. Before the initial section starts memory is reserved (Lines 3, 4) for the weather data. Line 2 specifies that the value of IDATE is an integer number. Two additional variables are included on Lines 2 and 4, again, to facilitate combining this module with those described in Chapters 4 and 5 (Listings 7-11). IDATE, an integer, is the truncated value of DATE, and ranges from 1 to 365. DATE is the sum of TIME elapsed since simulation started and DATEB, a parameter representing the Julian date at which simulation starts. The initial section (Lines 5-11) is followed by the dynamic section starting in Line 12. Three dummy variables are introduced (Lines 8-10) and used (Lines 40, 51, 97) to ease combining this with other modules (Section 4.4).

Actual weather data are read from tables using IDATE as input (Lines 103, 105, 106) and standard weather data are derived (Line 104). See Chapter 6 for further details. Ensure that manipulating DATE does not lead to values lower than 1 or higher than 365: CSMP does not reject an instruction to select data outside the TABLEs, but results will be nonsense.

The initial value of TIME (Line 113) should be 0.0. The FINish TIMe of 1000. is never reached: simulation always stops when either the FINISH conditions of maturity (DS = 2.0), or that of severe carbohydrate shortage (CELVN = 3.0) is reached.

For meaning and implications of the TITLE, PRINT, PRTPLOT and PAGE statements, for the TIMER variables DELT, PRDEL, OUTDEL and FINTIM, and for run control statements, such as FINISH, refer to the CSMP manual (IBM, 1975), or to Basstanie & van Laar (1982).

The last variables (Lines 120-125) are solely for convenient presentation of output.

3.4.3 Crop growth module with quarter-day time periods (L1Q)

Module L1Q (Listing 4) can be used for simulating crops in situations where temperature fluctuates a great deal, when the dynamics of plant reserves are important and when tillering and grain formation in cereals is under study. This module is organized in the same way as L1D. Module L1Q should not be combined with water balance modules L2SU and L2SS. To incorporate the module TIL (which simulates tillering and grain formation) into L1Q, substitute TIL for Line 36.

Formation and use of available carbohydrates (WAR) is programmed in Listing 4 Lines 14, 29, 30, 51. The rate of use is derived from the growth rates of the organs (Lines 52-56), these being equal to potential growth rates based on carbohydrate availability unless a reduction occurs, such as that resulting from sink size limitation (Line 35).

In each 24-hour cycle (i.e., sunrise to sunrise) the time period for integration (DELT) is equal twice to half the daytime (0.5 · daylength) and twice to half the nighttime (0.5 · (24h – daylength)). Though DELT itself remains equal to 0.25, all rates in integrals are multiplied with a correction factor for daylength, FADL (Line 144). This factor is larger than one, if the day part is longer than six hours, and vice versa. Simulation starts at 0.00 h, and output is printed at midnight (when PRDEL is a whole number). Printed rates shows their midnight values (so that photosynthesis is always 0.0). (In L1D, printing time corresponds with sunrise.) The variable DTIME indicates the starting time of the fraction of the day that the simulation has reached (first quarter: DTIME = 0., second quarter: DTIME = 0.25, etc.). The variable NIGHT signals whether it is day or night for calculating radiation intensity (Line 131).

Temperatures at different times during the day are reconstructed from the minimum and maximum temperatures by the function FUTP (Subsection 6.1.3).

The INTeGRaL function is used for various state variables to compute the running average of a variable (Line 47; Subsection 1.4.4) and to retain the maximum value that a variable reached (Line 46). The daily total of a rate can also be calculated with an INTeGRaL function; the content of such integrals is reset to 0.0 each new day (Lines 65, 66).

3.4.4 Functions and subroutines (module T12)

Module T12 (Appendix B) with special functions and subroutines completes the CSMP program. This terminal section can be used with the crop and soil modules at Production Levels 1 and 2. Several of its functions and subroutines are used in only one main program section and can be deleted if not used. Special remarks that are required to understand or use T12 are given here. Readers are advised not to change the contents of functions and subroutines as this can be intricate. Most variables inside these functions and subroutines are not defined in Listing 12. The functions (one output) and subroutines (several outputs) are carefully checked and can be used within the limits and conditions previously described.

END closes the DYNAMIC section with its model structure definitions; data for reruns can be entered after END (Basstanie & van Laar, 1982). END may be replaced by END CONTINUE to allow output specifications (such as PRDEL) to change during a simulation run. This can be useful for checking fluctuation of rates during a few interesting days without the burden of a huge output for the full growing season.

STOP terminates the section where reruns can be specified. It is followed by FUNCTIONS and SUBROUTINES, placed alphabetically. These program sections, written in FORTRAN, are placed here to maintain maximum lucidity in the main programs (CSMP permits their specification in the main program in several other ways). It is recommended that functions and subroutines be included in a subroutine-library (Subsection 1.4.4).

Calculations involving photosynthesis, growth and respiration are complex. Even experienced programmers easily make errors when rewriting a program or adapting functions and parameters. To avoid some of the most obvious errors a check on the C balance (FUCCHK function) is included. This consists of a comparison of the total net amount of C that entered and that is retained in the crop. The totals must be identical, but relative differences up to 1% are allowed for rounding off errors. In the models described here, the relative difference is usually less than 0.1%. If it does exceed 1%, check the consistency of the crop input data and of all statements that were changed, removed, or added. WIR represents the total reserves (starch) formed since simulation started and must be expressed as an integral function.

The FUPHOT function is generally used to compute the rate of gross photosynthesis of the canopy. The SUPHOL subroutine is an alternative to FUPHOT which permits computing photosynthesis of a canopy (PCGC) consisting of layers with different characteristics (Subsection 2.1.4).

The FUTP function approximates the fluctuation of air temperature during the day for the L1Q module. The FUVP function calculates the saturation vapour pressure (in kPa) that corresponds with a given temperature. Total daily radiation on fully clear days and daylength, both astronomical and photoperiodical, result from the SUASTR subroutine. Both FUPHOT and SUASTR call the subroutine for astronomical computations SUASTC.

The ENDJOB statement terminates the CSMP program.

3.4.5 Two examples of application of the modules

The first example addresses the potential growth of cowpea in Mali. Cowpea is an important crop in Sahelian countries, but hardly ever reaches its potential production level because of low soil phosphorus, water stress and insect problems. As a result, the potential yield of this crop is hardly known. The L1D + T12 modules can help provide an estimate of this crop's potential. The simulated growth curve (Figure 42) was obtained without calibrating the basic crop data. Simulation began at the date that the real crop started to grow. Data collected by P. Gosseye in an experiment in similar conditions, given in the same figure, show a similar pattern. This similarity provides some credibility for when simulating the potential yield of cowpea for other West African countries.

The second example simulated a field experiment with a silage maize crop (cv LG11) in Wageningen in 1985 using the module with quarter-day time periods (L1Q + T12). This was a cloudy year with a low final yield. Nutrients and water were always in ample supply. Accumulation of starch may have occurred on the few very bright days which could have resulted in reduced photosynthesis. Measuring with extensive equipment (Louwerse & Eikhoudt, 1975) of some crop plants within an enclosure of approximately 2x1x1 m provided a continuous record of CO_2 exchange. The measured CO_2 exchange rates were summarized over the same periods as simulation occurred (i.e., midnight till sun-up, sun-up till noon, etc.; adjusted for the position of Wageningen with respect to GMT and summertime). Corrections were also made for a 15% reduction of radiation by the enclosure. Figure 43 shows the results of the simulation and measurement for net CO_2 exchange of the above-ground dry matter for both day and night and the daily totals. Simulated nighttime

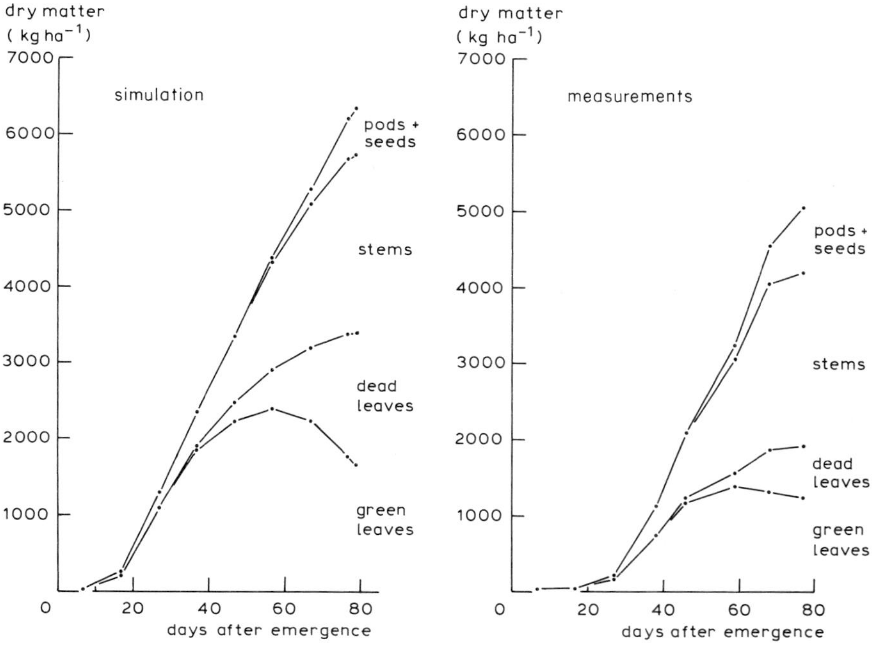

Figure 42. Simulated (left) and measured weight of a cowpea crop in Niono, Mali (Source: crop data from Gosseye, in Haverman (1986)).

respiration is higher than measured, but it is not certain whether the measured data are better than those simulated. Daily totals of net CO_2 exchange are fairly equal. High radiation caused the real crop plants to absorb a little more CO_2 than is simulated, possibly due to canopy disturbance by placing the enclosure. There is more variation in the real assimilation rates over the half-day periods, which may be due to an uneven distribution of radiation over the day not accounted for in the model. Reduced canopy photosynthesis due to carbohydrate buildup did not occur in the simulation. Photosynthesis of the field crop on bright days was not smaller than the simulated values and it is concluded that CO_2 assimilation was not limited by sink size.

Crop growth was established by periodic harvests (Sibma and Louwerse, CABO, personal communication). The final biomass was about 16.000 kg dry matter ha^{-1}, which was simulated closely (15.500 kg ha^{-1}) with a realistic distribution over the organs. This also supports the conclusion that photosynthesis was not restricted by the capacity to absorb its products.

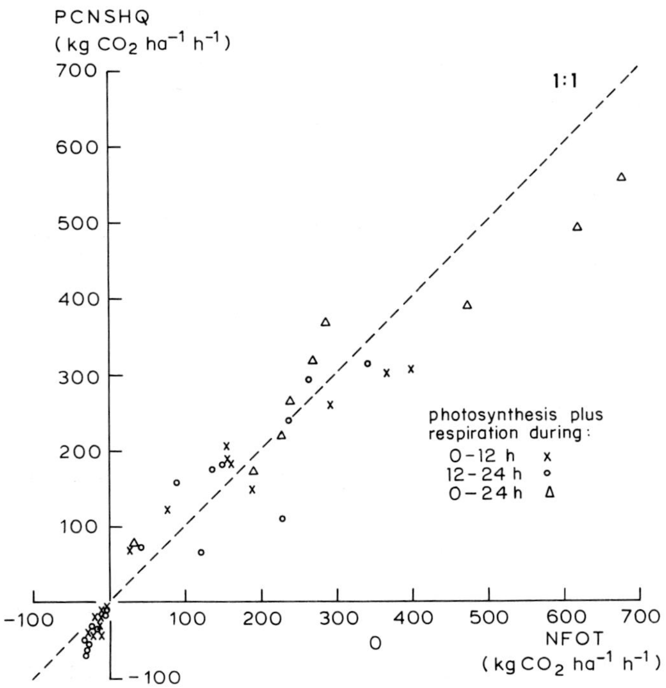

Figure 43. Simulated (PCNSHQ) and observed (NFOT) net photosynthesis of a maize crop in Wageningen on a few selected days of the 1985 growing season (crop data from Louwerse and Sibma, CABO personal communication).

3.5 Exercises

See Section 2.5 for an introduction to the exercises.

3.5.1 Morphological development

T1. Make a relational diagram of the processes and variables which determine morphological development.

T2. Estimate how much longer the growing season of IR36 lasts at 400 m elevation, as compared to sea level.

T3. How much earlier will potato cv Favorita be ready for harvest in comparison to cv Mara if planted at DS = 0.2 at 18 °C? How much at 26 °C?

T4. Are the responses in Table 13 of the average rice variety and Nipponbare to temperature really different?

S1. Plot daylength for photoperiodicity as a function of date for latitudes

−30°, 0°, 30° and 60°. What is the impact of daylength on the development rate of soya bean (cv Hawkeye) at day 80 at these latitudes? (Suggestion: make a MERGED PRTPLOT of DLP with L1D; eliminate the call for FUPHOT and the superfluous parts in the terminal section).

S2. Program a strong daylength sensitivity in rice during the development stage period 0.2-0.7 (multiplication factor 0.5 at 13 h and 0.33 at 14 h, such as cv BPI-76 in Vergara & Chang, 1985).

By how much is the growing season shortened due to the effect of daylength by transplanting at dates 250 and 350 as compared to day 150? And by how much in the northern Philippines (18°N) and in the southern Philippines (6°N), supposing all other things remain the same?

3.5.2 *Assimilate partitioning*

T5. Explain the difference between biomass and assimilate partitioning.

T6. Estimate how much senescing leaves contribute to crop yield in potato, rice and soya bean.

T7. In what environmental conditions do rice grains fail to reach their normal maximum weight?

T8. Make a relational diagram of the processes and variables which determine development of tillers, florets and grains (see Listing 2).

T9. Explain the meaning and formulation of CELVN and CELV in Listings 3 and 4.

T10. Assume the leaf area of a rice crop to be 7.5 $m^2 m^{-2}$ at flowering, 6.0 at DS = 1.5 and 1.5 at maturity. Propose a function for L1D to reproduce this reduction in leaf area.

S3. Determine the effect on grain weight and leaf area of a breeding program that leads to IR36 with the assimilate partitioning pattern of IR64 (Table 18). Did you expect a large effect? Propose a more effective assimilate partitioning for IR36 in Los Baños.

S4. Determine the effect on grain yield and its components of doubling and halving the time coefficient for tiller formation in rice (do not change other formation rates) in a sowing density experiment with 0.068, 0.2, 0.68, 2.0 and 6.8 million plants ha^{-1}. (Use modules TIL and L1Q). What is the optimum density? What do you notice about the number of grains per panicle? Is the pattern of tiller development and death realistic?

S5. Compare the effects of a heavily overcast period of three weeks before or after flowering on rice grain yield and yield components. Set radiation in overcast conditions to 17% of clear sky radiation (0.15 · RDTC, Section 6.1).

3.5.3 *Leaf area*

T11. How much is the rate of canopy photosynthesis with a leaf area of 3.0 $m^2\ m^{-2}$ and PLMX = 40.0 at a day that is 50% fully clear and 50% fully overcast, according to Figure 38? Does it correspond with Figure 22? Why is the fraction of the reference canopy photosynthesis in overcast conditions much higher than the fraction for clear sky conditions?

S6. Is the contribution of photosynthesis by soya bean pods significant? Estimate. Then evaluate the estimate with the model assuming that 2500 kg pod weight has a surface area of 1 ha (one side only).

3.5.4 *Simulation models*

T12. For what non-physiological reason is the simulation result of potato growth nonsense, if the response of development rate to temperature is specified as in Table 13 and the average temperature is between 30 and 35 °C? Is this a very special situation?

T13. What kind of C-balance errors are detected and which are not detected by FUCCHK?

3.6 Answers to exercises

3.6.1 *Morphological development*

T1. Carefully distinguish the types of variables and the relations involved. Use the symbols of Figure 5 to draw the diagram.

T2. The growing season lasts 1./0.013 + 1./0.028 = 113 d at sea level (Table 12). At 400 m elevation, the temperature is, on average, 2 °C lower. The growing season will then last about 7% = 8 days longer (Table 13)

T3. From DS = 0.2 to 1.0 in Mara at 18 °C takes 22.9 days and in Favorita 6.4 days less. From DS = 1.0 to 2.0 takes 66.7 days and 44.4 days respectively, so that Favorita is 28.7 days earlier. At 26 °C, interpolation in the functions of Tables 13 and 16 is necessary. Favorita can be harvested after 188 days and Mara only after 269 days.

T4. If relative values are calculated for the same temperatures the responses are almost the same in the range 22-32 °C; Nipponbare develops more slowly at 16-22 °C. The responses below 16 and above 32 °C cannot be compared. The assumption that the responses above 22 °C are identical will be difficult to prove wrong with an experiment.

S1. The result is a plot with three sinusoidal curves with different phases or amplitudes. The photoperiodic daylength at the equator is almost constant at 12.55 hour. The effects on the development rate at day 80 are 0.97978, 1.00, 1.00 and 0.86687, respectively.

S2. FUNCTION DRDT = 8.,1., 12.,1., 13.,0.5, 14.,0.33,...

24.,0.33 and DRV = ... as in Subsection 3.1.2.

In Los Baños, the growing season decreases from 135 to 102 and 93 days; in the north from 142 to 102 and 93 days, and in the south from 121 to 104 and 98. This is all due to daylength effects because the temperature for these exercises is kept constant. In reality the effects of daylength and temperature are intertwined.

3.6.2 Assimilate partitioning

T5. Assimilate partitioning implies that glucose distribution to organs is regulated and occurs according to a certain key. Biomass results from glucose after growth took place. Biomass partitioning implies that biomass allocation is regulated; and it suggests feedback from the organs to the distribution mechanism on the amount of glucose required. The first hypothesis seems more appropriate.

T6. Assuming 2500 kg ha^{-1} of leaves at flowering and only 50% of that still attached to the stems at maturity, the dropped leaves provided an equivalent of about 600 kg ha^{-1} of glucose. This corresponds with 460 kg ha^{-1} of potato tubers, 410 of rice grain and 280 of soya beans.

T7. When photosynthesis plus remobilization in the grain-filling period is much less than is anticipated during flowering, then there are too many grains to be completely filled. This can occur when the light level is very low (see exercise S4), during water stress, when diseases are present and at low temperatures.

T8. Distinguish carefully the types of variables and the relations involved in tiller, floret and grain development. Use the symbols of Figure 5 to draw the diagram.

T9. CELVN is the number of consecutive days without carbohydrate export from leaves and stems. The crop is assumed to die when the value reaches 3. Stems are included because they contribute to photosynthesis. CELVN increases by 1. each day that CELV is negative; it is reset to 0.0 if CELV becomes positive again (no after-effect). CELV in L1D is the gross photosynthesis minus maintenance of leaves and stems; the fraction of leaf maintenance that is contributed by excess energy from photosynthesis is subtracted. In L1Q, CELV is in principle the same, but the gross photosynthesis is to be accumulated during 24 hours and reset each day.

T10. The reproductive period lasts about 30 days. In the first 15 days, loss is 20%, in the second 15 days 75% of what remained. This is reproduced by:

FUNCTION LLVT = 0.,0., 1.,0., 1.5,0.03, 2.,0.15

S3. Grain yield (rough rice) of IR36 is 7691.4 kg ha^{-1} and 6358.2 of R36 x 64; leaf area at flowering is 9.9 m^2 m^{-2} and 11.4, respectively. The new cross is a crop with too much vegetative biomass. Lowering the share that goes to leaves by 0.1 and reducing the share to the stem to 0.1 and 0.0 at DS 1.0 and 1.25, respectively, decreases leaf area to 8.8 and boosts yield to 8200.1 kg ha^{-1}.

S4. The planting densities are obtained with WLVI = (0.68, 2., 6.8, 20., 68.) in kg ha^{-1}. WSO is in kg ha^{-1}, NTI in 1.E6 ha^{-1}, NGR in 1.E8 ha^{-1}. The results at maturity are:

TCFT	15			7.5			30		
WLVI	WSO	NTI	NGR	WSO	NTI	NGR	WSO	NTI	NGR
0.68	2326.2	1.245	0.983	2904.5	1.670	1.228	1712.1	0.842	0.724
2.0	4269.7	2.436	1.808	5202.4	3.231	2.202	3287.4	1.749	1.394
6.8	7574.1	4.571	3.237	7574.1	5.706	3.616	6192.0	3.409	2.616
20.0	7413.9	7.724	3.873	7413.9	9.299	3.957	7413.9	6.076	3.726
68.0	6917.9	9.754	3.680	6917.9	10.85	3.680	6917.9	8.482	3.655

The standard density (68 plants m^{-2}) appears to be slightly below optimal with the standard TCFT. The grain number is much more constant than the tiller number, so that the number of grains per tiller decreases substantially when the tiller number increases. The pattern of tiller development in a field experiment was roughly similar to this simulation for the standard case.

S5. Run the program first to find the date of flowering (DATEF, equals 56.). Then replace 'RDTMT(IDATE)' in Line 131 by LIGHT, and add:

```
PARAM DATEF = 56.
LIGHT  = FCNSW(CHOICE,LIGHTE,LIGHTN,LIGHTL)
LIGHTE = INSW((DATE – (DATEF – 21.)) * (DATEF – DATE),...
         RDTMT(IDATE),0.15 * RDTC)
LIGHTN = RDTMT(IDATE)
LIGHTL = INSW((DATE – DATEF) * ((DATEF + 21.) – DATE),...
         RDTMT(IDATE),0.15 * RDTC)
PARAM CHOICE = (-1.,0.,1.)
```

The grain yield drops from 7574.1 kg ha^{-1} to 3251.3 with early clouds, and to 3928.1 with late clouds. Late clouds do not reduce the number of tillers per plant 4.571, but early clouds reduce it considerably 2.828. The adjustment to low light levels due to late clouds causes tillers to carry much fewer grains, but the grains formed fill completely. Early clouds result in a low number of tillers, the few grains present are filled before maturity and the crop stops growing too early.

3.6.3 Leaf area

T11. Approximately $0.54 \cdot 1144. \cdot 0.5 + 0.77 \cdot 311. \cdot 0.5 = 429$ kg CO_2 $ha^{-1}d^{-1}$. This is similar to the rate in Figure 22. Canopy photosynthesis increases little at low radiation with PLMX beyond 25., but increases a lot in full light.

S6. The contribution to canopy photosynthesis by soya bean pods is probably small since the additional green area is small.

To simulate this, add the growth of pod area (= GSO / 2500 · 0.5) to ALV and adjust SLA as for stem material. In the standard conditions for soya bean, the yield increases by only 65.3 kg ha^{-1} with the extra green area.

3.6.4 Simulation models

T12. Because the temperature exceeds the specified range and CSMP extrapolates the relation from the last two data points, the development rate becomes negative. The problem is cured by adding 40.,0.01 to the functions DRVTT and DRRTT. Always take care that inputs do not exceed the specified range.

T13. Assimilated carbon remains in the plants or is lost by respiration. FUCCHK continuously checks whether all carbon is accounted for. Omitting one of the respiration rates causes an error. Another occurs when FCLV is unequal to CRGLV · 12. / 30. – CPGLV · 12. / 44. FUCCHK does not detect incorrectly calculated rates of photosynthesis, remobilization or respiration, nor incorrect carbohydrate partitioning (except when the sum of the fractions is unequal to 1.00).

4 Transpiration and water uptake

This chapter describes how to simulate transpiration and water uptake, the principal processes of crop water balance. Potential transpiration is discussed in Section 4.1, and water uptake, the limiting factor for transpiration during water shortage, in Section 4.2. Section 4.3 discusses the effects of water shortage on other physiological processes, such as phenological development.

4.1 Transpiration without water stress

4.1.1 Introduction

When there is sufficient soil water (as at Production Level 1) the photosynthesis rate largely determines the transpiration rate. When water is in short supply, the inverse is true, for the rate of water uptake from the soil is then of crucial importance. Both situations can be seen in Figure 6, Subsection 1.2.2, where with ample soil water, transpiration equals potential transpiration and photosynthesis equals potential photosynthesis; when there is water shortage, uptake is less than potential transpiration, and due to this stress, photosynthesis is below potential photosynthesis. This section is summarized in module L2C for crop processes at Production Level 2 (Listing 7).

In many cases it is important to consider crop water use. First, because it can be an important topic in itself in field or regional water balance studies, whether water is plentiful or not. Second, because water shortage is a very common phenomenon and water stress, brief or long, reduces the rate of crop growth and can affect dry matter distribution and hence the economic yield.

Water loss and uptake and the efficiency of water use by crops have been the topics of many studies, reports and reviews. The entire Volume 12B of the Encyclopedia of Plant Physiology (1982) is devoted to them. Other reports on transpiration and crop production modelling are: Slatyer (1967), Feddes et al. (1978), Doorenbos & Kassam (1979), Tanner & Sinclair (1983), and van Keulen & Seligman (1987).

The time period for simulating transpiration is 24 hours. The instantaneous effects of the driving forces of transpiration, irradiance and air-drying, are almost proportional to transpiration without water stress. Therefore, it is assumed that the daily total radiation and the average daytime drying power are proportional to daily transpiration. Amounts of transpiration per day are quite large in comparison to crop water content. Time periods even smaller than those for simulating transient carbohydrate levels (i.e., <6 h) would be required for a dynamic simulation of the plant water content.

'Transpiration' is the loss of water vapour by plants. 'Evaporation' is the loss of water vapour from the soil or from a free water surface. 'Evapotranspiration' is a broader term that covers both transpiration and evaporation.

Evaporation of a 1 mm layer of water requires as much as 2.4 MJ m^{-2} of energy. Transpiration by plants and evaporation from the soil or free water surfaces can therefore be considered as energy balance processes. The driving force for transpiration at any time is the gradient of water vapour pressure. Resistance to the transpiration and evaporation process is strongly related to wind speed. These two environmental variables, air humidity and wind speed, are sometimes, in combination, referred to as the 'evaporative demand' or 'drying power' of the air. Transpiration cools the evaporating surface so that a power source is required to maintain the surface temperature and hence the vapour pressure gradient. Solar radiation supplies the bulk of this power. Radiation is therefore the main environmental factor and driving force determining the transpiration rate. Net thermal (long wave) radiation is mostly a negative term in the heat balance. Heat carried by moving air is another power source, but this is usually a smaller source than radiation. Photosynthesis traps less than 5-8% of solar radiation and is disregarded here. Respiration yields an insignificant amount of energy. To simplify the treatment of transpiration, it is considered as resulting from two factors: radiation and evaporative demand.

When the crop has sufficient water the transpiration rate is the potential transpiration rate of the canopy. The term refers to the current weather and crop situation. Reduced photosynthesis due to sink size limitation, ageing, or even low air humidity increases stomatal resistance and lowers the potential transpiration rate. The actual transpiration rate is below the potential rate when stomatal resistance increases in response to water shortage.

4.1.2 Potential canopy transpiration

The Penman calculation is useful for calculating transpiration and evaporation rates for crop growth modelling. Penman calculated evapotranspiration from a low grass sward for 10-day periods. His method has been refined and extended and is now used for daily totals of canopy transpiration and for daily totals of soil evaporation (e.g., van Keulen, 1975; Feddes et al., 1978; Jarvis, 1981). The equations used are a combination of physical and micrometeorological terms. Physiological characteristics do not play a role, with the exception of leaf resistance. Physical processes and properties are averages in time (daytime) and space (upper leaf layers).

The transpiration rate is related partly to irradiance and partly to evaporative demand. Both parts are first calculated for a single, flat surface as a function of current weather; they are then adjusted for the actual leaf surface of the canopy. The calculations of both transpiration rate fractions are executed in the SUEVTR subroutine. The equations used are similar to those by Goudriaan (1982b) and Jansen & Gosseye (1986) (discussed in Subsection 5.4.5).

The equations are not in the main program because their physical background is generally accepted, their complexity exceeds that of other sections of the program, and because they are repeated several times.

Crop data required for calculating transpiration are: the canopy leaf area, the reflectance of the crop for solar radiation, and the diffusion resistances. The reflection coefficient for solar radiation is similar for all crop canopies and amounts to 0.2-0.3. The resistances for the movement of water vapour from the canopy to the open air (leaf resistance, boundary layer resistance and canopy resistance) are of moderate importance (Figure 44). They are considered in Subsection 4.1.4.

The rate of canopy transpiration in a humid environment is 0-5 mm d^{-1} and is 5-10 mm d^{-1} in a warm, dry climate (Figure 45). Its value is difficult to determine experimentally, but simulation readily provides it. (Figure 45 shows that measurements and simulations of daily transpiration are on average equal, but vary considerably; the simulated day-to-day values are probably more accurate than those measured: see Subsection 6.2.4).

4.1.3 Canopy transpiration, leaf area and environmental conditions

Leaf area is used in calculating both parts of transpiration: the drying power of the air is proportional to the area of the upper leaf layers, and absorption of solar radiation is exponentially related to leaf area (the average extinction coefficient for visible and near infrared radiation is about 0.5: Listing 7 Line 16). Figure 46 provides an example of the relation of leaf area and transpira-

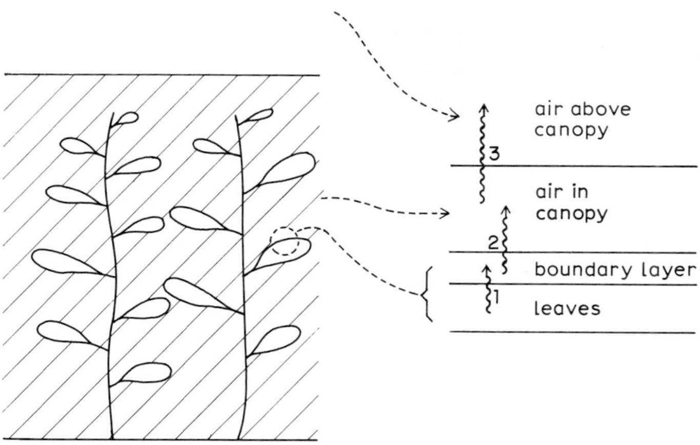

Figure 44. A diagram to locate the three principal resistances (arrows) to gas exchange of crops. 1 refers to the leaf resistance, 2 to the boundary layer resistance and 3 to the canopy resistance.

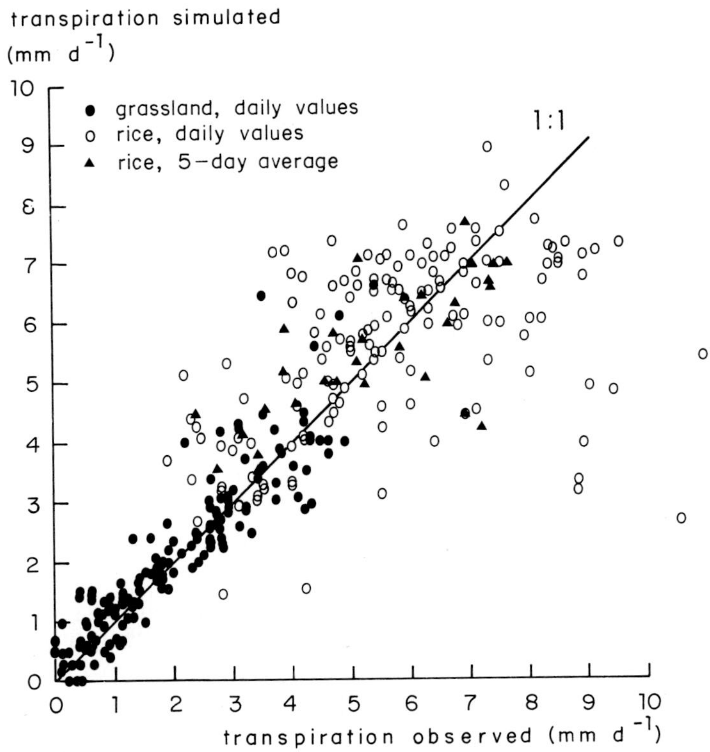

Figure 45. The simulated transpiration rate versus the measured value for all days of a growing season of a grassland crop in Wageningen, the Netherlands (black dots), and of transpiration plus evaporation of a rice crop in Los Baños, the Philippines (white dots, black triangles). (Source: Garrity, IRRI, personal communication).

tion rate: canopy transpiration in this case increases rapidly till about 4 mm d^{-1} at a leaf area of about 3 m^2 m^{-2}, and only a little beyond. The potential rate of transpiration can be twice as high in a dry tropical climate, whereas in overcast conditions transpiration is always much less.

Lower leaves do not contribute much to transpiration because little light penetrates deep into the canopy, their leaf resistance is higher, the air around them is more humid and wind speed is reduced. In the module presented here (L2C, Listing 7), only the upper leaves down to a cumulative leaf area of 2.5 m^2 m^{-2} contribute to the part of transpiration related to drying power. (Tanner & Sinclair (1983) proposed a similar construction; the value of 2.5 is somewhat arbitrarily chosen, but canopy transpiration rates are not sensitive to the size of this area when it is between 2 and 3).

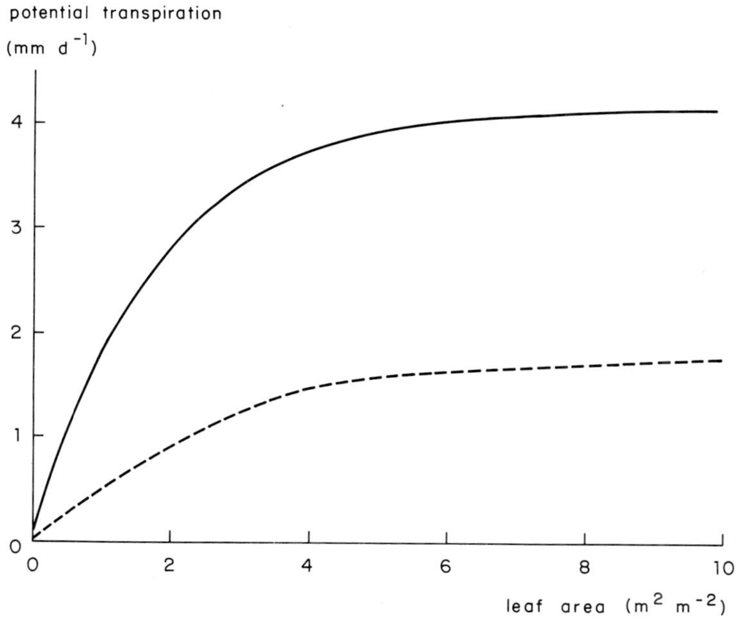

Figure 46. The transpiration rate of a Faba bean crop on a clear, warm dry day in the Netherlands as a function of the leaf area. The dashed line indicates the part attributed to the drying power of the air.

Radiation is often the most important weather variable (Figure 47). Figure 48 shows the effect of air humidity (via the vapour pressure gradient and not via stomatal resistance) on a clear and on an overcast warm day in the Netherlands. It shows that the drying power of air can sometimes be more important than radiation. The effect of wind is due to its influence on the resistance of the boundary layer and on the canopy resistance. The effect increases only up to windspeeds of about 2 m s^{-1} (Figure 49). Note that the air is often dry when radiation on a certain day is high, so that transpiration rates on successive days fluctuate more than Figures 47 and 48 separately suggest.

4.1.4 Resistances for water vapour movement

The resistance for water vapour movement consists of leaf, canopy and leaf boundary layer resistance in series. The first is always the most important. All resistances are computed for the upper layers only, because they contribute most to transpiration. The calculations yield a single value per day for each resistance.

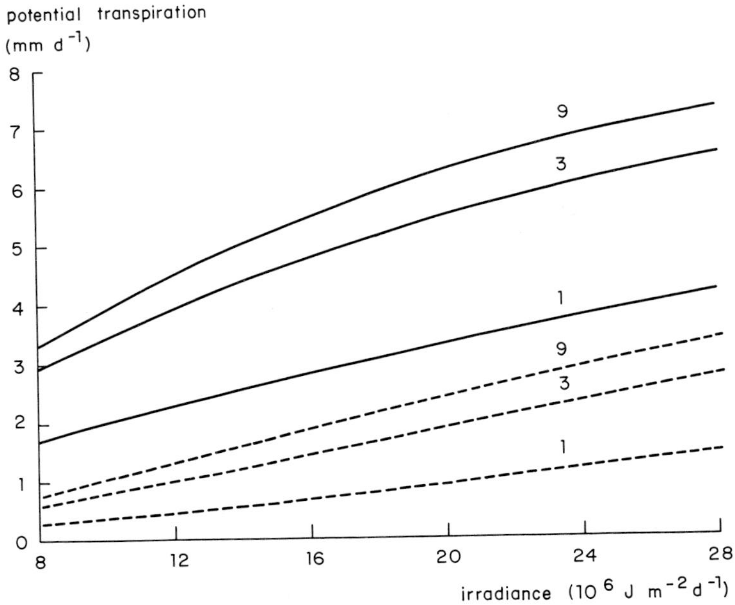

Figure 47. The transpiration rate of a crop as a function of radiation in clear, warm dry conditions for leaf areas 1, 3 and 9 m^2 m^{-2}. Full lines represent total transpiration, dashed lines the part attributed to the drying power of the air.

Leaf resistance

The leaf resistance for water vapour of unstressed plants ranges from a minimum of 50 to 300 s m^{-1}, to a maximum of about 2000 s m^{-1}. Its actual value is determined by the crop and it varies considerably from day to day in response to weather and crop conditions. To derive the value of the average leaf resistance, a procedure suggested by Goudriaan (1982b, p. 112) is used. The upper layer of the canopy is regarded as a single big leaf. Leaf resistance is determined by applying Ohm's law and dividing the net rate of photosynthesis by the CO_2 gradient across the stomata. The latter is quite predictable (Subsection 4.1.5). The daytime average of upper leaf resistance equals the CO_2 gradient divided by daily net photosynthesis per unit leaf area, minus the boundary layer and canopy resistance (Listing 7 Lines 20-21; some constants are used in this equation: 68.4 is the weight conversion, length and time dimensions; 24. is the number of hours per day; and 1.6 represents the ratio of the diffusion resistance for CO_2 to that of water vapour). The average net rate of photosynthesis per unit leaf area (Line 24) is equal to the total canopy net photosynthesis (average daytime rate of gross photosynthesis minus daytime maintenance respiration of the leaves; respiration due to metabolic activity need not be

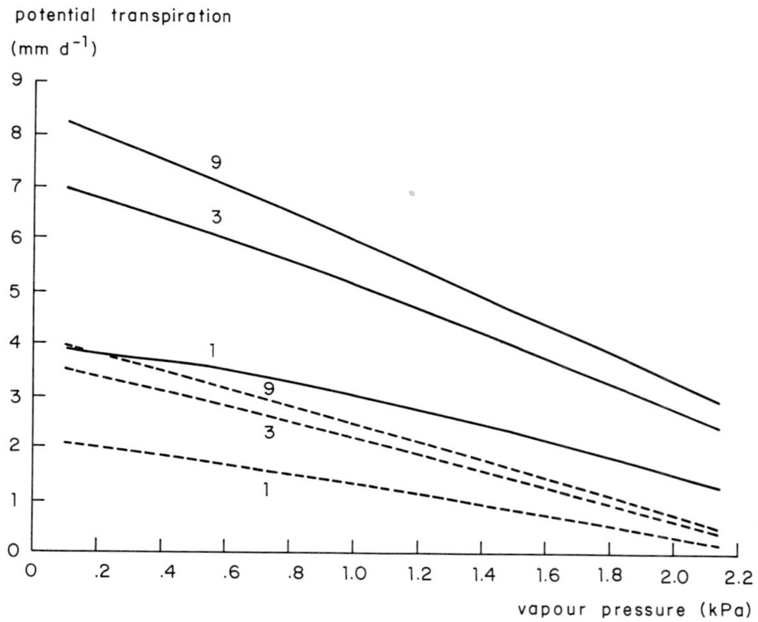

Figure 48. The transpiration rate of a crop as a function of air humidity in clear, warm dry conditions and a windspeed of 2 m s^{-1} for leaf areas 1, 3 and 9 m^2 m^{-2}. Full lines represent total transpiration, dashed lines the part attributed to the drying power of the air.

counted, see Subsection 2.3.3), divided by the leaf area (with a maximum of 2.5 m^2 m^{-2} because over 90% of photosynthesis occurs in the top layers).

Leaf resistance is limited between an upper and a lower limit. Minimum leaf resistance can be derived from the maximum leaf photosynthesis rate, determined in a ventilated leaf chamber with a known CO_2 gradient, respiration being 10% of gross photosynthesis, with a normal boundary layer resistance (10 s m^{-1}) and without a canopy resistance (Line 23). The maximum resistance of the leaf can be interpreted as the cuticle resistance. This maximum has been set at 2000 s m^{-1} for all crops (Line 20).

Calculating leaf resistance helps in understanding why the choice of 2.5 m^2 m^{-2} for upper leaves contributing most to transpiration is of little consequence for the rate of canopy transpiration. Supposing that a smaller area contributes most to photosynthesis implies a higher average photosynthesis rate per unit area. This implies a lower leaf resistance, a higher transpiration rate per unit leaf area, and a canopy transpiration rate that changed little. The range of values over which this compensation can take place, is limited.

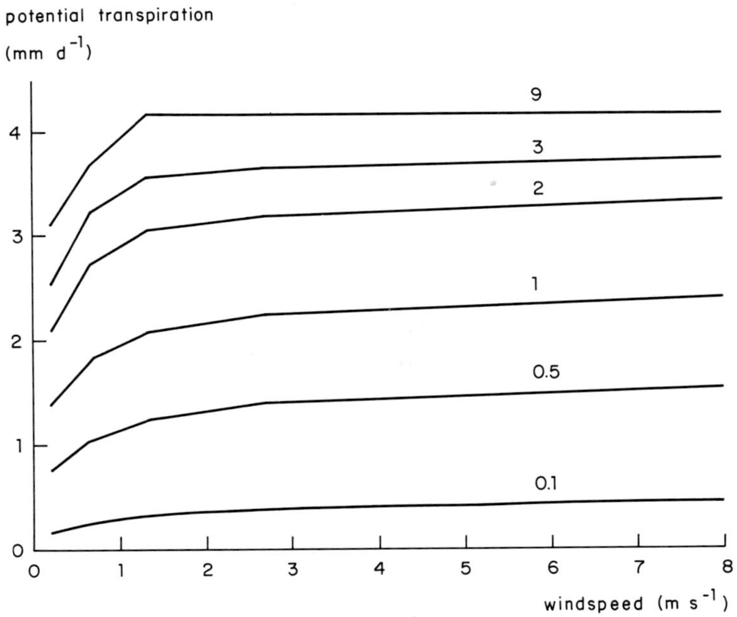

Figure 49. The transpiration rate of a crop as a function of the windspeed on a clear, warm dry day in the Netherlands for leaf areas of 0.1 to 9 m^2 m^{-2}.

Canopy resistance

Resistance for movement of water and heat from air spaces within the canopy to the air above it (to which the meteorological data refer) is called the turbulent or canopy resistance. Its value varies a great deal (from less than 10 to over 100 s m^{-1}) and is related to windspeed, crop height and canopy architecture in a fairly complex manner (Goudriaan, 1977; GELGAM, 1984). Its calculation for a whole canopy is simplified here and is found in the FURSC function Listing 7 Line 26 (FURSC is part of Appendix B). Crop height (m) as a function of development stage is a required input, but need not be very accurate. Table 21 gives a list of data for common cultivars of several crop species.

Leaf boundary layer resistance

The boundary layer resistance to diffusion results from a thin laminar air layer at the leaf surface. Its value depends on leaf width and windspeed (Goudriaan, 1977, p. 74; Listing 7 Line 25) and is typically 5-10 s m^{-1}. It relates to a surface from which vapour is released on both sides, since leaves of crop plants generally have stomata on both sides. The boundary layer resistance is twice as large for leaves with stomata on one side only. The wind velocity in the canopy

Table 21. Plant height as a function of development stage. Data given in CSMP style: development stage is the first and crop height (m) the second number of each pair.

Species	Development stage and crop height				
Barley	0.,0.,	1.,0.6,	2.1,0.6		
Chick pea	0.,0.,	1.,0.5,	2.1,0.5		
Cotton	0.,0.,	1.,1.5,	2.1,1.5		
Cowpea	0.,0.,	1.,0.75,	2.1,0.75		
Faba bean	0.,0.,	1.,0.5,	1.5,1.,	2.1,1.2	
Groundnut	0.,0.,	1.,0.6,	2.1,0.6		
Maize	0.,0.,	0.2,0.5,	0.5,1.5,	1.,2.,	2.1,2.
Millet	0.,0.,	1.,2.,	2.1,2.		
Potato	0.,0.,	1.,0.6,	2.1,0.6		
Rice, IR36	0.,0.,	1.,1.,	2.1,1.		
Rice, IR64	0.,0.,	1.,0.9,	2.1,0.9		
Sorghum	0.,0.,	0.5,1.5,	1.,2.,	2.1,2.	
Soya bean	0.,0.,	1.,1.,	2.1,1.		
Sugar-beet	0.,0.,	0.5,0.5,	2.1,0.5		
Sunflower	0.,0.,	1.,2.,	2.1,2.		
Sweet potato	0.,0.,	1.,0.3,	2.1,0.3		
Wheat	0.,0.,	1.,0.7,	2.1,0.7		

drops almost exponentially from the top downwards. To account for this, the average windspeed in the upper leaf layer is reduced to 60% of that above the canopy (Line 25).

4.1.5 Regulation of leaf resistance

Regulation of leaf resistance can be demonstrated by artificially decreasing or increasing the external CO_2 concentration, while keeping all other conditions constant: stomata will be opening or closing, respectively. From simultaneous CO_2 exchange and transpiration measurements it has been concluded that the CO_2 concentration in stomatal cavities tends to be constant in plants that grow in air with a normal CO_2 concentration (e.g., Raschke, 1979; Schulze & Hall, 1982). The CO_2 concentration is around 210 vppm in C_3 species and around 120 vppm in C_4 species (1 vppm = 1 cm^3 CO_2 per m^3 of space). It is not the absolute values of CO_2 concentrations that are constant, but the ratio of the concentration in the stomatal cavity and the ambient concentration. This internal/external ratio of CO_2 concentrations is typically 0.4 for C_4 and 0.7 for C_3

crops. The value of this ratio is independent of the light level and of the photosynthesis rate. To maintain such a constant ratio, leaf resistance can change over a 10-20-fold range.

The variation observed in the internal/external ratio between individual plants of field crops (wheat, potato, sunflower) without water stress in the Netherlands was only 10-20% (Teubner, 1985). The ratio is also unaffected by the photosynthesis rate, respiration, nutritional level, or leaf age (Tanner & Sinclair, 1983) (Figure 50). An increase was observed in old rice leaves (Makino et al., 1984; Fukai et al., 1985) and other crops. This may indicate a loss of the regulatory capacity when leaves senesce. It appears that the effect of temperature on the internal/external ratio has not been systematically studied. These observations confirm, by and large, that the regulation of stomatal resistance is effective in many situations and in many crops. The internal/external ratio is therefore introduced in the program as a characteristic of a species or cultivar (L2C, Listing 7 Line 22). Table 22 provides a list of these values

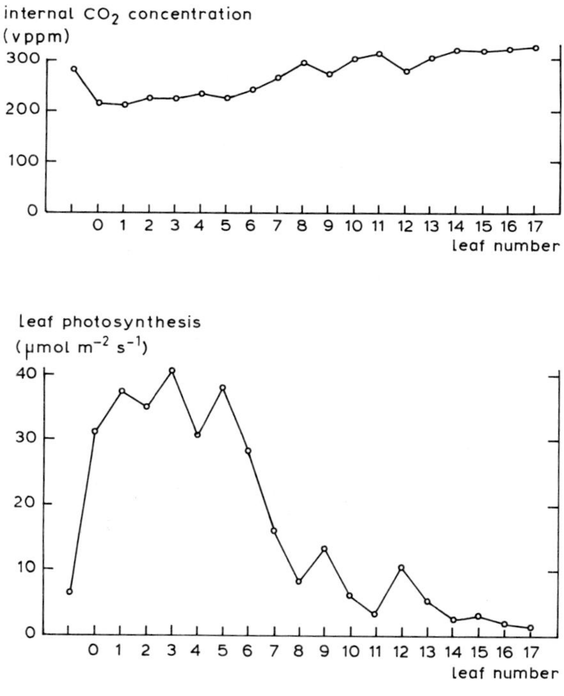

Figure 50. The internal CO_2 concentration and the photosynthesis rate in sunflower leaves as a function of the location in the canopy when counting from the top downwards. Conditions: mature leaf of a field-grown plant, full light, 26 °C (Source: Teubner, 1985).

Table 22. Values of the internal/external fraction of CO_2 concentration for several crops. Data refer to mature, healthy leaves, full light, no water stress, high relative humidity. Values are taken or derived from the references cited; values within parentheses refer to observations outside the common range. A value of 0.9 corresponds with no regulation.

Species	C_3/C_4	Fraction	Reference
Barley	C_3	0.65	Louwerse, 1980
		0.9	Bell, 1982
Cassava	C_3	(0.22)	El-Sharkawy et al., 1984a
		0.62	Veltkamp, 1985
Cotton	C_3	0.9	Dubbe et al., 1978; Bell, 1982
		(0.67)	Constable & Rawson, 1980
		(0.40)	El-Sharkawy & Hesketh, 1965b
Cowpea	C_3	0.8	Summerfield et al., 1983
Faba bean	C_3	0.7	Goudriaan & van Laar, 1978b; Bell, 1982
Groundnut	C_3	0.67-0.69	Bhagsari et al., 1976
Maize	C_4	(0.21)	Louwerse & v.d.Zweerde, 1977
		0.5	Dubbe et al., 1978
		0.2-0.5	Bell, 1982
Millet	C_4	0.4	Jansen & Gosseye, 1986
Potato	C_3	0.59-0.66	Teubner, 1985
Rice	C_3	0.55	El-Sharkawy et al., 1984b
		0.67-0.76	Fukai et al, 1985
		0.65	Dingkuhn, 1985 (IR20,IAC25 and Azucena)
Sorghum	C_4	(0.065)	El-Sharkawy et al., 1984b
		0.4	Schulze & Hall, 1982
Soya bean	C_3	0.62-0.66	Rawson et al., 1977
		(0.40)	Dornhoff & Shibles, 1970
Sugar-beet	C_3	0.39	Nevins & Loomis, 1970
Sugar-cane	C_4	0.36	Bull, 1969
Sunflower	C_3	0.64-0.68	Rawson et al., 1977
		0.9	Louwerse, 1980
Sweet potato	C_3	(0.33)	Hozyo et al., 1983
Wheat	C_3	0.74-0.77	Rawson et al., 1977 &
		0.5	Bell, 1982

for different species. However, this important characteristic of field crops is still insufficiently understood and deserves more attention.

In some cases, the internal/external ratio is high and does not appear to be controlled by the crop. The leaf resistance in daytime and without water stress is then continuously equal to a minimum value (Louwerse, 1980); at night the stomata are still closed. The absence of regulation causes a considerable increase in transpiration. It also leads to a higher photosynthesis rate, resembling the effect of a higher ambient CO_2 concentration (see Subsection 2.1.4). For instance, the observed growth rate of a well-irrigated maize crop in Mali (Penning de Vries, 1982) of about 350 kg ha^{-1} d^{-1} could only be correctly simulated with a maximum leaf photosynthesis rate of 100 kg CO_2 ha^{-1} h^{-1}. A leaf resistance of only 70 s m^{-1} has been observed (van Keulen, CABO, personal communication) when the CO_2 concentration in the stomata must have been about 210 vppm, or almost twice as high as normal. The same phenomenon was observed in sunflower plants.

Transpiration with fully open stomata can be simulated by setting the internal/external ratio close to unity. The concurrent effect on leaf photosynthesis in C_3 crops can be included in the program by adding the multiplication with the ratio of the high and normal fraction internal/external to Listing 3 Line 53. This demonstrates a trade-off between increasing photosynthesis and water use efficiency.

The evaporation rate of a wet canopy, such as after rain, can be quite high and depends strongly on windspeed and air humidity. The evaporation rate is so high that the leaves dry quickly and the high rate is no longer sustained. This shows how much stomata protect the crop from water loss. The simulated rate of transpiration is not corrected for water intercepted from rain because the amount is insignificant.

4.1.6 The transpiration/photosynthesis ratio

The transpiration coefficient can be defined as the total amount of water transpired, divided by the amount of above-ground dry biomass produced (kg kg^{-1}) (soil evaporation is not considered in this coefficient). The transpiration coefficient is often calculated at the end of the growing season. As was established many years ago (cf., de Wit, 1958, Encyclopedia of Plant Physiology, Vol 12B, 1982; Tanner & Sinclair, 1983), the coefficient determined during water shortage is equal to that measured without stress. This is due to the constancy of the ratio of the internal over the external CO_2 concentration at different stress levels. There are considerable, but predictable, differences in the transpiration coefficient between environments and species.

The transpiration coefficient is still a crude concept in crop physiological studies, so the water use coefficient of the crop, defined as the amount of water transpired per amount of gross photosynthesis (kg water kg^{-1} CO_2), is used instead. It is calculated on a day-to-day basis (Listing 7 Line 62). The value of the water use coefficient ranges from about 50 or less, to 200 or more kg H2O per kg CO_2 fixed (the lower values apply to C_4 crops in humid conditions and the high values to C_3 crops in dry climates). This corresponds with a transpiration coefficient of 150-600. The coefficient is strongly related to the internal/external fraction CO_2 concentration (Figure 51), because the transpiration rate depends a great deal on this fraction while the photosynthesis rate does not. There is a small difference between C_3 and C_4 crops over the entire range of internal CO_2 concentrations due to photosynthesis characteristics, but the fact that their internal/external fractions are different has the largest effect.

Van Keulen & Seligman (1987) present much experimental evidence indicating that the ratio of transpiration and photosynthesis of a canopy is independent of the water stress level. Yet, transpiration is largely a physical process and gross photosynthesis is largely a biological process. Respiration also

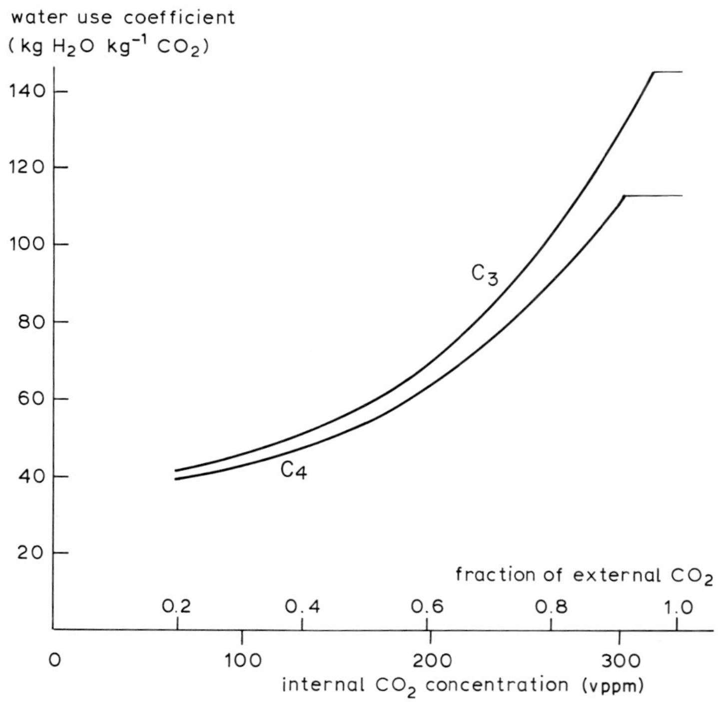

Figure 51. The water use coefficient of C_3 (PLMXP = 35.) and C_4 crops (PLMXP = 80.) as a function of the internal/external fraction of CO_2 concentration at an ambient concentration of 330 vppm. The maximum is attained with a leaf resistance equal to zero. The location of the labels C_3 and C_4 corresponds with common internal/external fractions for both groups.

plays a role in the transpiration coefficient and in the water use coefficient of a canopy; it is a biological process, but totally different from photosynthesis. These processes are all affected differently by water stress and stomatal closure and the extent to which their rates are reduced under stress vary. The direct and indirect (e.g., temperature rise) effects of stress are not the same in all environmental conditions. However, complications such as these are currently difficult to evaluate quantitatively. Since the experimental evidence suggests that the variability of the water use coefficient is small in practice, the constant ratio of transpiration to gross photosynthesis under water stress is also adopted here (Listing 7 Line 11).

Differences are often found in water use efficiency between genotypes. For instance, there are potato varieties with transpiration coefficients in the Netherlands from 125 to 180 (Bodlaender, 1986). Other internal/external fractions may be the basis for such differences. A moderate nitrogen shortage does not alter the water use efficiency (van Keulen & Seligman, 1987).

Altitude has an indirect effect on stomata; higher elevations stimulate transpiration. Lower ambient CO_2 concentration probably causes lower leaf resistance, though direct observations are not known. The CO_2 concentration at 1000 m above sea level is 290 vppm and 256 vppm at 2000 m elevation. Combined with the lower rate of assimilation, the increased transpiration leads to an increase in the water use coefficient of about 10% and 20%, in C_3 and C_4 crops respectively, from that at sea level, all other conditions being equal (Figure 52). Note that weather conditions change drastically with elevation (Koerner & Mayr, 1979; Oldeman & Frère, 1982).

4.1.7 Air humidity and stomatal resistance

A direct effect of low air humidity on stomatal resistance (i.e., not related to the crop water potential) has been reported at high vapour pressure deficits (Figure 53, cf., Schulze & Hall, 1982, pp. 192-196). It is significant in leaves of cassava, potato and of some other crop species (El-Sharkawy et al., 1984b; Teubner, 1985). This effect of air humidity on transpiration is probably insignificant in crops with well-developed leaf canopies, for humidity in the canopy builds up and lessens the direct effect. For instance, on a hot dry day (temperature 30 °C, relative air humidity 25%, windspeed 1-2 m s^{-1}) the vapour pressure difference between leaves and air above the canopy is about 3.0 kPa, but between the leaves and air surrounding them it is 2-2.5 kPa. More commonly, young plants may suffer from air that is too dry, because they still have little protection from neighbouring plants and are close to the soil surface which may be much warmer than the air at 2.0 m.

This humidity effect is mimicked by a direct relation between the maximum rate of leaf photosynthesis and either the vapour pressure deficit inside the canopy (Listing 7 Line 12, and add '* PLEH' to Listing 3 Line 53), or 0.75 times

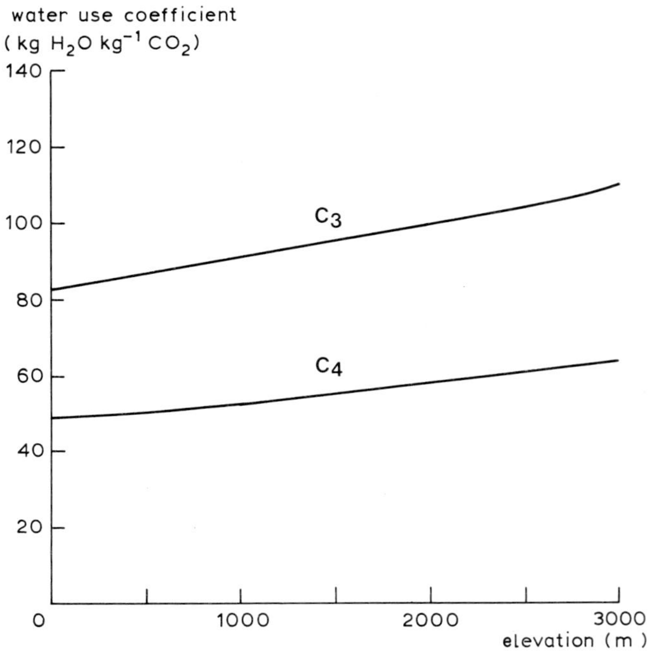

Figure 52. The water use coefficient of C_3 and C_4 crops in clear, warm dry weather conditions at elevations up to 3000 metres.

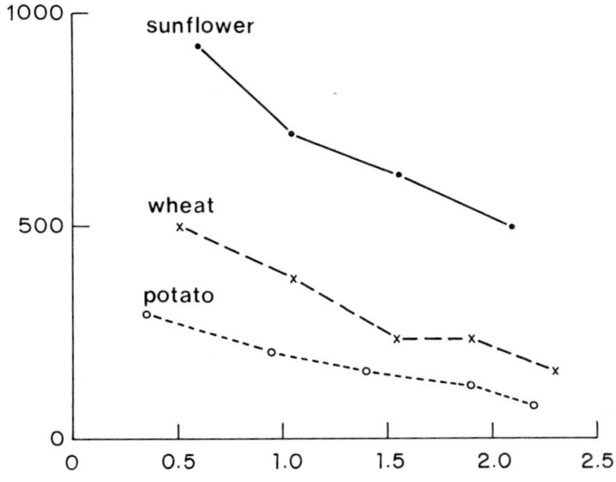

Figure 53. The effect of the humidity of the ambient air on leaf conductivity in field crops in the Netherlands (Source: Teubner, 1985).

the vapour pressure gradient when the canopy resistance is not computed (e.g., Listing 4 Line 75). Table 23 gives some data for several crops.

4.2 Water uptake

4.2.1 Introduction

When there is ample water, the rate of water uptake follows the transpiration rate very closely. If insufficient water is available, the plant lowers its water uptake, the stomata close to a certain degree, and actual transpiration becomes lower than potential transpiration. Transpiration then follows the rate of water uptake. Water in the crop provides only a small buffer between uptake and loss and their daily totals can be considered to be equal.

Water uptake occurs only where there are roots. The weight of roots at

Table 23. The relation between the multiplication factor of the maximum rate of leaf photosynthesis and the vapour pressure deficit (kPa) for some crop species. The originally reported effects are adjusted to account for less extreme situations in the morning and afternoon. The functions are given in CSMP style: vapour pressure is the first number of each pair and the factor is the second number.

Species	Relation	Reference
Cassava	1.0,1., 2.0,0.94,... 3.0,0.69, 4.0,0.42, 5.0,0.15	El-Sharkawy et al., 1984b
Cotton	1.3,1.0, 3.3,1., 5.3,0.86	Bierhuizen & Slatyer, 1964
Cowpea	2.0,1.0, 3.0,0.99, 4.0,0.93	Schulze & Hall, 1982
Faba bean	1.0,1., 2.0,1.,... 3.0,0.84, 4.0,0.52	El-Sharkawy et al., 1984b
Potato	0.5,1.0, 1.0,1.0, 1.5,0.90,... 2.0,0.65, 2.2,0.52, 4.0,0.52	Teubner, 1985
Rice	1.0,1.0, 2.0,0.99,... 3.0,0.86, 4.0,0.71	El-Sharkawy et al., 1984b
Sorghum	1.0,1.0, 2.0,0.98,... 3.0,0.88, 4.0,0.71	El-Sharkawy et al., 1984b
Soya bean	0.8,1.0, 1.2,1.0,... 1.4,0.99, 1.8,0.99	Rawson et al., 1977
Sunflower	0.5,0.96, 1.0,1.0,... 1.5,0.98, 2.0,0.89,2.2,0.84	Teubner, 1985
Wheat	0.8,1.0, 2.2,1.0	Rawson et al., 1977

certain depths and even their density do not reflect the water extraction pattern, for this occurs evenly over all rooted layers. To simulate water uptake in semi-arid regions, van Keulen (1975) assumed that the roots in each centimetre of rooted depth absorb the same amount of water, provided that all layers are equally moist. This implies that the resistance for water flow in the soil is larger than that in the roots. The independence of uptake and root density in annual grasses of a Sahelian rangeland provide an example (Figure 54). The water use of the grassland was well simulated when uniform uptake per unit of rooted depth was assumed; the distribution of root mass and root density decreased exponentially with depth in this situation (Penning de Vries, 1982). The total root mass of about 1500 kg ha^{-1} and the rooting density in the upper layers, exceeding 2 cm cm^{-3}, are common values for annual crops.

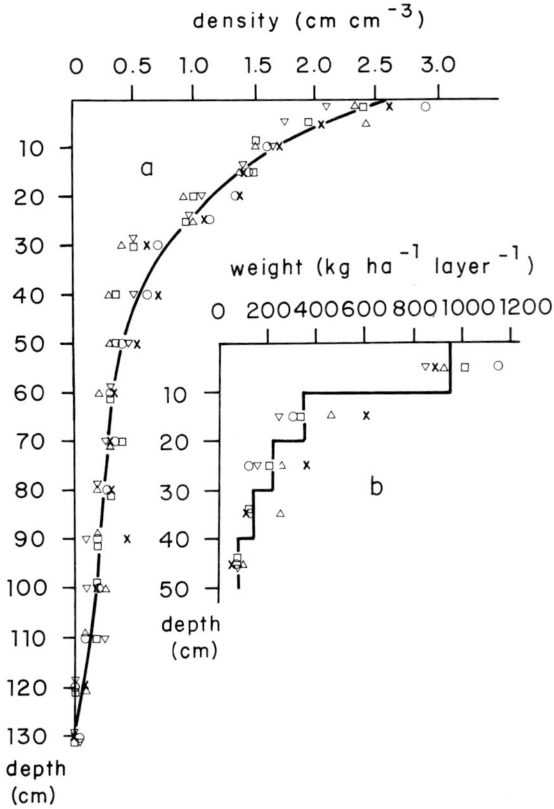

Figure 54. The density (a) and weight (b) of the root system of a vegetation of annual grasses in a Sahelian rangeland. Symbols represent different treatments (Source: Penning de Vries, 1982).

Many models of water uptake from the soil have been published. Detailed models exist of water movement in soil layers and of root uptake driven by water potential gradients. Some of these have two- and three-dimensional flow patterns (e.g., Lambert et al., 1976). Such detail in time and space is excessive for the crop growth models here, where van Keulen's approach (1975) is followed.

The root system is usually in contact with several parts of the soil profile that differ in texture, compaction and water content. Most soil water balance processes are more intensive near the surface. Hence, the soil profile is considered to consist of layers, and the conditions in each layer are treated separately. In the soil water balance modules, the profile is divided into a minimum of three horizontal layers, as in the simple soil water balance module L2SU for free-draining soils (Section 5.2), and into a maximum of 10 layers in the soil water balance module L2SS for soils with impeded drainage (Section 5.3). Water uptake simulation applies to both cases.

4.2.2 Uptake per rooted layer

The potential rate of water uptake per centimetre of rooted depth is an important variable. It is calculated by dividing the potential transpiration rate of the canopy by the total rooted depth (Listing 7 Line 28). There is no absolute maximum to the potential rate of uptake per unit of rooted depth in this module (but this might be a realistic addition if the root system is heavily damaged, such as after mechanical damage, insect attack, or advanced senescence).

The actual rate of water uptake is less than the potential rate when the soil water content is below a certain threshold. The uptake per layer is equal to the potential uptake rate per centimetre of rooted depth, multiplied by a stress factor and by the thickness of the layer. The calculation is repeated for all soil layers. The total water uptake is the sum of water withdrawn from the individual soil layers (Listing 8 Lines 11-14, Listing 9 Lines 63, 71-72).

The effect on uptake of low availability of soil water in a layer is represented by a multiplication factor, with a value between 0.0 and 1.0. Figure 55 schematically shows its relation to the relative soil water content. Plants do not suffer from water stress when the soil water content is at field capacity (−0.1 bar, or pF 2.0; see Subsection 5.1.3). No crop species absorb water from soils with a water content at −16 bar (pF 4.2, permanent wilting point), but there are differences between plant species. Hardy species do not reduce the rate of uptake until the soil water content is very low, while species sensitive to water stress reduce uptake at a water content a little below field capacity. The lowest soil water content where water uptake is unrestrained is a threshold value that lies between field capacity and permanent wilting point. Its exact location depends on the sensitivity of the species (characterized by a parameter), on the potential transpiration rate and on leaf area. The more resistant a species, the lower the threshold; the higher the potential transpiration, the higher the threshold. It is assumed that soil type in itself has no effect on this relation. This water stress effect is computed by the FUWS function (Listing 8 Lines 17-19, Listing 9 Line 68), which is similar to that used by van Keulen & Wolf (1986 p. 108). Table 24 provides a value for the sensitivity of some species to water stress. Note that it gives an impression only and that differences between varieties may be as large as 0.2 units of this scale or more.

Rice in rainfed lowlands is more sensitive to water stress than that calculated with the FUWS function. Rice yields drop significantly when the soil is not flooded for some time, even when the soil water content is still above field capacity. Irrigated and rainfed lowland rice have shallow root systems with almost all roots in the upper 0.2 m. The roots, grown in anaerobic conditions,

Table 24. The sensitivity of species to drought on a scale from 0.2 (non-resistant) to 1 (resistant) and to flooding from 0.2 (sensitive) to 1 (insensitive).

Species	Sensitivity to drought stress	Sensitivity to flooding
Bean	0.5	0.2
Cassava	0.65	
Cotton	0.65	0.2
Cowpea	0.8	0.2
Groundnut	0.65	0.2
Millet	0.8	0.6
Most cereals (wheat, maize)	0.65	0.6
Pea	0.5	0.2
Potato	0.5	0.2
Rice, upland	0.5	0.6
lowland	beyond scale	1.0
Sorghum	0.8	0.6
Soya bean	0.8	0.6
Sugar-cane	0.8	0.6
Sunflower	0.5	0.6
Sweet potato	0.65	0.2

Sources: most values for sensitivity to drought are derived from van Keulen and Wolf (1986), those for flooding are estimates partially based on Jackson & Drew (1984).

may differ from roots grown in aerobic soils and could be more sensitive to water stress (Subsection 4.2.4). Insufficient data for rice were available to derive an equation relating water stress, soil and atmospheric conditions. A simple solution for deriving water stress for rainfed and irrigated rice directly from the relative soil water content (RWCLQT) is by replacing Line 68 in Listing 9 by:

```
RWCLQT = (WCLQT(I) − WCWP(I)) / (WCST(I) − WCWP(I))
WSE    = AFGEN(WSET, RWCLQT)
FUNCTION WSET = −0.1,0., 0.,0., 1.,1., 1.1,1.0
```

The WSET function quantifies (in CSMP style) the relation between stress and relative water content.

No *a priori* preference is attributed to uptake from the upper soil layers. Yet, simulations often indicate that, because of more frequent wetting, more water is withdrawn from upper layers than from the lower layers of the same thickness. Preference may be given to water uptake from top layers, for example, to simulate the hypothesis that the resistance to water flow within the root system is high, such as when the xylem vessels are very narrow (cf., Encyclopedia of Plant Physiology, Vol 12B, 1982; Taylor & Klepper, 1978). The potential rate of water uptake per centimetre of rooted depth is then above average in the top and less than average in the lower layers. Quantifying these effects at the process level requires further study and experimental data for calibration.

A direct effect of temperature on water uptake has not been included here as it is probably not very important in most cases. High transpiration rates coincide with high radiation levels and usually with non-limiting soil temperatures. A temperature effect must be added when winter or spring crops are simulated (cf., van Keulen, 1975), because of the dependence of root permeability on temperature. Data for this are unavailable for most species.

4.2.3 Rooted depth

The rooted depth is defined as the depth from which the crop effectively extracts water. A density of 0.10 cm root length per cm^3 of soil volume may be adopted as a lower density limit. This is a low threshold value because water is mobile and flows relatively easily to roots. Rooted depth does not refer to the extreme depth where a few roots are still found.

The length of fibrous roots can vary enormously without much impact on root weight. Hence, simulation of rooted depth occurs independently of the growth of root mass. Rooted depth can increase at a rate of 3-5 cm d^{-1}, but soil physical, soil chemical and biological factors can reduce it (Taylor & Klepper, 1978). Table 25 presents a few data concerning the maximum rate of increase in rooted depth. Root growth generally stops around flowering. Water stress or low soil temperature reduces root growth (Listing 7 Line 34). Because of the

Table 25. Rates of increase in rooted depth in moist soil at an optimum temperature and the maximum effective rooted depth (i.e., root density >0.1 cm cm^{-3}) for different crop species.

Species	Rate of increase in rooted depth ($m\ d^{-1}$)	Maximum rooted depth (m)	Reference
Barley	0.03	1.5	Day et al., 1978
Cotton	0.025	1.8	Bassett et al., 1970
	0.03	1.8	Taylor & Klepper, 1974
Cowpea	0.028	1.0	Haverman, 1986
Faba bean	0.014	0.7	Grashoff et al., 1987
Maize	0.06	1.0	Sibma, 1987
	0.06	2.0	Taylor & Klepper, 1973
Millet	0.04	1.0	Jansen & Gosseye, 1986
	0.02	1.0	Gregory & Reddy, 1982
Potato	0.014	0.8-1.0	Vos & Groenwold, 1986
Rice(upland)	0.02	0.4->0.8	Yoshida & Hasegawa, 1982
Rice(lowland)	0.01	0.3	Sharma et al., 1987
Sorghum	0.05	1.4	Kaigama et al., 1977
Soya bean	0.035	1.7	Stone et al., 1976
Sugar-beet	0.02*	1.2	Brown & Biscoe, 1985
Tulip	0.02*	0.4	Benschop, 1986
Wheat winter	0.018	1.3	Gregory et al., 1978
spring	0.012	1.8	van Keulen&Seligman,1987

* estimate

lack of specific data, the effect of temperature on root extension is supposed to equal that of photosynthesis. The effect of water stress on the rate of increase in rooted depth is supposed to equal that of water uptake in the layer where the root tips are found (Listing 8 Line 27, Listing 9 Line 74). The effect of anaerobic conditions on root extension downwards is handled by setting the rooted depth increase to zero at depths below 0.2 m when there is less than 5% air in the soil (Listing 9 Line 73).

Roots grow down to a certain maximum depth if they are not restricted by soil conditions. The maximum depth depends on the plant species and ranges from 0.5-1.5 m or more. Table 25 gives some approximate values for the maximum rooted depth. Significant differences between cultivars for this characteristic are reported for upland rice (Gupta & O'Toole, 1986) and are also expected to exist within other species. Sensitivity analysis has established this as an important characteristic, though little is known about it in field crops. Maximum rooted depth should be determined around flowering in soil profile pits, either by using root observation tubes (Vos & Groenwold, 1983), or indirectly by monitoring (with neutron probes) the depths from which water is drawn when drainage is insignificant.

A very dense soil offers mechanical resistance which hampers the extension of roots downwards and reduces the maximum attainable depth. An obvious case is where shallow soil lies on bedrock. High soil densities can also be found at depths of 0.30-0.80 m in deep soils, particularly just below the ploughed layer. Its presence may be intentional, such as during soil preparation in irrigated rice where a hard pan is needed to reduce percolation of irrigation water. A compact layer can also develop unintentionally, such as when harvesting crops with heavy machinery. A physical limitation to rooted depth is approximated by specification of a maximum depth as a soil characteristic; the shallowest of the rooted depths set by the soil and by the crop is used (Listing 7 Line 33). Note that cracks, tunnels from animals or decayed roots, and other irregularities can make dense layers more penetrable for roots than the soil density measurement of a uniform piece of soil may indicate (Subsection 5.1.4).

Loss of rooted depth in a senescing root system may be added to a simulation model, but field data are needed to calibrate this effect.

4.2.4 Anaerobic conditions

Plants with roots in fully saturated soils generally suffer from stress. For an extensive review of physiological effects of excess water see Jackson & Drew (1984). Root systems of agricultural crops that are developed in aerobic soils do not have aerenchym and degenerate within several days when anaerobic conditions are imposed. Root permeability first decreases and uptake slows down. Root cells disintegrate and die when their metabolism no longer provides sufficient energy (i.e., O_2) for maintenance. Hence, wilting is sometimes, though not always, observed after flooding.

Flooding quickly depletes the O_2 in the soil and the supply is then almost nil. Anaerobic conditions occur on heavy soils following intensive rainfall and when the groundwater table is very high. Those conditions can be simulated with the module L2SS (described in Section 5.3).

Rice in irrigated or rainfed lowland soils has an effective root system in anaerobic conditions, because its roots develop aerenchym tissue that provides air channels (Yoshida, 1981). The rate of diffusion through the narrow channels provides sufficient O_2 to permit roots to extend to about 0.2 m. Several other crops develop roots with aerenchym, but not as extensively as rice.

The effect of flooding on water uptake is approximated with the FUWS function, in a similar manner to the effect of water shortage (Figure 55, right hand part of the graph). The effect is assumed to be proportional to the soil water content between field capacity and saturation, and independent of the transpiration rate. To mimick the non-water stress effect of flooding which occurs after a root system is established, a FINISH condition similar to CELVN (Subsection 2.3.4) can be added to the program, stating that crop death occurs if flooding lasts a certain number of days. This maximum flooding period is dependent on the species and its development stage. Some species grow a new root system with aerenchym in anaerobic conditions. Rice does so extensively, but other crops, including soybean, wheat and sunflower, also have this capacity. The regrowth rate of an effective root system after flooding also effects the degree of crop survival. This is not considered here.

Diseases are common after soils are flooded and cause much damage. This is due to the crop's physiological condition and the high humidity accompanying flooding. This is not considered here.

4.3 Non-stomatal effects of water stress

4.3.1 Introduction

This section discusses simulation of the effects of water stress on the physiological processes of plants which are not mediated by stomata. The degree of

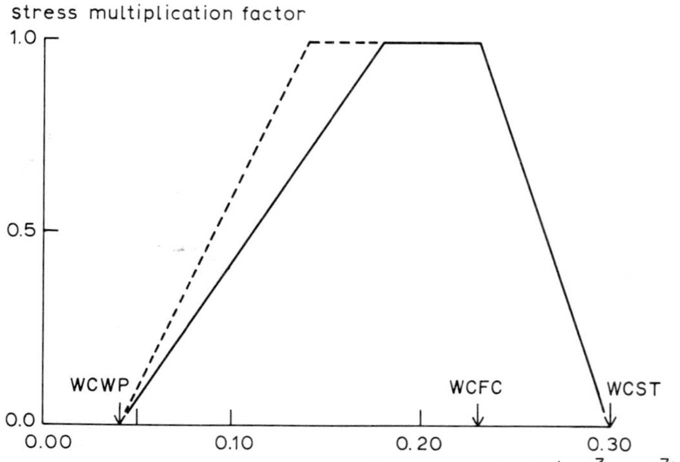

Figure 55. The relation between the soil water content and the stress multiplication factor on the rate of water uptake. WCWP, WCFC and WCST represent the soil water content at wilting, field capacity and saturation, respectively. The dashed line represents either a more drought resistant species under the same field conditions, or the same species under a lower evaporative demand.

stress can not be simulated explicitly because of the relatively long integration period. The ratio between actual transpiration (TRW) over potential tranpiration (TRC) is used to represent the degree of stress. When this ratio is above 0.5, the effects on physiological processes are usually small. The most significant influence of water stress is the indirect effect through reduced photosynthesis, particularly if the stress occurs during a sensitive period, such as during grain initiation in cereals (Subsection 3.2.5).

This definition of water stress implies that the effects of stress are over as soon as the soil is moist again. However, after prolonged stress this is not correct. To account for aftereffects, van Keulen (1982) computed a particular running average of the relative transpiration deficit (RTDA and RTD, respectively) to characterize water stress. This running average increases when the relative deficit exceeds 0.4 and decreases when it is less. Van Keulen chose a 10-day time coefficient for buildup and breakdown of the average value. This resembles the simulation of buildup and breakdown of hormone levels. In van Keulen's model, the running average, representing water stress, affects biomass partitioning, leaf photosynthesis characteristics and leaf senescence. This description of the effect of water stress can be achieved by replacing TRW / (TRC + 1.E-10) in Listing 7 Lines 10-11 by (1.0 − RTDA), where:

```
RTDA = INTGRL(0., INSW(RTD − 0.4, RTD * (1. − RTDA) / 10.,...
       −RTDA / 10.))
RTD  = (TRC − TRW) / (TRC + 1.E-10)
```

4.3.2 *Crop development during water stress*

From the limited amount of data available, it appears that a moderate level of water stress often has no direct effect on the rate of crop development (Halevy, 1985; Section 3.1). However, there are exceptions, for the physiological development of some crops slows under stress and this lengthens the vegetative period. In other crops water stress stimulates physiological development, for example, the development rate of Faba beans increases once the actual/potential transpiration ratio is below 0.7 and this increases up to twice the normal value to a ratio of 0.0 (Grashoff, CABO, personal communication). The development rate for rice in rainfed lowland and upland situations decreases under moderate stress so that flowering and maturity are postponed by 7-10 days (Buresh, IRRI, personal communication). Morphological development of sugarcane stops, or even reverses, under stress. Acceleration or deacceleration of development can be represented in the model as a relation between the stress level and a multiplication factor for the development rate (Listing 3 Line 97; Listing 7 Line 10). Very little data on this relation is reported in the literature. Without explicit data, the effect is assumed to be negligible.

A freely transpiring crop cools itself so much that the leaf temperature can be several degrees below the air temperature, even at high radiation intensities (Idso et al., 1981). The temperatures of water-stressed leaves can be several degrees above those from non-stressed leaves (Gupta & O'Toole, 1986). Reduced transpiration under nutrient stress can have the same result. Growing points do not have the same temperature as leaves, but have a temperature between that of leaves and the air. Increased temperature can accelerate flowering and shorten the crop's life cycle by several days (Seligman, ARO, personal communication). This temperature increase could be approximated from the reduced transpiration rate, but it is difficult to provide a fair level of accuracy. No general solution is proposed here.

Extreme levels of water stress may kill part of the plant. If new sprouts or tillers emerge, the crop rejuvenates and returns to an earlier development stage. No general solution is proposed here for simulating the effects of extreme stress levels.

4.3.3 Carbohydrate partitioning

Carbohydrate partitioning between shoot and root under water stress is altered in favour of the root biomass. Brouwer (in de Wit et al., 1978) described the biological principle of the mechanism; roots are formed in proportion to the demand from shoots for water. Yet it is difficult to quantify the growth stimulation of root biomass in response to water stress. It is assumed that up to a moderate stress level (actual/potential transpiration rate is 0.5) there is no significant effect on partitioning. At higher stress levels during the vegetative phase, the share that goes to the roots increases by up to 50% of the amount that otherwise would go to the shoot (Listing 7 Line 9, and Listing 3 Line 40). The flow of carbohydrates to storage organs in vegetative crops, such as sugarbeet, increases under stress at the expense of the flow to leaves. This is not included in the program. Other differences between species in this respect are not yet known. Although water stress stimulates root growth relative to shoot growth, water stress in the layer with root tips reduces the root extension rate to greater depth. Drought in upper layers does not stimulate Faba bean roots to grow deeper (Grashoff et al., 1987).

It is assumed that the relative partitioning of carbohydrates between leaves, stems and storage organs is not affected by water stress. The impact on formation of reserves is an indirect effect. Clearly, these are only approximations, though experience with field crops indicates that they are often acceptable (Section 3.2).

4.3.4 Leaf area

The effect of water stress on the growth of leaf area occurs through the effect of stress on root-shoot partitioning, and hence on an increase in leaf weight. Water stress can lead to higher values of the specific leaf weight, but this is not simulated here. An explicit way to deal with the effect of stress on the growth of leaf area is described by van Keulen & Seligman (1987) and is specific for the development of leaf weight in wheat.

Severe water stress can lead to progressive death and removal of leaf area. It is impossible to simulate this process dynamically, because the extreme values reached during the day are of critical importance. These are not obtained with the modules described here. Heterogeneity of soil environment is also important. Modellers should use experimental data to mimick leaf and stem death under severe stress for their crop and their situation. Van Keulen (1982) reduces biomass of wheat and grasslands by 0.1-0.2 d^{-1} when water stress exceeds a certain level and reduces leaf area correspondingly to mimick gradual crop death.

Some crops use leaf rolling as a means of reducing leaf area when water-stressed, such as rice (Gupta & O'Toole, 1986). Rolling increases diffusion resistances and reduces the area of exposed leaves. Other crops move or fold leaves when under stress to intercept less radiation. In these ways, they avoid drought, save some water and have a better chance of survival. But they are also less productive during the stressed period. These phenomena can be simulated by replacing the leaf area (ALV) by the effective leaf area (ALVE) in all appropriate rate equations and functions and by reducing the effective leaf area in relation to the stress level. For example:

```
ALVE = ALV * AFGEN(ALVRT, TRW / (TRC + 1.E-10))
FUNCTION ALVRT = 0.0,0.4, 0.4,0.4, 0.5,0.5, 0.6,0.8, 0.8,1., 1.0,...
                 1.0
```

This numeric example for leaf rolling is derived from O'Toole & Cruz (1980) for lowland and traditional upland rice cultivars. Reduced leaf area due to wilting can be similarly mimicked. (In this approximation, an 'implicit loop' creates a pitfall, for it is constructed by making the effective leaf area a function of the actual/potential ratio, because the actual transpiration itself is a function of the effective leaf area. A loop, such as this, can be solved in CSMP by putting in the DYNAMIC section 'FALVE =...' as the last line instead of 'ALVE =....' (as above), and 'ALVE = IMPL(ALV, 0.05, FALVE)' as the first line. CSMP will try each time period again with different values for ALVE until all rate equations are balanced, and only then proceeds with integration. For further information on the IMPL function, see IBM, 1975).

4.3.5 Leaf photosynthesis characteristics

Severe water stress can reduce the capacity of leaves to photosynthesize. This may cause damage to chloroplast structure and adaptation to lower actual photosynthesis rates. Van Keulen (1982) suggested considering the maximum leaf photosynthesis rate (read: amount of intact RudPCase) and the initial efficiency (read: amount of chlorophyl) as state variables. Both state variables decrease under severe stress by 0.0-0.05 d^{-1} and have only a limited recovery capacity. It is a good approach, but the parameters are not general and need to be defined for each case. Readers may want to explore this further.

If temperatures are so high that the photosynthesis apparatus might be damaged, attention should be directed to the most vulnerable leaf layers when calculating leaf temperature. The simulation of transpiration given here is not sufficiently detailed for this purpose.

4.3.6 Maintenance respiration

Maintenance respiration under water stress may intensify due to ion gradients increasing when the osmotic value of cytoplasma increases. On the other hand, the flux of excess energy increases when photosynthesis is reduced under stress (Subsection 2.3.3). Therefore, it is assumed that the carbohydrate requirement for maintenance respiration is unaffected by water stress. The energy for maintenance in leaf cells in daytime is still provided even when CO_2 assimilation has almost stopped.

4.4 Exercises

See Section 2.5 for an introduction to the exercises.

4.4.1 Transpiration without water stress

T1. Make a relational diagram of the processes and variables which determine canopy transpiration without water stress. Where is the water use coefficient? How is stomatal regulation indicated?

T2. Is it important to establish whether the maximum resistance of the leaf exceeds 2000 s m^{-1}? If so, suggest ways how to establish it.

T3. How is the transpiration coefficient (H_2O/CO_2) converted into the water use coefficient (water/dry matter)?

S1. Evaluate the importance on yield of leaf rolling in rice during the wet season on loamy soil, as discussed in Subsection 4.3.4. What happens if the precipitation in the beginning was slightly more (29 · 6.), but the downpour of 300 mm occurred only at day 260 and no rain fell at the end of the growing season (15 · 0.)? Explain the result.

S2. Suppose that soya bean stomata can be made to regulate at a CO_2-internal/CO_2-external fraction of 0.4. How is yield affected in the wet and in the dry season on a loamy soil, and how is the average water use efficiency and average stress level affected? What is the consequence of a disease that induces loss of stomatal control, leaving them fully open all the time? Why is the water use coefficient in both seasons similar? Do you expect a similar result on sandy soil?

4.4.2 *Water uptake*

T4. Make a relational diagram of the processes and variables which determine water uptake under water stress.

S3. Determine the numerical value of the effect of stress on water uptake for all soil water contents between air dry and saturation for values of TRC from 0.5 to 10 mm d^{-1}, for ALV from 0.5 to 10 $m^2 m^{-2}$, and for the water stress and flooding sensitivity coefficients between 0.0 and 1.0.

Is a value of −1.0 for WSSC or WFSC biologically meaningful?

4.4.3 *Non-stomatal effects of water stress*

T5. Explain the operation of the statements to compute a running average for the relative transpiration deficit (Section 4.3.1). What are the maximum and minimum value of RDTA?

T6. The effect of high or low values of the water stress sensitivity coefficient on crop production in monoculture is often not as large as one might expect. Why not? Why can this be different if weeds are present?

T7. The actual rooted depth attained in simulations is usually 0-3 cm more than the maximum specified. Explain. What are the consequences?

S4. Add to the rice model (Listings 3, 5, 7, Appendix B) that the development rate is unaffected down to a water stress level of 0.9, but reduced by 30% when the stress level reaches 0.5. Run it with the SAHEL water balance. Reduce the precipitation by replacing 29 · 4., 200. with 30 · 0.0. By how much is the crop delayed and is yield decreased? Explain.

S5. What would be the difference in rice yield if a breeding program could make rice absorb water more effectively compared to ineffectively (WSSC equals 1. and 0., respectively)? Assume that the development response to stress and precipitation are those from the previous exercise. How does it compare to increasing the maximum rooted depth by 0.1 m? Explain the results.

S6. Add the equations about the running average of transpiration deficit (Section 4.3.1) to the dry season rice program on loamy soil. What is the effect on yield and water use efficiency?

4.5 Answers to exercises

4.5.1 *Transpiration without water stress*

T1. Carefully analyse the processes and variables. Draw the diagram using the symbols presented in Figure 5. Compare your result with Figure 6. The water use coefficient is the ratio of actual transpiration and actual photosynthesis. The stomatal resistance is mainly determined by photosynthesis and the CO_2-internal/external fraction.

T2. It is unimportant because the absolute value of the transpiration rate is already very low when this maximum becomes effective.

T3. The conversion of carbohydrates into dry matter often yields about 0.5 g g^{-1}. CO_2 weighs 44/30 times as much as the carbohydrates. The water use coefficient is therefore about 3x as high, the actual ratio depending on conditions and the species.

S1. Replace ALV by ALVE in all rate equations and in Lines to calculate WSE1-3.

The rainy season yield is 6146.9 kg ha^{-1} at maturity with leaf rolling; without response to water stress it is 6186.0 on the same date. The stress on this soil was too mild to give the responsive crop an advantage in survival and it lost a little in productivity. The late rain in the 'dry' wet season is just in time for the leaf-rolling crop to survive the serious drought, so that it yields 4295.0 kg ha^{-1} at day 290. The non-adaptive crop had less water left in the soil when the late rain arrived. Net photosynthesis was negative from day 258 onwards and the crop died before the rain arrived at day 261. (Note: use FUPHOT without the restriction on ALV imposed by SUERRM to allow the IMPL function to perform preliminary computations over a wide range.)

S2. Use L1D, L2C, SAHEL and T12 with soyabean data for the wet and dry season, the standard sandy soil and short weather data set. FIEC = (0.64, 0.4, 0.95). The average stress level may be computed as the accumulated ratio of TRW and TRC divided by TIME.

The yields are 3452.0, 3446.4 and 2124.9 kg ha^{-1}, respectively, in the wet season, with average water use coefficients (kg H2O kg^{-1} CO_2) of 101.3, 72.2 and 212.9, and average stress levels of 0.994, 1.000 and 0.822. In the dry season, the numbers are: for yield 2418.4, 4620.8, 406.0 (early death); for water use coefficient 111.1, 78.2, 202.5, and for average stress levels 0.870, 0.990, 0.699, respectively. The water use efficiencies in both seasons are the same because there are almost no differences in seasons in the short data set. The effect of changing FIEC from 1.0 to 0.4 is much more pronounced when less water is available, such as on a sandy soil, or with less rain.

4.5.2 Water uptake

T4. Carefully analyse the processes and variables involved. Draw a diagram using the symbols of Figure 5.

S3. Use only Line 17 from Listing 7 and the FUWS function from Appendix B. Make reruns for combinations with multiple value parameters and with TRC = TIME. The output statements PRTPLOT WSE1 and PAGE MERGE make it easy to check results. WSE1 plotted versus TRC is a graph with curved lines at different positions.

Negative values of WSSC or WFSC imply that uptake is restricted by roots or soil even at field capacity. This could occur if the soil moisture contains salt or when roots are damaged.

T5. This running average is more complex than that in Subsection 1.4.3, because the rate of change of RDTA depends on another variable RTD. RDTA decreases when RTD is low, and increases when RTD increases. Multiplication with '1. – RDTA' prevents RDTA from exceeding 1.0. Without stress, RTD and RDTA are equal to 0.0.

T6. The WSSC and WFSC coefficients do not make more water available to the crop, but increase the rate at which it can be absorbed. The difference is significant only if a crop grows under moderate water stress, but not if it is in soil that dries and then recieves water again. When a crop is competing with weeds, crop water absorption may slow down while weeds continue to grow apace (cf., Lof, 1976). The crop's share of water is then smaller and crop yield is reduced.

T7. The final rooted depth exceeds the maximum specified because the daily increment is several centimetres and growth stops only when the maximum is exceeded. The time period for integration is relatively long for simulation of root growth. If the specified maximum rooted depth is equal to the profile depth, roots 'stick' out of the simulated system and these root tips cannot absorb water. As a result, the total water uptake is a few percent too low.

S4. The yield is 5706.5 ykg ha^{-1}, harvested at day 292. Without any development rate response to stress, the yield is 236.6 kg ha^{-1} less and maturity is reached 2 days earlier. The longer growth duration permits higher production because stress was not severe.

S5. The wet season differences are small. The difference between the WSSC extremes is 152.8 kg ha^{-1} in grain yield and 2 days in maturity date. This soil appears to provide a large buffer.

Increasing the rooted depth by 0.1 m raises the yield by 219.2 kg ha^{-1} and advances maturity by 2 days (reference: values of the previous exercise, delayed development). The advantage is small because the soil provided just enough water at a rooted depth of 0.7 m and stress develops only in the last days. Rooted depth of 0.8 m still has water left. The difference would be larger in a drier year.

S6. The formulation with RTDA leads to a reduced water use efficiency (WUDM = 76.0 versus 95.9 kg dry matter kg^{-1} water at TIME = 70.; WUPC remains almost the same). There is now a little more water stress and this leads to premature death (at DS = 1.598). Values of RTD below 0.4 have less effect on growth than the original equation, while those that exceed 0.4 have progressively more effect.

5 Soil water balance

Simulation of the soil water balance is discussed in this chapter. Though rooted in soil physics, the main simulation modules are relatively simple because the water balance is only considered in relation to crop growth and many details are hidden. Section 5.2 discusses a module (L2SU, Listing 8) to simulate soils that drain excess water freely and quickly and uses a one-day time period for integration. Section 5.3 discusses an alternative module (L2SS, Listing 9) simulating soils in which drainage is impeded, and in which saturated layers may develop. This second module is more complex and simulates fluxes with time periods much smaller than 24 hours, while crop simulation continues with one-day time periods. Section 5.4 presents details of the two modules and provides examples of their use as part of a full simulation program.

5.1 Soil physics for simulating crop growth

5.1.1 Introduction

Water in the rooted part of the soil is quite mobile. Its distribution over time and depth is important because it determines the amount of water available to the crop. Water enters the soil as rain or irrigation water and often by capillary rise from the groundwater table. It leaves the soil by evaporation, drainage, or is taken up by roots. Gravity and gradients in moisture suction cause water movement within the soil. Soil physical characteristics (e.g., hydraulic conductivity) codetermine the rate of water flow. Horizontal (lateral) inflow or outflow can be significant in hilly regions and in plots next to ditches and waterways.

A soil profile generally consists of layers of different soil types with distinct physical characteristics. To represent the vertical heterogeneity in water content, physical characteristics and root activity, the soil is divided into horizontal compartments or layers. The thickness and physical characteristics of each layer must be specified. The mathematical equations describing the soil processes are the same for all layers, but because the value of the variables and constants in the equations vary from layer to layer, the outcome is specific to each.

Spatial variability of the soil water content at given depth in the field is caused by spatial heterogeneity of physical characteristics of soil layers, irregularities at the surface, artificial drainage structures and by heterogeneous root distribution. This variability is not included in these models.

Two soil water balance modules are discussed: a simple module (L2SU,

Listing 8) for situations where the groundwater table is very deep (so that no water flows upwards to the rooted profile) and where the hydraulic conductivity of the soil does not limit downward water flow; and a more complex module (L2SS, Listing 9) for situations where hydraulic conductivity may limit water transport in the soil and where water can flow from the groundwater to the rooted profile. Neither module is suited to handle water infiltration into soils that crack when drying, as in vertisols (Eswaran, 1985); water uptake patterns in these soils are different, for roots prefer to grow along ped faces that offer little mechanical resistance. See also Table 1.

5.1.2 Soil texture

Soils differ in chemical and physical properties and in morphological characteristics and can be classified according to several systems. A widely used classification system is that described by the United States Department of Agriculture (1975). Another system uses hydrological conditions as the classification key (e.g., Kanno, 1956, 1962; cited by Moorman & van Breemen, 1978, p. 51). A classification based on physical characteristics is used here, since these are important for the soil water balance.

The number of pores and the distribution of pore sizes determine the hydraulic properties of a soil (e.g., Schuh & Bauder, 1986). To a large extent, the distribution of particle sizes affects the distribution of pore sizes. After removing soil particles larger than 2.0 mm in diameter from a soil sample, three particle classes are distinguished by size: sand (0.05-2.0 mm in diameter), silt (0.002-0.05 mm) and clay (<0.002 mm). Each textural class has a specific range of combinations of fractions of these particles (Figure 56). A soil consisting of 0.35 g clay per g of dry material and 0.30 g g^{-1} of silt (and therefore of 0.35 g g^{-1} of sand) is classified as a clay loam.

Equations and parameters to quantify the relation between the soil water content and soil water potential, and also the relation between the soil water content and soil hydraulic conductivity are discussed below. The parameter values are related to the textural class. These relations are necessary to simulate the soil water balance in a deterministic way. For parametric simulation of free-draining soils, only three specific points of the soil water content – water potential relation are needed: the water content at field capacity, at wilting point and when air dry. Those water contents may be computed, or obtained from measurements (Subsection 5.1.3).

For a soil of a given textural class, total porosity and pore-size distribution change with bulk density (i.e., the dry mass per volume unit of dry soil material, in kg m^{-3}). Soil samples of the same texture can differ in bulk density and thus have different hydraulic properties. The amount of organic matter in the soil also affects the physical characteristics (e.g., Schuh & Bauder, 1986; Saxton et al., 1986). Therefore numeric values cited for different textural classes should be used with caution for soils with a very high organic matter content

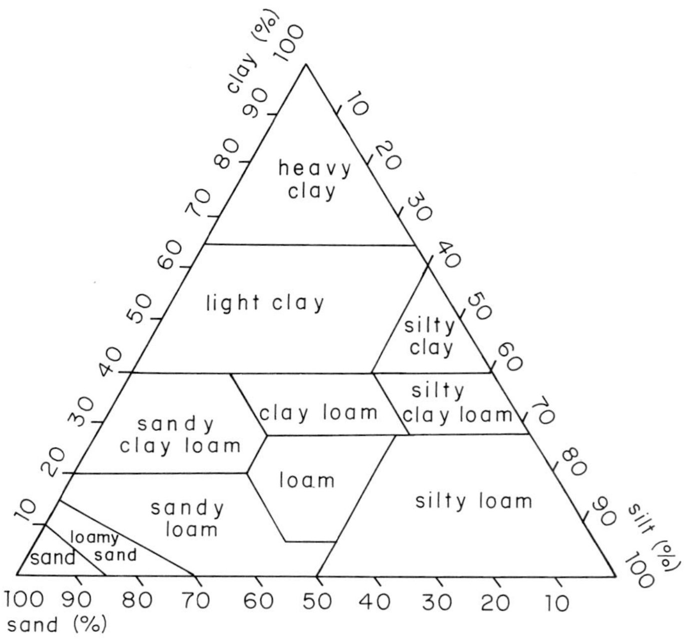

Figure 56. A graphical representation of the composition of mineral soil material as fractions sand (diameter 0.05-2.0 mm), silt (0.002-0.05 mm) and clay (<0.002 mm) (Source: Driessen, 1986a).

(more than 30%), for soils where swelling and shrinking clay minerals are dominant, or for compacted soils. Soils with a high content of gravel, stones, or cementing aluminum- and iron-oxides must also be regarded as exceptions. In these cases new observations are needed of the relations water content – water potential and water potential – hydraulic conductivity for the complex module (L2SS), and of the three characteristic soil water contents (field capacity, wilting point and air dry) for the simple module (L2SU).

The importance of accurate soil data for crop growth simulation varies greatly from one variable to the next, and can be strongly affected by soil hydrology and the crop's demand for water. Before starting a program to measure the variables, sensitivity analysis with a model is recommended.

5.1.3 Soil water content and soil water potential

Liquid water in the soil is subject to several forces. The extent to which water is attracted is expressed in energy terms. The energy status of the water is called a potential and the energy can be expressed in J per m^3 of water, or on the basis of mass (kg) or weight (N) units.

The part of the potential which is due to matric forces (or capillarity) is the matric potential. Gravity pulls water downwards. The gravity force is associated with gravitational potential. The actual force on the water is the sum of the matric forces and the gravity force. The forces may work in the same or in opposite directions. The matric potential and the gravity potential are sometimes combined and then called the 'total soil water potential'. Water may flow upwards or downwards, but always to the lower total potential. It is assumed that the gas pressure in the soil is equal to that in the atmosphere, that the matrix is rigid, and that osmosis plays no role.

Several names are used for the potential associated with matric forces: matric potential ($J\ kg^{-1}$), pressure potential ($J\ kg^{-1}$), soil water pressure ($J\ m^{-3}$ = Pa), tensiometer pressure (Pa), pressure head ($J\ N^{-1}$ = m) and soil moisture suction (Pa, or m). The names are different according to the units in which these potentials are expressed, but they all refer to the same concept, as long as soil gas pressure is atmospheric. Also the gravitational potential can be expressed in any of these units. The total soil water potential is also called hydraulic potential ($J\ kg^{-1}$) or hydraulic head (m). By definition, a pressure potential of 0.0 is assigned to free water under atmospheric pressure. Similarly, the reference level for the gravitational head is the soil surface. Pressure head h is negative in non-saturated soils. Suction $|h|$ has the opposite value of pressure head. The term suction is not used when the matric potential is positive, as may occur in saturated soils. All potentials are expressed here in head equivalents (cm). For more extensive treatment of this subject see Koorevaar et al. (1983).

The soil moisture content is preferably expressed on a volume basis (e.g., $cm^3\ cm^{-3}$). When the volumetric soil moisture content θ decreases below the saturated volumetric water content θ_s, the matric potential becomes negative and the soil moisture suction $|h|$ (cm) increases. A decrease in θ is not proportional to an increase in suction. The relation between suction and θ is mainly determined by pore geometry and is specific to soil type. A coarse sandy soil has many wide pores and loses much of its moisture when exposed to low or moderate suction. But a clay soil consists of fine particles and is finely porous. Narrow pores retain water with a higher force than wide pores and only release water to roots when the high forces are compensated for by high root suction.

Soil moisture suction during the growing season usually ranges between 0.0 and 20,000 centimetres, but can reach higher values. The logarithm of suction (in cm) is used to facilitate manipulation with this variable as it varies over a wide range. This logarithm is called the pF-value and the relation between θ and $\log|h|$ is the soil's pF-curve (Figure 57).

Empirical equations have been developed to describe the $\theta(h)$ relation for different soil types (e.g., Rawls et al.,1982; Saxton et al., 1986). Driessen (1986a) gave the following descriptive equation:

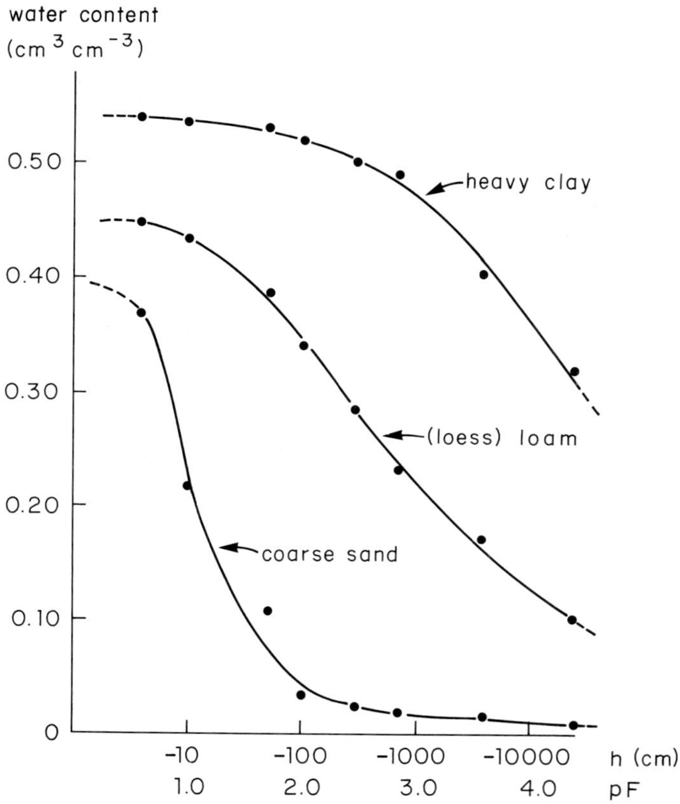

Figure 57. Some characteristic pF-curves for soils with different textures. *h* is the pressure head (Source: Driessen, 1986a).

$$\theta = \theta_s \cdot e^{-\gamma \, ln^2 \, | \, h \, |} \qquad \text{Equation 2}$$

θ_s and γ are parameters specific for texture and bulk density. The equation has provided satisfactory results for soils in the Netherlands. It is used in Listing 9, Lines 33-35 in the FUWCMS function. The indicative values for texture specific parameters of Table 26 can be used when direct measurements are not available.

As soil dries out, it becomes increasingly difficult for plants to extract water. At high soil water suctions (depending on environmental conditions), plants may wilt during the day and recover at night when evaporative demand is low. Beyond a certain value of moisture suction, plants do not recover at night and wilt permanently. The soil moisture suction then usually has a value of about 16,000 cm, or pF 4.2; the value differs between plant species. The volumetric

Table 26. Indicative values for soil physical characteristics and texture specific constants: th
content and hydraulic conductivity at saturation (WCST, KST) and four empirical constan

Soil type	WCST cm^3cm^{-3}	KST cmd^{-1}	KMSA1 cm^{-1}	KMSMX cm	KMSA2 $cm^{2.4}d^{-1}$	MSWCA –
1 Coarse sand (cs)	0.395	1120.	0.196	80	0.080	0.0853
2 Medium csd (mcs)	0.365	300.	0.1385	90	0.63	0.0450
3 Medium fine sand	0.350	110.	0.0821	125	3.30	0.0366
4 Fine sand	0.364	50.	0.0500	175	10.9	0.0255
5 Humous loamy mcs	0.470	1.0	0.0269	165	15.0	0.0135
6 Light loamy mcs	0.394	2.3	0.0562	100	5.26	0.0153
7 Loamy mcs	0.301	0.36	0.0378	135	2.10	0.0243
8 Loamy fine sand	0.439	26.5	0.0395	200	16.4	0.0299
9 Sandy loam	0.465	16.5	0.0750	150	0.24	0.0251
10 Loess loam	0.455	14.5	0.0490	130	22.6	0.0156
11 Fine sandy loam	0.504	12.0	0.0240	300	26.5	0.0186
12 Silt loam	0.509	6.5	0.0200	300	47.3	0.0165
13 Loam	0.503	5.0	0.0231	300	14.4	0.0164
14 Sandy clay loam	0.432	23.5	0.0353	200	33.6	0.0101
15 Silty clay loam	0.475	1.5	0.0237	300	3.6	0.0108
16 Clay loam	0.445	0.98	0.0248	300	1.69	0.0051
17 Light clay	0.453	3.5	0.0274	300	2.77	0.0085
18 Silty clay	0.507	1.3	0.0480	50	28.2	0.0059
19 Heavy clay	0.540	0.22	0.0380	80	4.86	0.0043
20 Peat	0.863	5.3	0.1045	50	6.82	0.0108

Sources: Driessen (1986a) and Rijtema (1969)

water content at this suction value is called the permanent wilting point (or simply wilting point) of the soil. Its value depends strongly on soil type. The water content at the permanent wilting point is an important variable both in the simple and complex soil water balance modules. The amount of water available for uptake by the crop is the total amount of water in the soil, minus the amount of water at permanent wilting point, minus loss terms.

'Field capacity' is the second point of the pF-curve that needs particular consideration before simulating the soil water balance in free-draining soils. Field capacity is the volumetric water content of the soil after wetting and initial redistribution. It is often treated as a soil characteristic (Section 5.2, van Keulen, 1975; Stroosnijder, 1982; Driessen, 1986a; Jansen & Gosseye, 1986). However, it is not really constant. The original definition of field capacity reads: 'the amount of water held in soil after excess water has drained away and the rate of downward movement has materially ceased, which usually takes place within 2-3 days after a rain or irrigation in previous soil of uniform structure and texture' (Veihmeyer & Hendrickson, 1949). Field capacity in the Netherlands is usually defined as the volumetric water content when the soil moisture has a suction of 100 cm (pF 2., Driessen, 1986a). This corresponds with a groundwater table at 100 cm depth. Other authors place field capacity at a lower or higher soil moisture suction (from 50 to 330 cm; McKeague et al., 1984). For the purpose of this book, field capacity is pF 2.

In reality the soil water content when air dry, θ_{ad}, is one third or less of the content at wilting point. This concept is not well-defined, but simulation results are not sensitive to its value. For simulation purposes, the soil moisture suction of an air dry soil is assumed to be 10^7 cm (pF 7).

5.1.4 *Hydraulic conductivity*

Moisture moves through soil pores at a rate proportional to the 'hydraulic head gradient'. Movement is inversely proportional to pore resistance. The hydraulic head gradient in the bulk soil can be approximated as the ratio of the difference in hydraulic head and the flow distance between two points. Soil resistance is expressed as its reciprocal value, i.e., the hydraulic conductivity. Conductivity depends on the soil type and soil water content. When water infiltrates into a dry soil, the small pores fill with water first. After more water infiltrates, bigger pores are also filled which have much higher conductivity. Hydraulic conductivity thus increases super-proportionally with increasing soil water content. The relation between hydraulic conductivity k (cm d^{-1}) and suction can be described by two empirical equations (Rijtema, 1969; Wösten et al., 1986; Figure 58) :

$$k = k_s \cdot e^{-\alpha |h|} \qquad (|h| \leq |h|_{max}) \qquad \text{Equation 3a}$$

$$k = a \cdot |h|^{-1.4} \qquad (|h| > |h|_{max}) \qquad \text{Equation 3b}$$

where

k_s is texture specific conductivity at saturation (cm d^{-1}), α is texture specific empirical constant (cm^{-1}), a is texture specific empirical constant ($cm^{2.4}$ d^{-1}), and $|h|_{max}$ is a texture specific upper limit for suction.

Values for these parameters are given in Table 26. The equations are used in the SUMSKM subroutine (Appendix B). If other equations are preferred (e.g., those described by Schuh & Bauder, 1986), the SUMSKM and SUMFLP subroutines in Appendix B should be adapted accordingly.

When vertical cracks occur in the soil, surface water infiltrates fast. However, little water is retained in the upper (cracked) part of the soil for this part is 'passed by'. Even when only a small part of a soil cross section is cracked, the

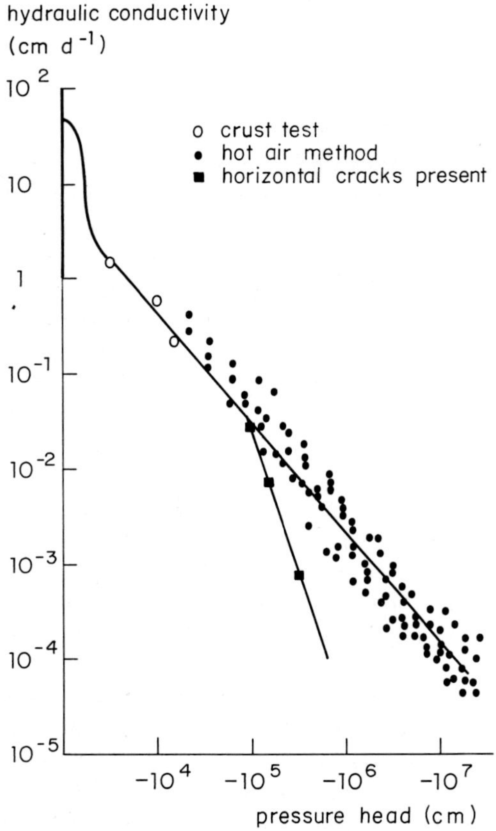

Figure 58. Hydraulic conductivity as a function of the pressure head (determined with a crust test and a hot air method) and for the same soil with horizontal cracks (Source: Wösten et al., 1986).

hydraulic conductivity is high at saturation and is usually low at small negative pressures (Bouma, 1985). No attempt is made here to simulate infiltration in such soils, for insufficient information is available about the *k(h)* relations in these situations. This does not imply that these phenomena are unimportant. Simulation of upward flow to the root zone is not really different between cracked and non-cracked soils, unless horizontal cracks develop as well. In that case, the *k(h)* relation must be adapted (Bouma, 1985; Figure 58).

5.2 A water balance module for free-draining soils

5.2.1 Introduction

The water balance of a sandy and loamy soil with a deep groundwater table (>1 m below the root zone) is relatively easy to simulate. This type of soil has high hydraulic conductivity when wet, permitting fast downward water transport, so that saturation of soil layers does not occur. Upward water flow is disregarded here. These soils and conditions are encountered in many areas in temperate, tropical semi-arid and subhumid regions. The water balance processes considered here are infiltration, percolation, evaporation, transpiration and downward redistribution. Lateral influx or outflux of water is always fully negligible in these situations. The approach can also be applied on clay soils with a deeper soil water table (>2 m below the root zone), but the dynamics of water infiltration into the soil, redistribution between layers, and evaporation are then simulated more crudely.

The simulation module for this situation is L2SU (i.e., Production Level 2, soil, unsaturated) which is presented in Listing 8. L2SU is based on models by van Keulen (1975), Stroosnijder (1982) and Jansen & Gosseye (1986). Its concept is described by the acronym SAHEL, for Soils in semi-Arid Habitats that Easily Leach. A brief description is given here. Figure 59 illustrates some of its basic features.

The inflow and outflow of water in separate layers is simulated on a daily basis. Inflow into the first layer is from rainfall. Field capacity is the highest water content that the soil can attain. The amount of water that cannot be stored in one layer, drains into the next layer or out of the profile. Water is extracted from layers by evaporation and transpiration. The soil profile is divided into three layers and each is considered to be homogeneous. Thickness and physical characteristics of each layer are inputs. The upper layer should be 0.10-0.20 m thick, the second 0.2-0.4 m, and the third 0.4-1.0 m. Their sum should slightly exceed the maximum rooted depth. The model can be extended to account for more heterogeneous situations by adding more layers.

The soil characteristics needed for L2SU are the volumetric water contents of the soil layers at field capacity, wilting point and when air dry (Subsection 5.1.3). Some values for common soil types are given in Table 27. The size of clods on the surface and the surface albedo must be included. Listing 10 gives an example of how soil data can be presented.

5.2.2 Infiltration

The infiltration rate (mm d^{-1}) is equal to rain minus interception and runoff. Irrigation can be treated in the same way as rain.

Not all water that reaches the surface infiltrates the soil, especially during heavy rain. Runoff from a field can be 0-20% of precipitation, and even more

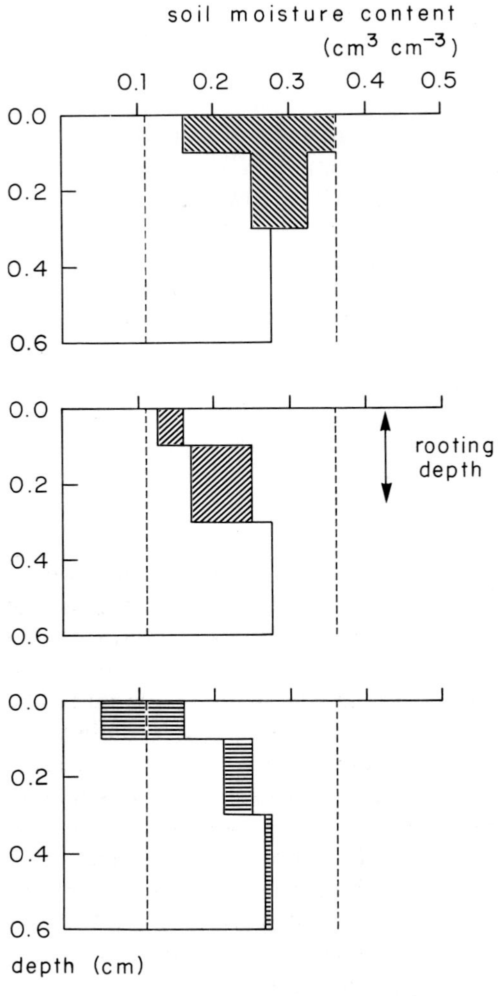

Figure 59. A graphical representation of changes in the soil water content of three layers due to infiltration (top) transpiration and evaporation (bottom). (Source: Stroosnijder, 1982).

Table 27. Typical soil water contents for different soil types (calculated from Table 26 with the equation of Subsection 5.1.3).

	Water content ($cm^3\ cm^{-3}$) at			
	Air dry	Wilting point	Field capacity	Saturation
Course sand	0.005	0.01	0.06	0.40
Fine sand	0.005	0.03	0.21	0.36
Loam	0.01	0.11	0.36	0.50
Light clay	0.05	0.24	0.38	0.45
Heavy clay	0.18	0.36	0.49	0.54

on unfavourable surfaces (Stroosnijder, 1982). It is negligible under proper irrigation and soil management conditions. Runoff occurs when the rate of water supply at the soil surface exceeds the maximum infiltration rate and the accumulated excess exceeds the surface storage capacity. In reality, the maximum infiltration rate is influenced by the water content at the surface. The simplest solution for calculating runoff is used in L2SU: it is assumed to be a constant fraction of precipitation (Listing 8 Line 40). This is not always satisfactory. The runoff fraction can be calculated as a function of daily precipitation (e.g., Jansen & Gosseye, 1986), and of the soil water content. However, the necessary data for calibrating will not often be available. Several other attempts have been made to describe the variation in runoff due to surface conditions (e.g., Davidoff & Selim, 1986; see also Subsection 5.3.5). A detailed consideration of runoff is warranted when its quantification is crucially important, such as for establishing a crop in semi-arid zones.

Run-on of water from adjacent fields may also occur and can be added to precipitation. However, it is difficult to quantify or measure this term in the field.

5.2.3 Soil water movement

When a soil layer is filled beyond field capacity, water percolates into the next lower layer. Most drainage occurs within 24 hours (except in heavy soils), and as one-day time periods are used in L2SU, it is only a small over-simplification to assume that all drainage occurs within one day. Simulation is therefore straightforward: if on any day more water infiltrates a soil layer than can be held by that layer, then the excess water drains into the next layer (Lines 41-

43). If more water enters the deepest layer than can be retained, the excess is lost as deep percolation. Some upward flow is implicitly simulated as the contribution of each layer to soil evaporation.

Soil layers can become no wetter than field capacity. Limiting the infiltration rate into the second layer is a simple method of mimicking a water-logged top layer. Runoff should then be increased by the water that is in excess of the saturated top layer, plus maximum surface storage. Choose the value of maximum infiltration rate such that realistic periods of waterlogging result. Dynamic simulation of waterlogging can be undertaken with the module L2SS (Listing 9) when the saturated conductivity of the impermeable layer is known.

5.2.4 Evaporation

Surface evaporation is important for bare soils, but it is much less than transpiration under a well-developed crop canopy. Water can evaporate until the soil is air dry. The water content is then only about one third, or less of the permanent wilting point (see Table 27, Subsection 5.2.1). The amount of water held between wilting point and air dry can be lost by evaporation, but is inaccessible to crops (see Figure 59, Subsection 5.2.1). Similarly, rain or irrigation must first wet the soil till the wilting point is reached and then provide the water required by the crop.

The potential soil evaporation rate is determined by the same energy balance processes as that of leaf transpiration. Therefore, the potential evaporation rate is also computed using the Penman approach (Section 4.1) in the SUEVTR subroutine. The first step of this calculation does not consider canopy shading, but includes the effect of the crop on windspeed near the surface (Listing 7 Lines 41-47). Shading is accounted for in the next step; the extinction coefficient for shortwave radiation together with near infrared radiation is about 0.5 (Listing 7 Line 40). This potential rate applies for the day at which water infiltrates into the soil; a maximum is the extent to which the upper layer can be depleted (Listing 8 Lines 48, 49).

Surface roughness, characterized by clod height, affects the resistance of the boundary layer. It is used to calculate soil evaporation in the same way as leaf width (Listing 7 Line 45). Windspeed near the soil surface is less than at canopy or screen height; atmospheric resistance is related to crop height and density. Sensitivity of crop growth for these variables is low. Effects of soil tillage and formation of mulches or ridges on evaporation are not considered.

Reflectivity of the soil surface for solar radiation affects its energy balance. Its value depends on the surface colour and moisture content of the upper layer (Menenti, 1984). The values for the albedo or whiteness of a dry soil surface runs from 0.15-0.4 (Table 28). The dependence on soil humidity is represented simply as a negative proportionality (Listing 7 Line 44). Although not entirely correct, the average water content of the upper soil layer is used to calculate reflectivity. The surface emissivity for long wave radiation is between 0.9 and

Table 28. Albedo values for wet and dry soils.

Surface type	Wet	Dry
Dune sand	0.24	0.37
Sandy loam	0.10-0.19	0.17-0.33
Clay loam	0.10-0.14	0.20-0.23
Clay	0.08	0.14

Source: ten Berge (1986)

1.0 for all moisture contents (ten Berge, 1986). In SUEVTR its value is set equal to 1.00. The sensitivity of crop growth to both surface characteristics is fairly low.

The evaporation rate diminishes as soon as the topsoil starts drying. In L2SU, this is assumed to happen the day after the last rainy day, the latter being defined as a day with at least 0.5 mm of precipitation (Listing 7 Lines 57-58). The reduction of the evaporation rate over time is mimicked by using the observation that cumulative evaporation is proportional to the square root of time (Stroosnijder, 1982; see also Subsection 5.3.6; Listing 8 Lines 50-51). The evaporation proportionality factor (the rate on the first day of this sequence) is assumed to be equal to 60% of the potential soil evaporation. Rains too small to trigger resetting of days since the last rain are added to the evaporation, since they are assumed to be lost the same day.

All soil layers contribute to evaporation, but the top layer considerably more than the bottom layer (see Figure 59, Subsection 5.2.1). The actual partitioning at any moment depends on the depth and thickness of layers, their water content, and a soil specific extinction coefficient that is used to mimick the upward flow. Partitioning is calculated in Listing 8 Lines 52-61. The extinction coefficient is approximately 10 m^{-1} for heavy and 30 m^{-1} for light soils.

The calculation of the potential rate of soil evaporation from incident energy is not correct if the soil is an important net source or sink of energy during a one-day period. This is rarely the case in the tropics, but in temperate climates the soil is a net sink of heat during spring and releases heat in the autumn or fall. The soil may also act as a source or sink if abrupt changes occur in air temperature or radiation intensity, such as those associated with the passing of large scale meteorological systems. This is significant to soil surface temperature and to the interpretation of thermal remote sensing imagery, but it is generally insignificant for crop growth modelling.

5.2.5 *Water uptake and transpiration*

Water uptake by the root system and partitioning over the rooted layers is described in Section 4.2 and depicted in Figure 59 (Subsection 5.2.1). The actual water uptake from soil layers is calculated in Listing 8 Lines 11-15, the effect of water stress on uptake in Lines 17-19, and the root depth in each layer in Lines 21-23.

Maximum available water (i.e., all water held between field capacity and wilting point) is from 0.5-2.5 mm water per cm of rooted depth. If evaporation could be avoided, a C_3 crop could produce 170-800 kg ha^{-1} total dry matter on the water stored in each 10 cm of rooted depth and a C_4 crop about twice as much (Subsection 4.1.6). Obviously, water stored in the soil provides an important buffer in periods of little rain. Dry season cropping is, in fact, possible in many climates, provided that at the start there is a saturated soil profile and at least 0.5-0.7 m of rootable profile (see Subsection 5.4.6 for an example).

A crop dies from water stress even before the lower soil layer reaches wilting point. The rate at which water is extracted near wilting point is so low that photosynthesis provides insufficient energy for maintenance respiration, and the crop dies.

In simulation, water is always withdrawn from the entire soil layer as soon as roots enter it. The withdrawal rate is proportional to the root length in the layer. When the maximum rooted depth is reached, but roots do not extend to the bottom of the soil profile, water is slowly extracted from below the rooting zone.

5.2.6 *Soil temperature*

Soil temperature can affect the downward extension rate of roots. It is approximated by a simple running average of air temperature (Listing 7 Line 60). A more sophisticated treatment requires consideration of thermal soil properties, which are usually unknown. An example of a deterministic analysis is given by de Wit & Goudriaan (1978). A detailed simulation study is presented by ten Berge (1986).

5.3 A water balance module for soils with impeded drainage

5.3.1 *Introduction*

There are many poorly drained soils and many situations where the groundwater table is between 0.0-1.5 m below the surface during part of, or the entire growing season (e.g., Marletto & van Keulen, 1984). In soils where lowland rice is grown, downward water flow is usually impeded by a hardpan formed by puddling. A perched water table may then develop near the root zone. Capillary rise provides a significant water supply during dry spells.

When downward water flow is limited or groundwater is perched near the root zone, the L2SU module for water balance simulations should not be used. In these cases, the Simulation Algorithm for Water flow in Aquic Habitats (SAWAH) is more appropriate. The SAWAH subroutines are included in Listing B. The main subroutine, SUSAWA, makes use of several other subroutines to compute the rates of change of water contents in soil layers. These rates are integrated and their interaction with the crop is simulated in the soil water balance module L2SS (Listing 9). This module simulates the water balance for crop growth at Production Level 2 with partially saturated soils. All subroutines are presented in Appendix B.

Technically, L2SS and L2SU differ: in the number of layers simulated, in the complexity of the flow computations, in the time period used and, consequently, in running time on the computer. L2SS plus SAWAH is more deterministic, but also requires more input data. L2SS is to be combined with L2C and L1D (Table 1).

Simulation of the soil water balance processes becomes more accurate when the simulated soil layers are made thinner, and more layers are considered. The shape of the soil moisture profile is then approximated more precisely and soil and root properties can be defined in more detail. A disadvantage of thin layers is the associated short integration period in explicit numerical solution schemes (Subsection 1.4.2). This is due to the low capacity of thin layers (i.e., the amount of water that can be absorbed or released to reach a given change in potential). The deterministic simulation of soil water flow requires integration periods very much smaller than those for crop processes or parameteric simulation of waterflow as in the L2SU module. Redistribution of water in the soil during any one day resulting from sudden changes is simulated with short time periods.

The L2SS module may therefore be used for a crop-soil system where such sudden changes occur once every 24 hours due to extraction of water from the soil by roots, due to precipitation or evaporation, and to a possible change in groundwater depth, all of which are simulated as instantaneous events.

5.3.2 Integration periods

L2SS with the SAWAH subroutines simulates water redistribution induced by root uptake, rainfall, evaporation and capillary rise. These processes can occur relatively fast. Simulation of water flow requires time periods of 0.001-0.1 day, depending on the thickness and hydraulic properties of the soil layers. Thus, SAWAH covers every one-day period of the crop in many small steps. The resulting changes are the rates computed in L2SS. Interaction between the crop and the soil occurs every 24 hours at sunrise when the simulation day begins.

The integration period in SAWAH can be explicitly chosen (parameter DTFX), but it is recommended that it is established by SAWAH. Minimum and maximum time periods should be specified (the minimum DTMIN and maximum DTMX, Listing 10 Line 15). The actual integration period is the smallest of those specified and computed. When a fixed time period is chosen, SAWAH does not check whether the time period is larger than the allowed time coefficient. However, it does sometimes reduces the prescibed time period because it cannot exceed the time required to saturate a layer or to remove all its available water. A maximum time period is specified to ensure that at least a minimum number of repetitions occur each day. The maximum time period overrides the fixed time period if the latter is too long. The last integration period of each day completes the 24 hour cycle.

5.3.3 Soil water movement: unsaturated flow

The daily rate of change of water content of each soil compartment is an output of the SAWAH submodel.

Water exchange between soil layers is governed by the hydraulic conductivity of the soil and the local gradient of the hydraulic head. Water flows towards the location with the lowest hydraulic head. To keep the calculation time within acceptable limits, the number of layers for a soil profile of 0.7-1.5 m is limited to 10. The upper layer should be at least 0.05-0.10 m thick (Subsection 5.3.6); the thickness of other layers may vary. All layers have their own pF-curve and hydraulic conductivity function. The use of only 10 soil layers implies that these layers are relatively thick and that the simulation of the soil water profile is somewhat crude. This can be viewed as a price for faster computation, but in practice the soil data needed to characterize many soil layers individually are often not available.

The inaccuracy associated with the use of thick layers is reduced by applying the concept of 'matric flux potential'. Consider the flux density or Darcy equation, which expresses the flux q (cm d^{-1}) as a function of hydraulic conductivity k (cm d^{-1}) and the gradient of the hydraulic head H (cm):

$$q = -k \cdot \frac{dH}{dz} = -k \cdot \frac{dh}{dz} + k \qquad \textit{Equation 4}$$

where z (cm) is the space coordinate, positive downwards, $-h$ (cm) is the matric suction, $-k$ (dh / dz) is the matric component of the flow, and k is the gravity component.

In numerical simulation of flow processes, differentials such as dh, are replaced by differences, such as $h(i) - h(i\text{-}1)$, the index referring to a layer number. A difference such as this is called a 'finite difference' to stress the contrast with 'differential', an infinitely small increment. A finite difference is represented here by the symbol Δ. This distinction is the basis for several 'tricks' applied in numerical simulation. One of these is the use of the matric flux potential instead of hydraulic conductivity. The present context does not allow elaboration on the theory and only the most relevant equations are presented (for theory and application of matric flux potential, see Klute, 1952; Gardner, 1958; Raats, 1970; Shaykewich & Stroosnijder, 1977; for measurements see ten Berge et al., 1987).

The matric flux potential is defined as:

$$\Phi(h) = -\int_h^o k(h)dh \qquad \textit{Equation 5}$$

It can be substituted into the flux density equation:

$$q = -\frac{d\Phi}{dz} + k \qquad \textit{Equation 6}$$

Because it is assumed that the flux between the centers of two adjacent compartments is constant with depth, the integrated form of Equation 6 is:

$$q\Delta z = -\Delta\Phi + \int_{z_{i-1}}^{z_i} k(z)dz \qquad \textit{Equation 7}$$

The second term on the right-hand side is the gravity term. Its value can be approximated by assuming a linear course of the matric suction between the centres of adjacent compartments i and i-1:

$$\int_{z_{i-1}}^{z_i} k(z)dz = \frac{dz}{dh} \int_{h_{i-1}}^{h_i} k(h)dh \qquad \textit{Equation 8}$$

Since this only applies to the gravity term it is not a severe simplification of the overall flow process. The flux density equation is then rewritten as:

$$q = \left(\frac{1}{\Delta h} - \frac{1}{\Delta z}\right) \int_{h_{i-1}}^{h_i} k(h)dh \qquad \textit{Equation 9}$$

The integral on the right-hand side is identical to $\Delta\Phi$. The term $1 / \Delta h$ represents the gravity component and $1 / \Delta z$ the matric component of the flow. The matric flux potential 'weighing' therefore also applies to the gravity term.

A complication arises when intrinsic soil properties, such as hydraulic conductivity and pF-curve, change with depth, as is often the case. The gradient of the matric flux potential has a straightforward meaning (i.e., it is a flux density) only if the soil material is homogeneous. An averaging procedure with values of two compartments is used for layered soil. The integral term of Equation 9 is then replaced by:

$$\int_{h_{i-1}}^{h_i} k(h)dh = \left[\int_{h_{i-1}}^{h_i} k_1(h)dh \cdot \int_{h_{i-1}}^{h_i} k_2(h)dh \right]^{1/2} \qquad \textit{Equation 10}$$

k_1 and k_2 are the conductivity functions of the two adjacent layers. This is analogous to the procedure recommended by Vauclin et al. (1979), with an additional weighted averaging over the matric suction.

After each time period the new water content in each layer is obtained by subtracting the outflow from the inflow during that period and adding the resulting change to the previous value of the water content. The matric suction in each compartment is assessed again for the new moisture content; the whole procedure is repeated in the next time period.

The small simulation program SWD (short for: soil water dynamics, Listing 6) combines the basic processes of water flow in unsaturated soils, infiltration, capillary rise and redistribution. As a deterministic model, the program is based on the flux density equation and the mass conservation equation. The equations are solved for each set of two adjacent compartments and for each short time period. Soil water movement is the result of a gradient in total water potential. When flow across the upper and lower boundaries of the profile is permitted, as in Listing 6, water continues to flow as long as the difference between suction and gravitational head is not constant throughout the profile. With fixed suctions at both boundaries, a steady state profile finally develops, characterized by the same flux across all compartments. SWD can be modified to block drainage or evaporation (set fluxes at bottom or top of the profile to zero), or to extract water from layers (e.g., to mimick transpiration). SWD cannot handle non-homogeneous profiles (and this cannot be overcome by defining the parameter values KST, WCST, KMSA and MSWCAfor each layer separately). Since it simulates only unsaturated conditions, saturation of one or more soil compartments cannot be handled. However, the program can be used for evaluating complex models when these are applied for simplified sets of inputs. The program is the basis for exercises at the end of this chapter.

To prevent problems with rounding off when using Personal Computers, Line 33 resets K(I) to zero when it approaches that value.

5.3.4 *Soil water movement: saturated flow*

Saturated soil sections (a section consists of one or more layers) have different solutions for the flow equation from those with unsaturated flow. In a way, this situation is less complex, because the dependence of the transport coefficient on moisture content has vanished and at saturation only hydraulic conductivity is relevant. On the other hand, the pressure gradient at layer interfaces can no longer be calculated from the moisture contents of neighbouring layers. The entire set of saturated layers is now involved, because the weight of the overlaying water contributes to the local water potential in every layer. Moreover, different soil layers may have different values for saturated conductivity, so that the conductivity of an individual layer no longer determines the saturated flow through that layer. The product of local pressure gradient and local conductivity must now be equal at every point in the saturated section, since no changes in moisture content can occur. Only at the edges of the saturated section may moisture content change, related to expansion or contraction of the particular saturated section of the soil profile.

To simulate saturated flow, first the saturated sections of the profile are identified (Figure 60). At the saturated – non-saturated transitions, the matric suction is supposed to be zero. This usually closely approximates the real situation. Suction is not always zero when the boundary of a saturated layer coincides with the top or bottom of the soil profile. The pressure may be positive at the top of the profile, for water may be ponded on the surface. At the bottom of the profile the pressure depends on the depth of the groundwater table. If groundwater is present within the modelled part of the soil profile, the water pressure at the lower boundary is positive. For deeper groundwater tables, the pressure at the base of the profile is negative.

The total change in hydraulic head over a saturated section is calculated by adding the drop in gravitational head to the difference in pressure head between two edges. For n saturated layers within a section, a total of $n + 1$ unknowns must be solved: n times the difference in hydraulic head over a layer, plus the flux through the saturated 'package'. There are $n + 1$ independent linear equations; n times the flux density equation over a layer and the

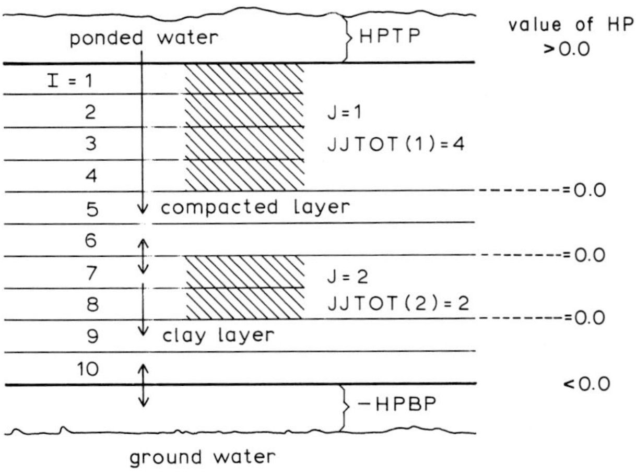

Figure 60. An example of water redistribution in a layered profile as simulated with the L2SS + SAWAH modules. Hatched zones indicate saturation. Low conductivity in Layers 5 and 9 cause a double perched water table and ponded water on the surface. The pressure head HP at the top of the profile is then positive (*HPTP*), at the bottom it is negative (*HPBP*) and it is zero at the other edges of saturated sections. Arrows indicate the flow direction (single-headed arrows) or possible flow directions (double-headed arrows).

summation of changes in hydraulic head over each layer, which yield the total difference in hydraulic head over the entire saturated section. This set of $n + 1$ linear equations is written in matrix form and solved for each saturated section of the profile. In most cases there will be only one saturated section, though separate saturated sections can occur after intermittent heavy showers or when the soil has several slightly permeable layers.

Groundwater depth is a boundary condition and an essential input to the model. Its value should be obtained by direct observations in the field for which simulations are made (see also Subsection 5.3.8).

5.3.5 Infiltration and runoff

Infiltration depends on rainfall intensity, soil properties and surface storage capacity. The amount of rain (or irrigation water) that cannot infiltrate or be stored on the surface is lost as runoff.

Usually, only daily precipitation total, not the intensity, is known. Rainfall can be constant for 24 hours, or may come in one short but heavy rainstorm. Where water is normally received over brief periods, e.g., when short showers predominate and on irrigated fields, the extreme of instant supply comes closer to reality than the other extreme of a constant supply.

In SAWAH it is assumed that all daily precipitation or irrigation water is usually received instantaneously at the start of the day. A water layer corresponding to the daily total of rainfall is then ponded on the surface. The rate with which the thickness of this ponded water layer changes is an output of SAWAH; the thickness of the layer is a state variable in L2SS (WL0QT, Listing 9 Line 50). Infiltration is simulated with the equations for unsaturated waterflow (Subsection 5.3.3) including both matric and gravity forces, as long as the surface compartment is not saturated. The effect of positive water pressure is taken into account. This hydrostatic pressure head is due to the weight of the free water layer and is equal to the depth of that layer. When the surface layer becomes saturated, the model switches to the equations for saturated flow, again taking into account the positive surface pressure.

If, after a day of infiltration, not all rain has entered the soil, the remainder is left ponded on the surface. When the amount of ponded water exceeds the surface storage capacity, the excess is considered as runoff. The maximum amount of water that can be stored on the surface (WL0MX, m) can be computed from surface characteristics (Driessen, 1986a):

$$WL0MX = 0.005 \cdot d \cdot \frac{\sin^2 (\sigma - \varphi)}{\sin \sigma} \cdot \frac{\cot (\sigma + \varphi) + \cot (\sigma - \varphi)}{2 \cdot \cos \sigma \cdot \cos \varphi}$$

Equation 11

where d is surface roughness (cm), σ is the clod/furrow angle (degree), φ is the average slope angle of the land (degree).
These characteristics are illustrated in Figure 61. The clod/furrow angle is normally between 30 and 45 degrees, the slope angle of the field in general does not exceed 17 degrees (= 30%), while the surface roughness or furrow depth depends on the way the land is cultivated: about 0.2 m for contour plowing, 0.06-0.08 m for tillage with light equipment, and 0.01-0.02 m for untilled soil (Driessen, 1986a). Equation 11 is not relevant for bunded fields. For rice, the effective bund height is numerically equal to the surface storage capacity.

The surface storage capacity is not calculated in the L2SS module, but is given as a situation specific parameter. Infiltration is computed with rain as input; runoff is an output (Listing 10 Lines 54-58).

5.3.6 *Evaporation*

Computing the actual soil evaporation is complicated in SAWAH in comparison with the 'square-root-of-time relation' used in L2SU (Subsection 5.2.4). The simple equations cannot cope with significant capillary supply from groundwater. Moreover, they do not deal with root water uptake. The formulation below implicitly incorporates both processes.

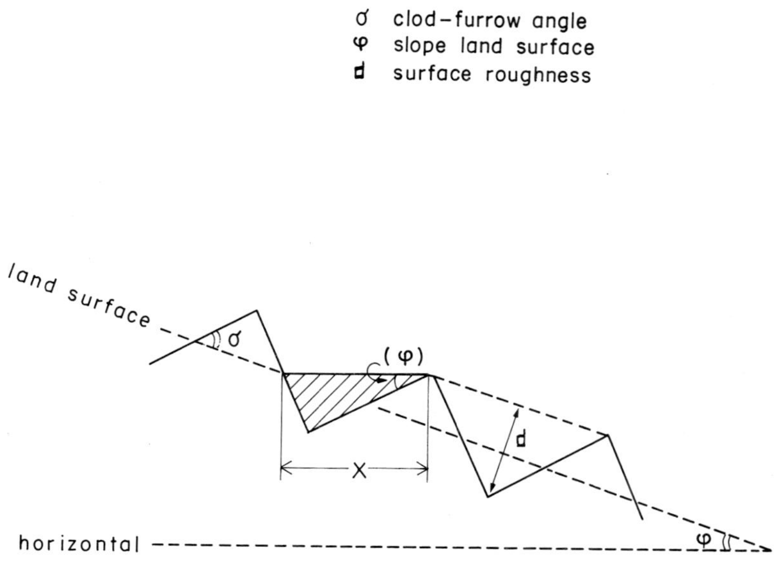

Figure 61. A schematic representation of the surface storage capacity of a field, WL0MX. WL0MX is equal to the shaded area divided by the length X. (Source: Driessen, 1986a).

The water loss rate is equal to the potential evaporation rate in saturated soil (Subsection 5.2.4). The evaporation rate does not depend on soil properties or soil condition. Topsoil drying is associated with a decreased evaporation rate, for water then evaporates below the surface and water vapour moves upward by diffusion. Formation of a dry surface layer gradually impedes further water loss. The rate of topsoil drying, and hence the rate at which evaporation decreases, depends on soil characteristics, water content and on the potential evaporation rate. Calculating the development of a dry surface layer is crucial for an accurate determination of soil evaporation.

There are two solutions for calculating soil evaporation in a deterministic way. The first is a numerical solution by simulating the flow processes involved; the second is an analytical solution. A purely numerical solution requires simulation of very thin surface layers to account for the steep gradient in the soil moisture content; this is impractical. Analytical solutions, however, are only available for homogeneous soils with uniform initial conditions. Therefore, a combined approach is necessary.

A basic assumption is that an 'evaporation front' exists at the surface or at some depth z_E in the soil. This is a sharp transition where liquid water is transformed into vapour. All transport is in the liquid phase below the front, while only vapour transport occurs above the front. The evaporation rate at any moment is a function of the depth of the evaporation front and of the rate of vapour diffusion through the dry layer. The balance of liquid supply to the front, and vapour diffusion away from it, determines the movement of the front itself.

At a front depth z_E (cm), the rate of water vapour loss from the soil (e, cm d^{-1}) can be expressed as:

$$e = \frac{c_2}{z_E} \qquad \textit{Equation 12}$$

c_2 is taken to be a soil-specific constant, though its value depends somewhat on air humidity, wind speed and soil temperature.

To compute the supply of liquid water to the evaporation front, some assumptions made by Parlange in describing infiltration can be applied to the description of evaporation (Giraldez, personal communication; for details, see Parlange (1971) and for applications, Smith & Parlange (1978); Giraldez & Sposito (1985)). The amount of water that has evaporated after a given period of time (E, cm) is given by:

$$E = z_E \cdot (\theta_i - \theta_{ad}) + \frac{c_1}{e} \qquad \textit{Equation 13}$$

where
$z_E \cdot (\theta_i - \theta_{ad})$ is the water evaporated from above the front, c_1 / e is the amount of water extracted from below the front, c_1 is a function of the initial moisture content θ_i, the final moisture content θ_{ad}, (air dry) of the top layer and the (liquid) soil water diffusivity D (which is a function of moisture content θ itself):

$$c_1(\theta_i) = \int_{\theta_{ad}}^{\theta_i} D(\theta)(\theta_i - \theta)d\theta \qquad \textit{Equation 14}$$

('Initial' in this paragraph refers to the water content at the start of the day, and not to the value at the start of the simulation). Combining Equations 12-14 and further developing the terms leads to an expression of the rate at which the evaporation front sinks into the soil:

$$\frac{dz_E}{dt} = \frac{c_3}{z_E} \qquad \textit{Equation 15}$$

The front depth is regarded as a state variable (Listing 10, Line 52); its change over a day is calculated in SAWAH. c_3 is defined as:

$$c_3(\theta_i) = \frac{c_2}{(\theta_i - \theta_{ad}) + c_1 / c_2} \qquad \textit{Equation 16}$$

z_E increases rapidly if c_1 is small compared to c_2, i.e. when vapour diffusion is rapid compared to the supply of liquid water from the subsoil. z_E also increases rapidly when the difference $(\theta_i - \theta_{ad})$ is small. A high initial moisture content also increases the value of c_1, as it increases the upper boundary of the integral in Equation 14. A high initial moisture content, therefore, results in low c_3 values and in the slow progress of z_E. The value of z_E is reset to zero when the topsoil is saturated by rain or rising groundwater.

The integral in Equation 14 is calculated as an explicit exponential function of θ_i:

$$c_1(\theta_i) = A \cdot (\mathrm{e}^{B \cdot (\theta_i / \theta_s - c_4)} - 1) \qquad \textit{Equation 17}$$

The parameters A and B are constants for the soil, c_4 is about 0.5 (dimensionless) for all soils, e is the base of natural logarithm, and θ_s is the water content at saturation.

The above concept was developed for an 'ideal' soil, i.e., deep, with uniform initial moisture content. It can also be used as an approximation for a layered soil provided that the top layer is at least 0.05-0.10 m thick. c_1 and c_3 are kept constant during each day, with θ en θ_i referring to the toplayer.

The Equations 12, 15-17 are incorporated in the module T12 (Appendix B, Subroutine SUZECA), and the parameters used are (Listing 10 Line 19): c_2 (about 0.1 $cm^2\ d^{-1}$), A (0.005-0.5 $cm^2\ d^{-1}$), B (5-15, dimensionless). Realistic combinations of A and B should give c_1 values ranging from 5.0 to 50.0 $cm^2\ d^{-1}$ with exceptions up to 400 $cm^2\ d^{-1}$. Coarse soils tend to have high B values combined with low A values. Low B and intermediate A values are recommended for silty soils, and low A and low B values for heavy soils. (Note that the units of z_E and e in Equations 12-17 are not equal to the units of the corresponding variables ZEQT and EVSW in Listing 12.)

5.3.7 *Transpiration*

Actual water uptake from the rooted soil is almost the same as described in Sections 4.2 and 5.2 (Listing 9 Lines 64-72) though its appearance is different (see Subsection 5.4.4). The same relation of water stress and water uptake as in the module L2SU applies here (Line 68), but there is a difference in the possibility of anaerobic conditions developing in layers. The extent of anaerobic conditions is assumed to be proportional to the water content between field capacity and saturation. Anaerobic conditions reduce the capacity of the root system to extract water from the soil in most species (Subsection 4.2.4).

5.3.8 *Flow to and from the groundwater*

The pressure head at the bottom of the profile is equal to the distance between the bottom of the profile and the water table. The pressure head is negative when the groundwater table is below the profile. The soil below the profile is then assumed to be in equilibrium with the water table. The corresponding matric suction at the lower boundary is then used to calculate the matric flux potential at this depth; the flux (upwards or downwards) is subsequently calculated from the difference in matric flux potential between the lower boundary and the centre of the lower layer.

Using pressure head at the bottom of the profile as a boundary condition has an advantage over introducing the ground water level directly. This is because the depth of the water table is often measured in piezometer tubes where water transport is faster than in the soil, particularly in soils with low conductivity. Thus, a soil section below the ground water table may be non-saturated for a while after the level in the piezometer has moved upwards to ‘pass’ that section. The reverse may also occur, leaving the soil saturated, although the piezometer readings show that the ground water level dropped. However, such data are rarely available for crop growth studies.

The profile simulated need not be deeper than the bottom of the tube in which the groundwater level was measured. A concious choice of soil profile depth saves much computer time and the results will be almost unaffected.

5.3.9 *Lateral flow*

Lateral inflow or outflow of water can be significant for the water balance of small fields. The extent of lateral flow depends on the position of the field in the landscape and its distance from ditches and canals.

Lateral flow occurs mainly in saturated soil layers. In contrast to rainfall, lateral flow is very difficult to measure in experiments. Angus & Zandstra (1980) demonstrated the principle of simulating lateral flow between a cascade of rice fields, but soil data are lacking to apply it in actual situations (Whisler, 1983).

Net lateral flow leads to a change in the groundwater table. Since the water table is an input into the water balance module, the net contribution of lateral flow to the soil water balance is implicitly accounted for and needs no further consideration.

5.4 Simulation modules for water balance and crop growth

5.4.1 Simulating the effects of water stress (module L2C)

The effects of water stress on crop growth and development can be simulated with the module L2C (Listing 7) for crop-related calculations at Production Level 2. The scientific basis for the module was presented in Chapter 4. Abbreviations used are explained in Listing 12. L2C is to be combined with the basic module (L1D), a soil module (L2SU or L2SS), and crop, soil, and weather data (such as that in Listings 5, 10, and 11 respectively), and with the terminal module T12 (Appendix B), as indicated in Table 1 in the Reader's guide.

The L2C module provides the soil module with data on the potential transpiration rate, rooted depth, potential transpiration per cm rooted depth, leaf area and crop height. The L2C module requires from the soil module the rate of water uptake (actual transpiration) and the effect of soil temperature and water stress on root elongation.

Lines 2-4 of Listing 7 are to be placed in the INITIAL of the full simulation program.

The water stress level is quantified by the ratio of actual transpiration and potential transpiration. Water stress always affects photosynthesis (Line 11), carbohydrate partitioning once the ratio is less than 0.5 (Line 9), and crop development depending on its sensitivity (Line 10). Air humidity may reduce photosynthesis independently of water stress (Line 12).

Weather data (discussed in Chapter 6) are incorporated in Lines 51-59. Soil temperature is approximated by a running average of air temperature (Line 60). The DSLR variable (Line 57, 58) keeps track of the number of days since the last day with at least 0.5 mm of rain (or irrigation, if applicable). It is used as an indicator of potential or less than potential soil evaporation in the L2SU module. If no rain was received DSLR is also used to calculate the actual evaporation rate for free-draining soil (Listing 8 Lines 47-51).

Statements to compute potential soil evaporation (Lines 40-47) are added to this crop module because they apply to both soil modules. To compute potential soil evaporation, the SUEVTR subroutine is called (Line 41, 42). (The 'leaf resistance' in this statement has no equivalent for soil and is set to 0.0. Soil evaporation continues at night, so that the fraction of the day is set to 1.00 (in contrast to canopy transpiration) and the average air temperature is used. The SUEVTR subroutine and FURSC function are explained in Subsection 5.4.5.)

Two auxillary variables are created to ease checking results, or to produce characteristic values for water use efficiency in Lines 62-63.

The following crop data are required (see Listing 5):

– for water uptake: the sensitivity of the species for water stress and flooding (Line 42);

– for transpiration: leaf width, crop height (as a function of development stage) and the internal/external CO_2 fraction (Lines 37, 39, 42);

– for crop development: the relation between stress and development rate (Line 36);

– for root growth: the maximum depth and growth rate (Line 43). The effect of temperature on root growth is assumed to equal that on photosynthesis.

5.4.2 Two soil water balance modules

The L2SU module to simulate the water balance of free-draining soils for crop growth at Production Level 2 is an alternative to the L2SS module for soils with impeded drainage. The choice of module depends on the objective of the study, on environmental conditions and on the availability of soil data. Sections 5.1, 5.2 and 5.3 provide the background of the processes. Abbreviations used in the water balance modules are defined in Appendix A.

Both soil modules use the potential rate of transpiration, rooting depth (and the ratio of both), leaf area and crop height, calculated in the crop module L2C as inputs. As with the check on the carbon balance (Subsection 3.4.4), there is a check on the water balance at each time period. Both soil water balance modules contain a number of statements that must be placed before or in the INITIAL section of the total CSMP program before running it. They are presented here, together with the DYNAMIC parts for reasons of clarity.

5.4.3 Module for free-draining soil (L2SU, SAHEL concept)

The water balance module of a soil with three layers, freely draining at the bottom is contained in L2SU, Listing 8. The numbers of layers can be extended by repeating the separate statements for new layers (the calculations of CKWFL and CKWIN must then be adapted).

The upper soil layer should be between 0.05 and 0.3 m thick and the other layers between 0.2 and 0.5 m. The simulated soil layers may correspond with actual layers in thickness and characteristics. For accurate calculations it is a disadvantage to include soil layers below the rooted depth. Three soil characteristics per layer are specified: water content at field capacity, wilting point and air dry (Subsection 5.1.3). An example is given in Listing 10.

The degree of water stress is calculated per layer with the FUWS function (see Figure 55, Subsection 4.2.2). Stress in any layer affects water uptake

(Lines 12-14); but only stress in the deepest layer where root tips are located affects the rate of root growth (Line 27).

Rooted depth is calculated per layer by repeated subtractions (Lines 21-23). A limit condition ensures that roots do not grow beyond their maximum depth (Listing 8 Line 32).

The actual daily evaporation, obtained with Lines 47-51, is equal to the potential rate on days when there is at least 0.5 mm of rain. After rain, evaporation diminishes proportionally with the square root of time (Lines 50-51). The partitioning of actual evaporation over layers is according to an exponential extinction with depth, assuming the centre of the layer to be halfway between its middle and upper boundaries (Lines 55-61).

Water percolating from the deepest layer (Line 43) is only used to check the consistency of the water balance (Line 66).

Factors 0.1, 10. and 1000. in the module are used to maintain proper units. Dividing by DELT converts amounts into rates (Subsection 1.4.3). The total amount of water in the soil profile is an auxillary variable (Line 39); not all the water is available to the crop.

5.4.4 Module for soil with impeded drainage (L2SS, SAWAH concept)

The water balance simulation module for soil with impeded drainage (L2SS) is presented in Listing 9. Rates of change of the water content of layers are computed by the subroutines that together represent SAWAH. In L2SS, these subroutines are all invoked through the subroutine SUSAWA (Lines 37-39, 54-56). The SAWAH subroutines are part of Appendix B.

The L2SS module looks different from the L2SU module in three respects: some variables are indexed (they refer to arrays of similar variables for individual layers); some are within DO-loop (the computations are performed for I, the DO-loop runner, with values increasing from 1 to NL, the number of layers); and, the sequence of the statements is not affected by CSMP, but determined by the modeller. FORTRAN is used because repeating lines for 10 layers (as when using CSMP) is clumsy and makes the program too large to conveniently read or modify. The L2SS module also looks different from the L2SU module because most computations are performed by a single large subroutine (SUSAWA) which calls many other subroutines. Indexed variables must be placed within PROCEDURES. CSMP sorts the contents of the entire PROCEDURE as a single statement, assuming that the arguments to the left of the equal sign are outputs and to the right of the equal sign are inputs. For further implications, check the CSMP manual (IBM, 1975).

Place statements related to organization of the module (STORAGE, FIXED) in the INITIAL of the crop part before running the program; those beginning with '/' (in column 1) link subroutines and the main program with each other and must be placed immediately below the STORAGE statements.

The number of layers (NL, often 10) is explicitly given in the calls for the subroutine SUSAWA. The module can be used with a smaller number of compartments by lowering the value of NL. No changes are needed in the STORAGE statement (Line 4, 5) in the call for the COMMON blocks (Lines 6-9), or in the dimensioning of variables in the subroutines. NL should be two or more. Increasing the number of layers above 10 will rarely be necessary and cannot be done without adapting subroutines and dimensions of variables.

The soil water contents of layers are initialized in SUSAWA in the INITIAL. This is identified by the first input being 1 (Line 38). Many arguments in the subroutine are not now used, so that the list of input arguments for sorting the PROCEDURE that carries SUSAWA has been shortened. When a switch parameter (WCLIS) has a value –1.0, initial water contents are set at values in equilibrium with the soil water table (Line 17). (If user specified water contents are preferred, set the switch to 1.0 and the initial soil moisture contents from TABLE WCLMQI(1-10) are then effective).

The DYNAMIC section seems short. The rate of water uptake is defined here as in the L2SU module (Subsection 5.2.5), but in FORTRAN style (Lines 64-72). No more water than is available is extracted for transpiration in any one time period (Lines 70, 71). The effect of water stress on root growth is calculated in Lines 73 and 74.

The majority of the soil water balance calculations are performed in the SAWAH subroutines, i.e., the unsaturated and saturated flow in all layers. Changes in water contents of the layers, in the level of ponded water, and in the evaporation front depth, are all rate variables that are integrated in L2SS to yield the new values of state variables after every day.

Data characterizing the soil are indicated by specifying the soil type. Indicative values for common types are given in tables KMSA1T, KMSA2T, KMSMXT, KSTT, MSWCAT and WCSTT of Listing 11. The numbers in table TYL (type of layer) specify the soil type for subsequent layers. Other soil types can also be used by replacing some of the present values.

The module requires the observed (or estimated) depth of the groundwater table as an input and forcing function (Lines 22, 57). The water table depth has a negative value when above the surface, and positive below it. The groundwater table is interpreted in SAWAH as a piezometer pressure at the lower profile boundary.

5.4.5 Functions and subroutines

A number of functions and subroutines for handling the soil water balance are given in Appendix B. They are described here briefly, in alphabetical order. Some are relatively complex, but crop growth modelling does not require understanding of the exact procedure of computations. All subroutines beginning with SUST are only used for saturated soil compartments. Most names in the first lines of the subroutines and functions are defined in Appendix A. Note that the L2SU module uses fewer subroutines than L2SS.

User defined functions look like other CSMP statements in the main program. In defining them, however, the output name does not appear, only the name of the function itself. Calls for subroutines in the main program are statements with several inputs and outputs (left of the equal sign). In defining the subroutines, all outputs follow the inputs on the right-hand side. The order of arguments should not be changed.

The FUCCHK function is explained in Subsection 3.4.4.

The FUPHOT function is explained in Subsection 3.4.4.

The FURSC function calculates canopy resistance. It is simple and will be adequate in many situations. Inputs are windspeed, leaf area, crop height and the height of windspeed measurement (Subsection 6.1.6). When more emphasis is put on micrometeorology, another function may be needed (cf., Goudriaan, 1977).

The FUVP function is explained in Subsection 3.4.4.

The FUWCHK function checks whether the integral of all fluxes into and out of the soil, corresponds to the change in water content of the entire profile since the beginning. When the relative difference between both exceeds 1%, a warning is printed. The program should then be checked for errors made while modifying the program.

The FUWCMS function calculates the volumetric water content at a specified value of the moisture suction.

The FUWRED function calculates windspeed near the soil surface under a leaf canopy, using equations based on Goudriaan (1977). Inputs are leaf width, total leaf area, crop height and windspeed at reference height.

The FUWS function quantifies the degree of reduced water uptake in a layer due to stress. It requires as inputs (in the following order), the potential transpiration rate of the canopy, leaf area, water content of the soil layer, sensitivity coefficients for water stress and excess, and the soil water content at wilting point, field capacity and saturation. FUWS can be used for all soil water contents; its value runs from 1.0 (no stress) till 0.0 (see Figure 55, Subsection 4.2.2). Stress is generally due to water shortage, but can also result from excess water.

The SUASTC and SUASTR subroutines are explained in Subsection 3.4.4.

The SUCONV subroutine converts units of variables between the main program and subroutines. The direction of the conversion is governed by the first argument, all others are variables to be converted.

The SUERRM subroutine is used in many subroutines to check whether variable X has a value between a reasonable minimum and maximum value. If not, an error message is written to a file with unit number NUNIT (6 for FOR06.DAT in CSMP) and the program is stopped. If XMIN or XMAX equals −99.0, no minimum or maximum is effective. Appendix A provides a list of the error messages.

The SUEVTR subroutine computes two evaporation rates of a canopy or soil surface from a few surface characteristics and weather data. The first rate is related to the daily total radiation, the second to the drying power of the air. Both refer to a single surface. The subroutine is used to compute potential canopy transpiration and potential soil evaporation. Inputs (in this order) are: radiation on a fully clear day, the measured radiation for that day, the reflection coefficient of the surface for solar radiation, the fraction of the day that evaporation is significant, the average temperature of the evaporating surface, the air humidity, and the leaf, boundary and canopy resistance (see also Subsection 6.2.4). The incoming long wave radiation is calculated from the average daytime temperature. Crops only transpire during the day, so only the daytime fraction of thermal radiation need consideration.

The SUGRHD subroutine determines the gravitational head at the top and bottom of layers. Only thickness and the number of layers are required as inputs.

The SUINTG subroutine performs rectangular integrations to update the water contents for small increments of time during the one-day time period. It also determines the size of these increments (DT). Inputs are the rates of change of water content in each layer, soil moisture contents and parameters for the minimum and maximum size of the time period. Outputs to SUSAWA are updated state variables (WCL, WL0); SUSAWA then returns the total change of each state variable over that day to the L2SS module for real integration.

The SUMFLP subroutine calculates the matric flux potential for a layer with a given water content and soil properties. SWICH3 = 1 computes the matric flux potential using analytical integration. The equation used must be adjusted if the hydraulic *k(h)* relation in SUMSKM is modified. A numerical Gaussian integration can be chosen (SWICH3 = 2) as an alternative if no analytical solution exists or is feasible. See ten Berge & Jansen (1989) for further details.

The SUMSKM subroutine calculates the hydraulic conductivity for a layer from its suction. SWICH3 has the same meaning as in SUMFLP. The *k(h)* relations in SUMSKM and SUMFLP must always correspond.

The SUPHOL subroutine is explained in Subsection 3.4.4.

The SUSAWA subroutine is the key subroutine for soil water balance. It is used in the INITIAL (first input argument is 1) to determine initial water contents, and in the DYNAMIC (first argument is 2) to compute rates of water redistribution for saturated and unsaturated layers, for change in the thickness of the surface water layer and for change in the evaporation front depth. Many inputs are not used in the first call; these are therefore not included in the PROCEDURE statement (Listing 9 Line 14). In the second call, all variable inputs are used (Lines 54-56). SUSAWA causes rates to be integrated with short time periods (in SUINTG); it maintains its own time scale, TIMTOT, which goes from 0.0 till DELT for each time period of the simulation program. Rain is ponded on the surface immediately after the subroutine is called, and root water uptake and evaporation are then extracted. Water is subsequently redistributed between layers for the remainder of the 24 hours. SUSAWA uses

SUZECA to compute the depth of the evaporation front and derives from it the maximum allowable rate of soil evaporation. Inputs to SUSAWA are: a switch argument; water contents of the layers and ponded water; number of layers; variables: crop water uptake, potential evaporation, rain and the soil water table depth; and constants: layer thicknesses and soil types, integration period of crop simulation (DELT, here always one day), the minimum, maximum, and fixed time period for SUSAWA, the maximum level of ponded water and three evaporation constants. Outputs are: the change in water content of the soil layers, of the ponded water level, the actual soil evaporation rate, runoff, daily drainage or capillary supply to the bottom of the profile, the rate of change of all water in the profile and of evaporation front depth, and the total profile depth.

The SUSEFL subroutine selects between fluxes calculated at the boundaries of saturated sets of layers in the saturated and non-saturated soil. Inputs are indexes of saturated layers and potential fluxes. Output is the array of net flows into layers (FLX). The flux through a saturated set (from SUSTFL) is usually not equal to the flux at the boundaries of the set (from SUUNST). It is assumed that the fluxes across the outflow end of the saturated set and across the internal interfaces are those computed by SUSTFL.

The SUSLIN subroutine is called in SUSAWA and calculates certain soil characteristics from the basic data. Inputs to the subroutine are soil types and layer thickness, number of layers and groundwater table depth. Outputs are the soil water contents of layers in equilibrium with the water table and the depth of the upper boundary of each layer. The subroutine checks whether the soil characteristics are between reasonable extremes, and writes them into the CSMP output for inspection by the modeller.

The SUSTCH subroutine searches for saturated layers in the soil profile. It organizes the saturated layers into JTOT sets of 'continuous' saturation, each consisting of JJTOT(J) layers. Each set is sandwiched between two unsaturated layers, or between an unsaturated layer and the top or bottom of the profile, or between the top and bottom of the profile. SUSTCH provides the indexes of saturated layers in the array INXSAT. The only inputs are the water content and number of the layers.

The SUSTFL subroutine calculates a tentative flux through each set of saturated layers, as allowed by the total hydraulic head jump over the set and the saturated conductivities of each layer. This is achieved by solving a system of JJTOT(J) + 1 linear equations for set J of JJTOT(J) layers. These consist of JJTOT(J) flux equations and one summation of hydraulic head intervals to the total jump (DHH). The solution is obtained by invoking the matrix decomposition subroutine SUSTMD and subsequently solving it with SUSTMS. The output array FLXSTT contains the computed saturated fluxes for SUSEFL. These values are still tentative, since gradients and conductivities in neighboring unsaturated layers must permit these rates to occur.

The SUSTHH subroutine identifies the difference in total hydraulic head

(DHH) between top and bottom of that particular set for each continuous set of saturated layers. If a saturated set is sandwiched between unsaturated layers, the matric suctions are defined as 0.0 at the saturated-unsaturated boundaries. If one or both boundaries are not free, the matric suction is replaced by the pressure heads imposed at the relevant interfaces. This occurs when a saturated set extends up to the surface and water is ponded (WL0 positive), or with a true groundwater level in the profile (HPBP positive). Other inputs are the array INXSAT indicating the saturated JTOT sets of JJTOT layers, number of layers, gravitational heads at the bottom and top of the layer, and water standing on the surface.

The SUSTMD subroutine decomposes the matrix A into upper and lower triangles to allow a rapid solution by the SUSTMS subroutine. The initial elements of A are the coefficients of the original set of linear equations defined in SUSTFL, but are replaced by transformed coefficients for SUSTMS. (Source: LUDCMP subroutine by Press et al. (1986) p. 35-36.)

The SUSTMS subroutine solves the set of N linear equations Ax = B. Here A is input, not as the matrix A, but as its upper-lower triangular decomposition determined by the SUSTMD subroutine. INDX is input as the permutation vector returned by SUSTMD. B is input as the right-hand side vector B, and returns with the solution vector x. A, N and INDX are not modified by this routine and can be left in place for successive calls with different right-hand sides B. This routine takes into account the possibility that B will begin with many zero elements (LUBKSB subroutine by Press et al. (1986) p. 37.).

The complex SUUNST subroutine calculates tentative fluxes into and out of unsaturated layers. The calculation is based on the matric flux potential concept. Pressure is assumed to be 0.0 at the interface of an unsaturated and a saturated layer (see also SUSTFL) and only the matric flux potential of the unsaturated compartment is required. Fluxes over the interface between two layers are determined by both gravity and matric terms. Inputs are: the choice of equations constituting the *k(h)* relation (SWICH3 = 1 or = 2); the indication that suction is to be computed from soil water content, or vice versa (SWICH4 = 1 or = 2); the water content of layers; thickness and number of layers; the level of ponded water; and the hydraulic head. Outputs are the fluxes between unsaturated compartments and the moisture suction of layers. If the difference in matric suction and matric flux potential across a layer have the same sign, the product is set to zero to suppress rounding errors in the Gauss integration.

The SUWCMS subroutine calculates the matric suction for a layer from its volumetric water content with Equation 2 (for its limitations, see Subsection 5.1.3) if SWICH4 = 1. Otherwise, the reverse calculation is performed.

The SUZECA subroutine determines by how much the depth of the evaporation front (ZEQT) changes in a day. This front sinks into the soil when the soil dries, and rises when capillary rise provides more water than the amount lost by evaporation. Its value is reset to 0.0 when rain exceeds potential evaporation. Inputs are: the soil water contents of the upper layer, the potential evaporation rate, precipitation, the net fluxes into and out of the first compartment, the amount of ponded water, the time period in the crop module (DELT) and three soil constants.

5.4.6 Two examples using SAHEL and SAWAH

An example using SAHEL is from an evaluation of the potential benefit of soya bean as a dry season crop after rice (Pandey, IRRI, personal communication). Experiments over the last few years show that soya bean can grow well on stored moisture plus incidental rainfall in Los Baños, the Philippines, and responds favourably to supplemental irrigation. To judge whether the crop could be introduced to local farmers, two agronomical questions were raised. What average yield (grain, fodder) may be expected and what is the yield variability due to erratic rainfall? Do soya beans respond well to irrigation, and if so, what is the optimal time or soil condition?

To draw conclusions for an upland site, SAHEL was applied (combine L1D + L2C + L2SU + data + T12) and run for 23 years of actual weather data (sowing data and initial soil moisture were fixed for this example). Figure 62 shows that a mean grain yield of 1.1 t ha^{-1} can be expected. The CV is as high as 30% (assuming a random distribution of yields) due to erratic rainfall. With full irrigation the mean goes up to 3.5 t ha^{-1} and, more importantly, the CV decreases to 10%. Irrigation increases fodder yield from 3.7 to 4.4 t ha^{-1} and the CV decreases from 9 to 8%. Irrigation almost exclusively benefits the grain yield. (Economic evaluation should follow these results before recommending any such procedure to farmers.)

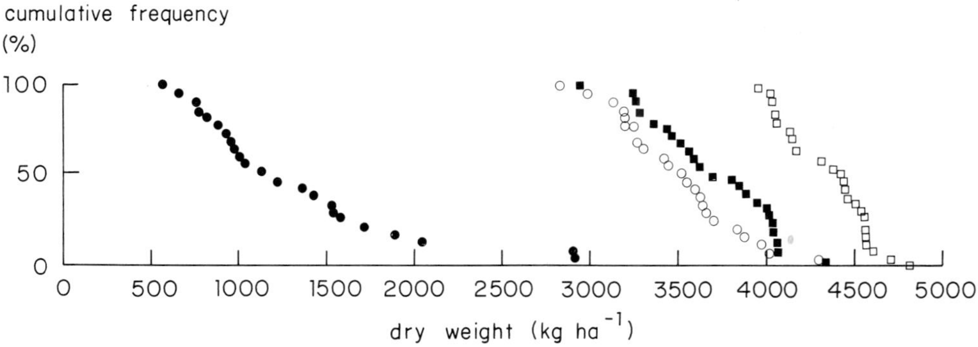

Figure 62. The cumulative frequency of pod + grain yields (filled symbols) and leaves + stems (open symbols) of a dry season soya bean crop in Los Baños, the Philippines, without irrigation (circles) and with irrigation (squares).

The second example uses SAWAH for simulating the contribution of water from a soil water table to the root zone. Capillary rise is assumed to be important in many subhumid areas in the dry season, e.g., in large areas of northeastern Thailand where rice and peanuts are produced in these conditions. The model consisted of the modules L1D + L2C + L2SS + data + T12. The soil and climatic data in this example apply to a large part of the Korat plateau near Khon Kaen. Its landscape undulates smoothly; a laterite pan at variable depths impedes water flow, allowing a shallow groundwater reservoir to develop during the wet season. In the dry season, water is supplied by capillary rise from this reservoir. The soil texture is relatively coarse, so the distance between the root zone and the soil water table is crucial to crop performance. The simulation starts November 1 with soil at field capacity, or wetter when the water table is less than 1.0 m deep. The soil profile chosen is typical for the area: a loamy fine sand on top (0.2 m) and a subsoil of a sandy clay loam (0.2-2.0) (Suraphol, Khon Kaen University, personal communication). The physical data corresponding with these soil types are given in Table 26. Different, but constant depths for the groundwater table are imposed. Weather data from the Khon Kaen station for 1986 show that there was no rain except on two days in November. Results are shown in Figures 63 and 64. The grain yield is almost 7000 kg ha^{-1} for a water table at the surface, drops quickly to 4000 kg ha^{-1} for a water table depth of about 0.5 m and decreases gradually to almost zero at water table depths of 3.0 m and more (Figure 63). This reflects the requirement of ample water for lowland rice. Rooting depth increases to 0.7 m when the water table decreases below 0.9 m. The yield – water table response curve is not smooth because of the interaction of rooting depth, different rates of evaporation and transpiration and the effect of water stress on uptake. The high value at 2.0 m occurs when the water table is exactly at the bottom of the simulated soil profile, which causes a deviation from the trend. No water from the groundwater reservoir reaches the root zone if the depth exceeds 3.0 m. A gradual drying of the topsoil is associated with reduced evaporation, but substantially less so for shallow, than for deep water tables. The water flux at the bottom of the profile can be substantial and varies with the water table (Figure 64). No hard evidence is available to confirm these results; indeed, they are very hard to obtain other than by simulation. The range of simulated rice yields agrees with observations in the region on well-fertilized rice fields with minimum disease and pest levels.

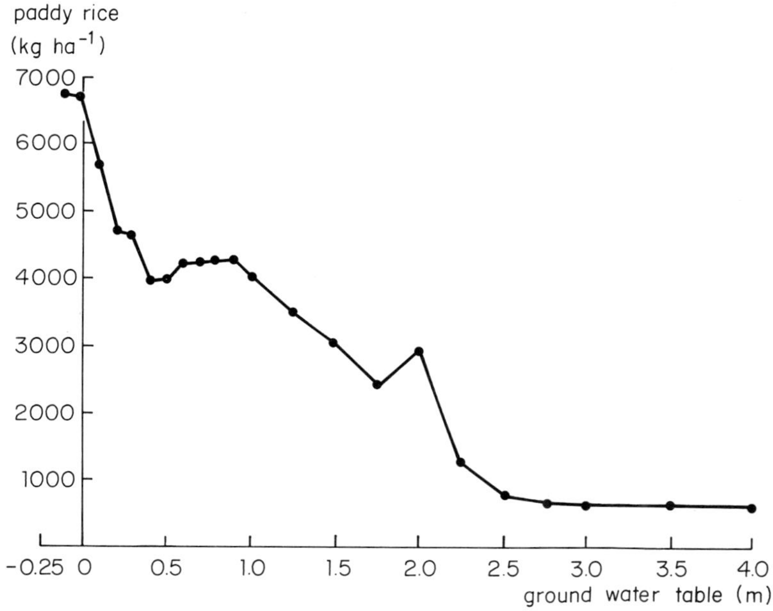

Figure 63. The simulated response of the grain yield (paddy rice, zero moisture) of a dry season rice crop in northeastern Thailand to a constant depth of the ground water table on a sandy clay loam below a loamy fine sand top soil. Almost no rain is received during the growing season.

5.5 Exercises

Use the data presented in the introduction to the excercises of Chapter 2.

5.5.1 Soil characteristics

T1. Classify the texture of the following soils: 38% sand, 20% silt and 42% clay; 10% sand, 70% silt and 20% clay; 60% sand, 30% silt and 10% clay.

T2. Draw the pF-curve for medium-fine sand and for clay loam. Use data from Table 26.

T3. How much is the pressure head that corresponds with a suction of +100 mb, +2 kPa and +1 bar?

T4. Estimate the amount of available water that can be stored between wilting point and field capacity in a profile 1 m thick of (a) fine sand, and (b) clay loam. Use Figure 57 and assume field capacity to be at pF = 2. Which soil has highest amount of water available if field capacity is at pF = 2.5?

T5. Draw the equilibrium volumetric water content versus depth, for a clay

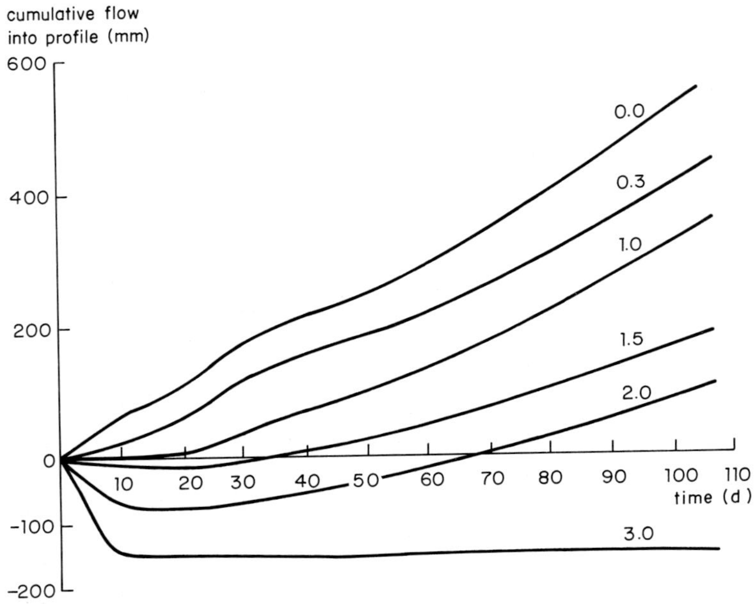

Figure 64. The capillary rise of water into the soil profile to a dry season rice crop in northeastern Thailand, as a function of soil water table depth of 0.0-3.0 m.

loam with the water table at 1.0 m and at 2.5 m. Do the same for a fine sandy loam. In which soil is the largest amount of water stored above the groundwater table? (Remember in equilibrium the suction gradient is 1 cm/cm and the suction is zero at the groundwater level.)

5.5.2 Free-draining soil

T6. Make a relational diagram of the processes and variables which determine the water balance of free-draining soil. Indicate the interactions between crop and soil.

T7. Write CSMP statements to make the runoff fraction dependent on rainfall intensity (mm d^{-1}) and on volumetric water content in the first layer. Assume that an amount of 10 mm d^{-1} can infiltrate at field capacity and beyond. This amount increases linearly with decreasing relative water content to 50 mm d^{-1} when the soil is air dry.

T8. Calculate the cumulative actual evaporation of a bare soil for 10 consecutive days after heavy rain. Plot the results versus time and versus the square root of time. What do you observe? Assume the potential soil evaporation to be constant at 5 mm d^{-1} and use the proportionality factor as in Listing 8.

T9. What is the effect of a larger coefficient for distributing evaporation over soil layers? In what type of soils should this coefficient be relatively high and in what soil types should it be small?

S1. Run L1D + L2C + L2SU for upland rice in the wet season on a loamy soil. Compute total transpiration, evaporation and deep drainage. What is the yield loss due to water stress? Notice how water infiltrates into the soil on a rainy day. What is the water use coefficient in these cases?

S2. Rerun L1D + L2C + L2SU for upland rice on a loamy soil in the dry season. What is the yield, and what is the yield loss, due to stress? How much would the yield improve if it were possible to lower the internal/external fraction in stomata to 0.4? What is the penalty of no regulation in this case? What is the water use coefficient in these cases?

S3. Rerun L1D + L2C + L2SU for upland rice for the wet season on a sandy, a loamy and a clay soil. How much water is lost by evaporation for each soil type during the season? Explain the difference in yield. How much more water is retained in soils that remain fallow?

S4. To the program created in exercise S1, add that roots preferentially draw water from the top layer; all other conditions being equal: 1 cm root in the top layer absorbs 1.2x the average, in the middle 1.0x the average and in the bottom layer 0.8x the average? How do yields change? How do the water contents of the upper and lower layers change?

S5. Although the SAHEL submodel was designed for well drained soils, it may be adapted to simulate in a rough but simple way a situation with impeded drainage. This may be useful if the information required by SAWAH is not available. Assume to this purpose that the interface of layers 1 and 2 limits percolation to a maximum of 2 mm d^{-1}. This causes waterlogging in the upper layer and water that exceeds the storage capacity of this layer (i.e., filling till the layer is saturated) runs off the field. Evaporation continues at the potential rate as long as the water content of the upper layer exceeds 90% of the saturated content. By how much is soya bean yield in the wet season on a loamy soil reduced, compared to a soil without this low-percolation layer? Why? How much water runs off or drains out of the soil profile?

5.5.3 Soil with impeded drainage

T10. How does a relational diagram of L2SS differ from a relational diagram of L2SU?

T11. Calculate the flow rate between an upper layer with WCL(1) = 0.25 cm^3cm^{-3} and a lower layer with WCL(2) = 0.05 cm^3cm^{-3}. Assume for simplicity that the total amount of water in each layer remains contant, and also that conductivity k is independent of moisture content and has a value of 5 cm d^{-1} (hypothetical!). The thickness of both layers is 0.1 m (case 1a). Use the pF curve of coarse sand to derive suctions.

What would be the flow rate if WCL(1) = 0.05 cm^3cm^{-3} and WCL(2) = 0.25

cm^3cm^{-3} (case 1b)? And what if the layer thickness were 0.2 m (cases 2a, 2b)?

S6. Consider the program SWD (Listing 6). Run it for a loess loam soil with groundwater at 0.8 m, starting with a moisture content of 0.08 cm^3 cm^{-3} throughout the profile. Choose the top boundary condition such that equilibrium with groundwater will finally be reached. Depict the moisture profile (WCL vs depth) for each day and note capillary rise. Why does water enter from the top and bottom of the profile? After how long is the equilibrium situation reached? How do you recognize this state? Why does the water content in the top layer exceed field capacity (pF = 2). Repeat this for light clay (soil Type 17, Table 26) and for fine sand (soil Type 4) with DELT = 0.001. Compare the equilibrium profile on loam with one computed from Figure 57.

S7. Rerun the SWD program as in exercise S6, but suppose that the soil profile does not permit water to enter at the top layer. How do results change? Does it take longer to reach equilibrium when starting with a soil at 99% of saturation?

Plot WCL(1) and WCL(10) versus time, and also the fluxes at the bottom of the first and last layers, FLX(2) and FLX(11). Plot WCL versus depth every day (manually) and the final matric suction (MS) versus depth.

Repeat this exercise for impeded drainage: set the flux at the bottom (FLX (11)) to zero (remove the old FLX(NL + 1) statement). Use an initial moisture content of 50% of saturation. What happens?

S8. Use a simple approach to include transpiration (water uptake) of potentially 2 mm d^{-1} to each of the top three layers of the SWD model (ground water at 0.8 m, MSBT = 0., MSTP = 80., FLX(1) = 0.0). To mimick stress, introduce a reduced uptake proportional to the relative water content in that layer. Be cautious with dimensions and units. What changes are due to water uptake?

S9. Run L1D + L2C + L2SS and data for lowland rice, wet and dry season, on a loamy soil. Use the equations to quantify water stress of Subsection 4.2.2. Assume the water table to be at 0.4 m, and a dense soil layer at 0.3 m that roots cannot penetrate. Total the relevant water fluxes. First use 10 layers, and then rerun it for 5 layers (NL = 5, TKL(1 − 5) = 5 * 0.2). What is the cause of the difference? Is it significant?

S10. Determine the grain yield and the capillary rise to the root system using the program from Exercise S9 for a constant water table at depth of 3.0, 1.8, 1.2, 0.8, 0.5, 0.3 and −0.01 m. What is the difference when the dense soil layer is removed and roots grow to a depth of 0.7 m? Are simulation results sensitive to the soil water table input?

S11. Use the program of Exercise 9 with a water table at 0.2 m at day 50., 0.5 m at day 100 and 1.0 m at day 150. What are the yields for transplanting dates 52, 62, 72 and 82? How much water is supplied from the ground water table and by how much is the soil water depleted? Is the rooting depth the same on both soil types?

5.5.4 *Soil water balance subprograms*

T12. Determine the tree of subroutines and functions used for L2SS and L2SU. Which subroutines and functions can be deleted if L2SU is used?

S12. Run both water balance simulation modules with soya bean in the wet season for a sandy, a loamy and a clay soil, with the water table at 3.0 m and all layers at field capacity at the start. Are results of the modules identical? Should they be? Does yield increase if the soil water table is at 1.5 m on these soils?

5.6 Answers to exercises

5.6.1 *Soil characteristics*

T1. The soil textures are those of light clay, silty loam and sandy loam.

T2. See Figure 57.

T3. The pressure head is −100. cm, −20. cm, and −1000. cm, respectively.

T4. In a 1.0 m deep soil of medium fine sand or clay loam, the maximum available water (mm) amounts to:

	WCWP	WCFC		Available water	
	pF 4.2	pF 2.0	pF 2.5	pF 2.0	pF 2.5
Medium fine sand	0.011	0.161	0.104	150.	93.
Clay loam	0.276	0.399	0.376	123.	100.

If field capacity is at pF = 2, the medium fine sand soil has most water available for a crop. When field capacity is at pF = 2.5, the clay loam can store more water (derived with the equation in Subsection 5.1.3).

T5. In both situations the clay loam has more water in the profile above the groundwater table. Close to the groundwater table, volumetric water content in the fine sandy loam is a little higher, but this is insufficient to compensate for the lower water content near the surface.

5.6.2 *Free-draining soil*

T6. Carefully distinguish the types of variables and relations involved. Use the symbols of Figure 5 to draw the diagram. See Figure 6.

T7. Equations can be used for linear relationships:

```
RUNOF = RAIN * FRUNOF
FRUNOF = AMAX1(0., (RAIN – MAXINF) / RAIN)
MAXINF = 50. – (50. – 10.) * RWCL1
RWCL1 = AMIN1(1., (WCL1 – WCAD1) / (WCFC1 – WCAD1))
```

A more flexible solution is to use the TWOVAR function (IBM, 1975). It allows linear interpolation in two dimensions to make the runoff fraction a function of rainfall intensity (the independent variable in the functions) and relative water content of the upper layer (RWCL1, the value after the function name). For instance:

```
RUNOF = RAIN * FRUNOF
FRUNOF = TWOVAR(RUNOFT, RAIN, RWCL1)
RWCL1 = AMIN1(1., (WCL1 - WCAD1) / (WCFC1 - WCAD1))
FUNCTION RUNOFT,0. = 0.,0., 10.,0., 20.,0., 30.,0.,...
   40.,0., 50.,0., 60.,.17, 70.,.29, 80.,.38,...
   90.,.44, 100.,0.5, 1000.,.95
FUNCTION RUNOFT,.5 = 0.,0., 10.,0., 20.,0., 30.,0,...
   40.,.25, 50.,.40, 60.,.50, 70.,.57, 80.,.63,...
   90.,.67, 100.,.7, 1000.,.97
FUNCTION RUNOFT,1. = 0.,0., 10.,0., 20.,.50, 30.,.67,...
   40.,.75, 50.,.80, 60.,.83, 70.,.86, 80.,.88,...
   90.,.89, 100.,.90, 1000.,.99
```

T8. On the first day actual evaporation equals potential evaporation. Cumulative evaporation has a linear relation with the square root of time since the last rain. After 10 days, it equals $5 + 0.6 \cdot 5 \cdot 3 = 14$ mm.

T9. With a larger partitioning coefficient, the contribution of deeper layers to evaporation decreases. In soils with a fast decrease of hydraulic conductivity with increasing suction the deeper layers tend to contribute less to evaporation. In general, coarse textured soils such as sands have a high partitioning coefficient; fine textured soils with a high silt content have a low coefficient.

S1. The totals of transpiration, evaporation and deep drainage are 332.24, 169.38 and 486.15 mm, respectively. The grain yield is 6186.0 kg ha^{-1}, which is only 324.6 kg less than with no water stress. The water use coefficient in these two cases is 95.5 (stressed) and 95.3 kg H_2O kg^{-1} CO_2.

S2. The yield without irrigation is 949.80 kg ha^{-1} and the crop dies early; with plenty of water, the yield is 7691.4 kg ha^{-1}. A crop with a lower internal/external ratio (0.4) produces 1457.2 kg ha^{-1}; a crop without stomatal regulation dies earlier and only 223.88 kg ha^{-1} is formed. The water use coefficient in these cases is 95.5 (95.3 no stress), 72.7 and 141.9 kg H_2O kg^{-1} CO_2.

S3. The yields on sandy, loamy and clay soils are 5615.5, 6186.0 and 5385.1 kg ha^{-1}, respectively. Cumulative soil evaporation is 179.45, 169.38 and 183.61 mm, respectively. The small difference in evaporation does not explain the different yields. The crop yields least from the soil with the smallest difference between field capacity and wilting point.

Simulate a fallow soil by breaking the interactions between crop and soil (the crop related statements can then be removed): redefine TRW to be equal to TRC and ZRT = 0., replace ALV by 0.0 in FUWRED and in EVSC =....

There is 139.01, 288.23 and 309.45 mm of water in the soil profile on sand, loam and clay at harvest time. Without a crop, there is more water left at the same date: 193.66, 343.42 and 363.25 mm, respectively.

S4. This can be simulated by modifying four statements:

```
TRRM = TRC / (ZRT1 * 1.2 + ZRT2 + ZRT3 * 0.8 + 1.E-10)
TRWL1 = TRRM * 1.2 * ZRT1 * WSE1
TRWL2 = TRRM *       ZRT2 * WSE2
TRWL3 = TRRM * 0.8 * ZRT3 * WSE3
```

The yield decreases by 165.3 kg ha^{-1}; 0.0083 cm^3cm^{-3} more water is left in the lower layer and 0.0202 less in the upper layer.

S5. The waterlogged crop yields 1681.2 kg ha^{-1} and the crop in free-draining soil 2477.0 kg ha^{-1} of storage organ. The total runoff in the first case is 600.5 mm and drainage is 8.2 mm, while runoff in the free-draining soil is zero and drainage is 509.7 mm. Therefore, more water is available and less stress occurs in the upper layer in the second case.

5.6.3 Soil with impeded drainage

T10. More layers are distinguished in L2SS and there are two additional state variables: WL0QT and ZEQT. The soil water table is an input. Runoff and infiltration are computed and are no longer a constant fraction of precipitation. Soil evaporation is computed in a different way.

T11. The different cases are:

1a: WCL(1) = 0.25;WCL(2) = 0.05;TKL(1) = 10 cm;TKL(2) =10 cm
1b: WCL(1) = 0.05;WCL(2) = 0.25;TKL(1) = 10 cm;TKL(2) =10 cm
2a: WCL(1) = 0.25;WCL(2) = 0.05;TKL(1) = 20 cm;TKL(2) =20 cm
2b: WCL(1) = 0.05;WCL(2) = 0.25;TKL(1) = 20 cm;TKL(2) =20 cm

From Figure 57 read suctions of 10 cm and 100 cm at WCL = 0.25 and WCL = 0.05, respectively. Using Equation 4, the flow rates q are calculated as:

1a.: $q = (+90 / 10) \cdot 5 + 5 = +50$ cm d^{-1}
1b.: $q = (-90 / 10) \cdot 5 + 5 = -40$ cm d^{-1}
2a.: $q = (+90 / 20) \cdot 5 + 5 = +27.5$ cm d^{-1}
2b.: $q = (-90 / 20) \cdot 5 + 5 = -17.5$ cm d^{-1}
(negative flows are directed upward)

S6. This model cannot deal with saturated soil; with 10 layers, layer thickness must be adjusted to 0.08 m. At the groundwater, MSBT = 0.; equilibrium can only be reached if boundary condition FLX(1) = 0. or MSTP = 80. is imposed. Choose the latter. Water initially enters from the top and bottom of the profile because hydraulic head decreases from the top downward and from the bottom upward. Water contents differ less than 1% from their equilibrium values (0.38826 cm^3 cm^{-3} in Layer 4) after three days; Layer 4 is the last to reach equilibrium. Water contents exceed field capacity according to the pF = 2 definition (0.36 $cm^3 cm^{-3}$, see Table 27) because the water table is less than 1.0 m deep. This near-equilibrium is reached after seven days on clay (in Layer 4, 0.39610 $cm^3 cm^{-3}$) and after only two days on sand (0.23638 $cm^3 cm^{-3}$, Layer 4). This state is recognized by almost zero fluxes and zero hydraulic head gradients.

The equilibrium profile for loam, reading Figure 57, starts with WCL = 0.365 $cm^3 cm^{-3}$ in the upper layer for an average pressure head of –76 cm, and WCL = 0.390 $cm^3 cm^{-3}$ in Layer 5. The difference with the numeric solution is small.

S7. This situation is obtained by replacing Listing 6 Line 18 by FLX(1) = 0. Water flows into the soil only by capillary rise. It takes longer to reach equilibrium: after 11 days on loam, 19 on clay and 4 days on sand, the top layer is always the last. Equilibrium is reached faster from wet than from dry situations because the conductivity of wetter soil is higher: in 3 days on loam, 2 days on clay and 1 day on sand. The implicit assumption in the L2SU module (that after a soil is saturated, most water held above field capacity drains within approximately one day) appears to be a fair approximation.

Impeded drainage leads to redistribution of water but no water enters or leaves the soil. On loam and clay there is no redistribution; on sand it takes about 10 days. Redistribution is slow because conductivities are low.

S8. In the computation of the rate of change DWCLDT(I) for the top three layers add a sink term equal to 0.2 cm d^{-1} / TKL(I) cm * WCL(I) / WCST(I). An actual equilibrium will not be reached, since the roots continue to extract water. To approach the equilibrium state, it takes longer with than without water uptake: 12 days on loam and 4 on sand. The message 'logarithm of negative argument' warns when the stress function is not properly designed and as a result non realistic water contents develop.

S9. Simulation stops at Julian date 290 when the crop is mature and the storage organs weigh 5341.6 kg ha^{-1}. Rooting depth is 0.3 m. Total transpiration is 262.4 mm, total evaporation is 258.3 mm, total drainage is 33.8 mm, and while total rainfall is 916.0 mm, runoff amounted to 361.9 mm.

The same case simulated with five layers yields 5157.5 kg ha^{-1} of storage organ, rooting depth 0.23 m, and 254.0 mm tranpiration, 264.7 mm evaporation, 38.9 mm drainage and runoff is 358.7 mm. The lesser rooting depth is the result of the above-threshold water content in the second of the five layers. The summed fluxes are different because gradients of pressure head are taken over larger distances, leading to a loss in accuracy. These differences are usually insignificant and in the same order of magnitude as differences due to inaccuracies in input data.

S10. At harvest time, the results for panicle weight (WSO), leaf area (ALV), total transpiration (TRWT, mm) and total evaporation (EVSWT) and total drainage (DRSLT) are:

DATEB = 52	WSO	ALV	TRWT	EVSWT	DRSLT
ZW = 3.0	209.0	0.51	28.4	43.3	30.0
ZW = 1.8	826.4	1.28	75.1	89.9	−105.4
ZW = 1.2	2738.0	2.07	140.2	282.3	−421.0
ZW = 0.8	4721.8	3.68	234.7	257.4	−487.9
ZW = 0.5	5371.9	4.53	274.7	246.3	−504.8
ZW = 0.3	5927.3	5.37	312.3	246.8	−546.1
ZW = −0.01	7688.3	8.38	439.2	201.8	−641.2

The wet season results are:

DATEB = 197	WSO	ALV	TRWT	EVSWT	DRSLT
ZW = 3.0	1034.9	0.72	56.4	326.7	377.0
ZW = 1.8	1708.3	1.17	88.2	320.2	304.1
ZW = 1.2	2924.0	2.00	140.8	352.3	157.6
ZW = 0.8	3924.0	2.84	190.1	310.1	137.7
ZW = 0.5	4421.6	3.44	219.3	285.0	131.4
ZW = 0.3	4902.3	4.10	249.6	264.5	66.2
ZW = −0.01	6509.0	6.33	351.2	219.3	−301.7

Without the dense soil layer, the values at the highest and lowest soil water tables are the same or quite similar. With the water table at 1.2 and 1.8 m, roots have access to more water and yields are higher by 1000-1500 kg ha^{-1}. Water table depth is an important input variable in both cases and simulation is consequently sensitive to this.

(The yield with the water table at 0.4 m (Exercise S9) is higher than that at 0.3 or 0.5 m in this exercise. The difference develops largely between days 72 and 80. The water content of layer 3 is then not fully saturated (which it should be) for the 0.3 m water table depth, while it is for the 0.4 m depth. This is a small error due to the small effective number of layers (3) combined with the constant values of the water tables.)

S11. Results are as follows:

	DATEB	WSO	dWCUM	DRSLT	ZRT
loam	52	5010.7	−89.95	−454.29	0.42
	62	5217.0	−97.74	−442.23	0.58
	72	5175.1	−91.70	−437.40	0.62
	82	4550.3	−81.72	−410.10	0.72
sand	52	4747.6	−76.07	−429.45	0.61
	62	5007.2	−60.71	−480.63	0.63
	72	5041.5	−57.62	−462.82	0.62
	82	4667.5	−62.32	−429.83	0.71

WSO is grain yield, dWCUM is the change in available soil water since the beginning, DRSLT is total drainage (negative values indicate capillary rise into the profile) and ZRT is rooting depth.

5.6.4 *Soil water balance subprograms*

T12. L2SU + L1D + L2C uses FUCCHK, FURSC, FUVP, FUWCHK, FUWRED, FUWS, FUPHOT (calls SUERRM, SUASTC) or SUPHOL (calls SUERRM, SUASTC), SUEVTR (calls SUERRM), SUASTR (calls SUASTC).

L2SS + L1D + L2C uses the same functions and subroutines as above, as well as the SAWAH subroutine assembly. It is invoked by calling SUSAWA, which employs the subroutines SUCONV, SUGRHD, SUSLIN (calls SUERRM), SUSTCH, SUSTHH, SUSTFL (calls SUSTMD, SUSTMS), SUUNST (calls SUWCMS (calls SUERRM), SUMFLP (calls SUMSKM), SUMSKM (calls SUERRM)), SUSEFL, SUINTG (calls SUWCMS (calls SUERRM)), and SUZECA.

S12. The results from both modules are not identical, but similar, as expected. Yields according to the L2SU module for loam, sand and clay are 3452.0, 3426.5 and 3046.6 kg ha^{-1}, respectively, and the available water in the profile at the end of the season is 210.4, 62.6 and 244.2 mm. For L2SS, the results are: 3309.9, 3157.6 and 2956.1 kg ha^{-1} and 205.3, 60.8 and 239.0 mm, respectively.

With a water table at 1.5 m, yields in L2SS increase slightly to 3359.2, 3212.7 and 2938.9, and the available water increases to 329.6, 177.2 and 337.0 mm, respectively.

6 Weather data

Section 6.1 discusses the nature and precision of weather variables required for crop growth modelling. These variables are:
- at Production Level 1: daily values of solar radiation, maximum and minimum temperature and air humidity;
- at Production Level 2: in addition to variables at Production Level 1: daily precipitation values, air humidity and windspeed.

Section 6.1 includes some hints on pitfalls of data collection.

A few derived weather variables are computed in Section 6.2: radiation from clear and overcast skies, net radiation, daylength and potential evapotranspiration. Some are needed in the simulations, others provide reference values for comparison of growth conditions. Generating data can be helpful when sufficient original observations are unavailable. This is discussed briefly.

6.1 Historic weather data

6.1.1 Introduction

Solar radiation is by far the most important weather variable for crop growth simulation at Production Level 1, but air temperature can also be crucial. Air humidity is important in very dry weather for some crops. For simulation at Production Level 2, precipitation is an essential input; solar radiation, temperature, and air humidity are also important, but windspeed has little impact on transpiration. Three full sets of historical weather data are shown in Figure 65 as examples of different climatic types. An example of how such data can actually be used in the form of CSMP tables is shown in Listing 11.

Weather data can be obtained from national meteorological services, from the Food and Agricultural Organization (FAO) (Frère, 1987) and from the institutes within the Consultative Group on International Agricultural Research (CGIAR) (e.g., Oldeman et al., 1987). Inspect all data carefully for definitions and units of variables. If data have been obtained locally or from small stations, also inspect the measuring conditions. Information about interpreting weather data and environmental physics can be found in Rose (1966), Monteith (1973), Campbell (1977), Doorenbos & Kassam (1979), and Oldeman & Frère (1982). The World Meteorological Organization recently produced a hardware-software-training package for a data base management system for climatological data (CLICOM). Weather data for modelling are often difficult to obtain, particularly if sets for 5 to 25 years of historical data for all six variables without missing values are required.

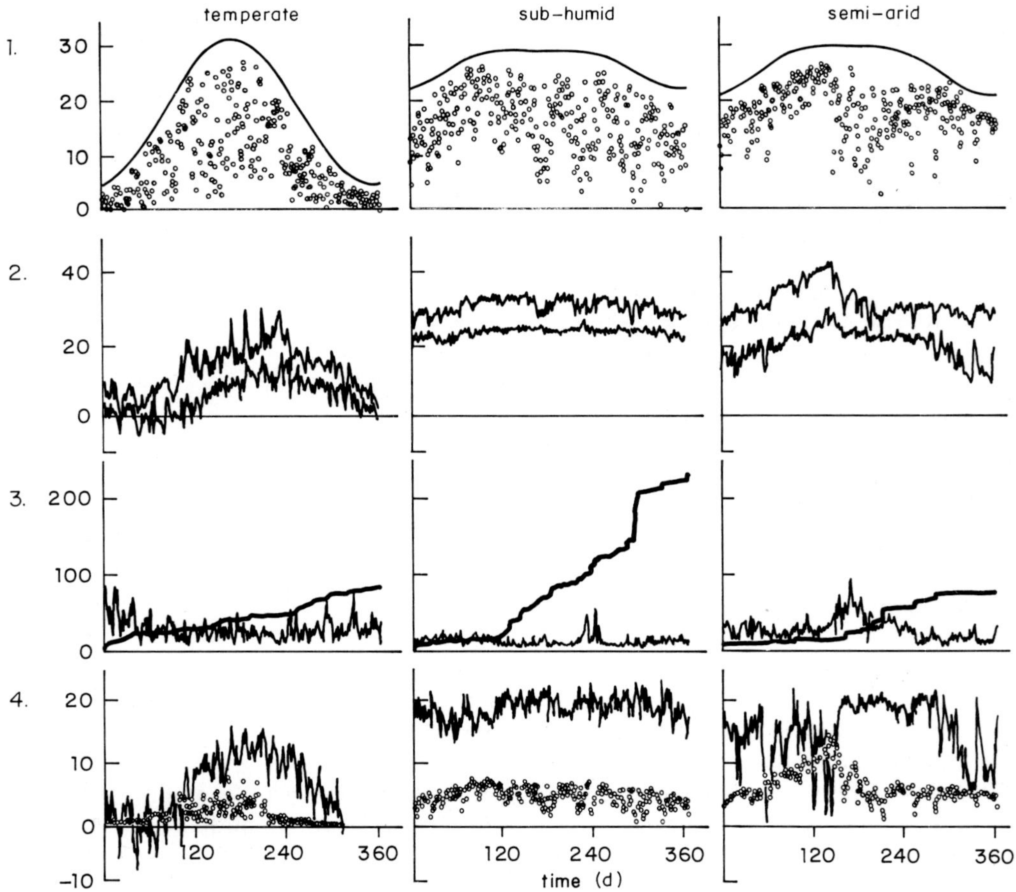

Figure 65. Patterns of weather variables in 1984 in a temperate climate (Wageningen, the Netherlands, left), a humid tropical climate (Los Baños, the Philippines, middle), and a semi-arid tropical climate (Hyderabad, India). 1: observed total global radiation (circles) and clear sky radiation (MJ m^{-2} d^{-1}); 2: maximum and minimum temperatures (°C); 3: average windspeed (m s^{-1} x 10) and cumulative precipitation (cm); 4: dew point temperature (line, °C) and the evaporation rate of a standard grass sward (mm d^{-1}).

The one-day time period used in crop growth simulation corresponds well with the frequency with which weather data are often recorded. Average values can be used if the basic data available are only weekly or monthly means. Obviously, the impact of short deviations from the mean can then not be evaluated. The use of averages for radiation, temperature, windspeed and humidity is appropriate for many purposes. Values of precipitation per day, either observed or generated, are essential for simulating water-limited production (van Keulen & Wolf, 1986). Daily values of radiation are required if sensitive phases (such as tillering and panicle initiation in cereals) fall in periods of variable cloudiness.

When the simulation time period is about six hours (as in module L1Q) intermediate values of weather parameters are estimated from the daily values. These may be replaced by actual data when available.

Weather data are given in the form of CSMP tables: a set of 365 values per variable. It is practical to always supply a full year of data, rather than only a growing season, so that crop growth or derived weather variables can easily be obtained outside the main season.

Two other input parameters characterize a site: latitude and elevation above sea level. Latitude affects the maximum amount of solar radiation and daylength, and elevation affects the CO_2 concentration of the air.

6.1.2 Solar radiation

Solar radiation is a key meteorological variable and its values should be obtained as accurately as possible. Daily values of the 'total global radiation' should be obtained, if possible, from a properly calibrated RIMCO pyranometer, or a Gunn Bellani integrator (Oldeman et al., 1987), or similar instruments. 'Total' refers to the sum of visible and near infrared radiation (400-1300 nm) and 'global' refers to radiation coming from all directions. Readings given in sunshine hours (Campbell-Stokes method) must be converted into J m^{-2} d^{-1} (see van Keulen & Wolf (1986) p. 64 for how to do this). Radiation on fully overcast days is, by convention, 20% of the value on fully clear days, though in reality, days with even less radiation occur. The radiation unit conversion factor in the modules L1D and L1Q (Listing 3 Line 103, Listing 4 Line 131) maintains proper calculating units.

Solar radiation can be partitioned in two ways: according to wavelength in photosynthetically active radiation (PAR, 400-700 nm wavelength) and near infrared (700-1300 nm), and according to direction in direct (from a point source) and diffuse radiation. PAR is always about 50% of the total solar radiation and this fraction varies little with radiation intensity (Monteith, 1973). But the fraction diffuse of the total radiation depends strongly on the daily total. The relation between the fraction diffuse and the daily total radiation relative to the extraterrestrial radiation at that location and date appears to be constant in temperate regions (Figure 66; Spitters et al., 1986). The relation is built into the SUASTC subroutine (Appendix B). The difference between direct and diffuse radiation is important for canopy photosynthesis and SUASTC always accounts for this. Recordings in a humid climate showed almost the same relation, though with much scatter and higher values on clear days (Figure 66). A higher minimum of the fraction diffuse radiation in the humid tropics can be included by replacing 0.23 by 0.35 in the SUASTC sub-

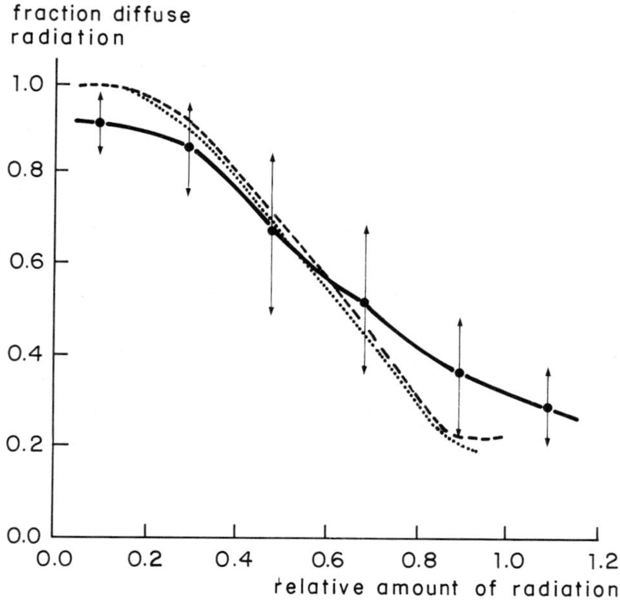

Figure 66. The ratio of diffuse radiation over total global radiation as a function of the ratio of total global radiation over its normal maximum at sea level (0.75 times extraterrestrial radiation). The dotted line is from Spitters et al. (1986), the dashed line from subroutine SUASTR. The drawn line and the standard deviations at some points are based on five years of recording by the Philippine-German Solar Energy Project (a joint project of the Republic of the Philippines and the Federal Republic of Germany, implemented by the Office of Energy Affairs and the Deutsche Geselschaft für Technische Zusammenarbeit).

routine (Appendix B).

In the L1Q module with quarter-day time periods, radiation is supposed to be partitioned equally over morning and afternoon, and is zero at night.

6.1.3 Minimum and maximum temperatures

Temperature affects the rate of most physiological processes. Maximum and minimum temperatures are not used as such in simulation, but are replaced by an effective temperature which is calculated from them. How this is done depends on the process and the time period. The effective temperature for processes that continue during the complete 24-hour time period is the average of the maximum and minimum temperatures. The effective temperature for photosynthesis is assumed to be the average day temperature, calculated as the mean of the 24 hour average and the maximum temperature (Figure 67, Listing 3 Lines 105, 106). Only in the quarter-day time period module (L1Q) is the effective temperature for each time period and for all processes calculated from a fixed, asymmetric pattern over the whole day, by using the nearest maximum and minimum temperatures (Listing 4 Line 133 plus the FUTP function, Appendix B). The four fractions in the call of the FUTP function were determined for Wageningen and also approximate a humid tropical location (Oldeman & Frère, 1982). However, the pattern could be different at other locations.

Maximum and minimum temperatures (or values at 14.00 h and sunrise, respectively) can easily be accurately measured. Values used here are based on observations at the standard screen height of 1.5 m above the soil surface.

The temperature near the soil surface can differ from the air temperature at screen height. Small plants may have a higher temperature during the day and a lower one at night. This is not accounted for here, but it may be worthwhile to

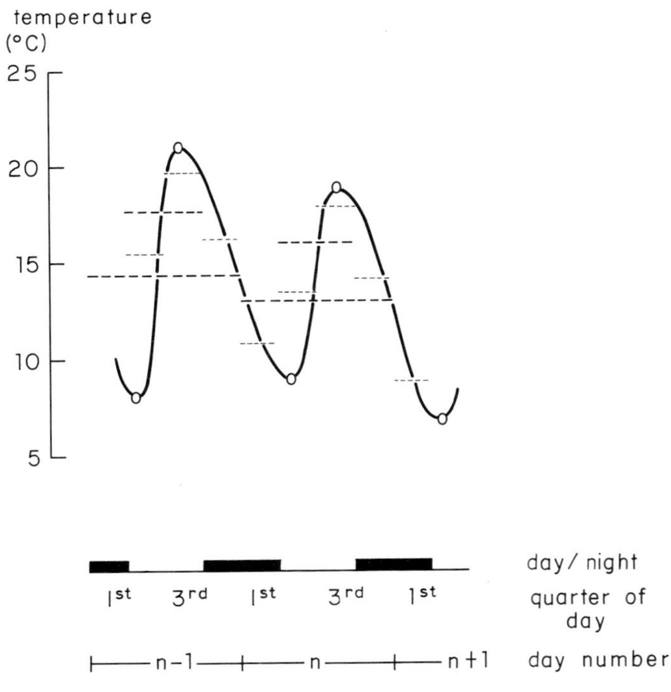

Figure 67. A graphical representation of the procedures to determine the effective temperature in the modules L1D (dashed line: 24-hour average and daytime average) and L1Q (fine dashes: effective temperatures in 6-hour periods). Circles indicate input data.

explicitly consider near surface temperatures when studying crop emergence. Leaf and growing point temperatures can also deviate several degrees from the air temperature (Subsection 4.3.2); the higher the transpiration rate the lower the leaf temperature relative to air temperature. Such deviations mean that using standard meteorological temperatures is not always totally satisfactory. However, as yet they are the best available.

6.1.4 Precipitation

Rain is a particularly important driving variable in the semi-arid and subhumid tropics, but is also important in temperate zones during in dry periods. Its value can change more from day to day than any other meteorological variable. On wet days its value is often 2-20 mm, but can reach 100 mm or more in intensive tropical storms. Determinating its value deserves proper attention and this is more difficult than may appear. Measurements generally underestimate the real value (TNO, 1977). An accuracy of more than 5-10% for absolute values of precipitation is difficult to attain. The wind profile around the rain gauge is very important, for the height of the rain gauge above the surface and its exposure have repercussions of 10-20% and more on the amount of rain caught in the gauge. The standard rain gauge at a meteorological station in the Netherlands has an opening of 200 cm^2, with its rim 40 cm from the soil surface; it underestimates reality by a few percent (TNO, 1977). The imprecision in precipitation measurements is significant for simulation.

The spatial variability of precipitation is also quite large. Its value must be determined at the field for which the study is undertaken whenever precipitation is a key variable for simulation. Though the quality of data of the nearest official meteorological station may be better, less accurate data from the field for which the simulation is performed can be more relevant.

Rainfall generally occurs over periods much shorter than 24 hours, the time period used in simulation. Observations of rainfall intensity with recording rain gauges are rarely made routinely. This is a serious handicap for runoff calculations. In the module L2SS for soils with impeded drainage, it is assumed that all precipitation from a single day is received in the first soil water balance time period.

Interception of precipitation by leaves, stems and fruits is 10-20% of the fresh weight of leaves, or almost equivalent to their dry weight. These amounts are equivalent to a water layer of less than 1 mm per occasion, which is disregarded here. Interception should be taken into account in environments receiving a large number of small rain showers.

Dew rarely amounts to more than 0.1 mm d^{-1} and is difficult to measure. Dew is partly condensed soil evaporation. Though the amounts of dew are small and negligible for water balance studies, wetness of the surface can be crucial for other processes. The leaf-wet period is important in simulating crop damage by pathogens, because diseases develop quicker on wet than on dry surfaces.

6.1.5 Humidity

Air humidity affects transpiration and evaporation, and reduces photosynthesis in some crops when its value is very low. Good air humidity measurements are not easy to obtain, but are not of overriding importance in crop simulation.

Air humidity can be measured in several ways and expressed in different units. The absolute concentration, expressed as the water vapour pressure in kPa (1 kPa=10 mbar), is preferred (its value generally changes little during the day, so that the time at which the reading is taken is less important). Humidity expressed in other units can be converted to kPa by equations shown in Table 29 in Subsection 5.4.5. Relative humidity changes a lot during the day and should be avoided as a basic measurement of humidity.

If reliable values for air humidity are unavailable, they may be approximated by assuming that the air is saturated with vapour at dawn when the daily minimum temperature is reached. The vapour pressure can then be obtained by calling the FUVP function for this temperature. This is a good approximation when there is dew, but it provides values which are too high for the dry season in arid and semi-arid environments.

Table 29. Equations to convert air humidity data (HUAA) into vapour pressure (VPA, in kPa). TPA is air temperature (°C) at the time the wet bulb temperature or relative humidity was taken, or average day temperature; FUVP(temp) is:

$$0.611 * e^{(17.47 * temp / (temp + 239).)}$$

If HUAA is dewpoint temperature: VPA = FUVP(HUAA)
If HUAA is wet bulb
temperature: VPA = FUVP(HUAA) – 0.0623 * (TPA – HUAA)
If HUAA is in mbar: VPA = 0.10 * HUAA
If HUAA is in mm Hg: VPA = 1.33 * HUAA
If HUAA is in percent relative
humidity: VPA = 0.01 * HUAA * FUVP(TPA)

6.1.6 *Wind speed*

Canopy transpiration is only sensitive to windspeeds up to 1-2 m s^{-1} (see Figure 49 in Subsection 4.1.3). Plant lodging due to high wind speeds and gusts is not considered here.

Windspeed is measured directly as a rate and averaged over 24 hours, or obtained as a daily windrun and expressed in m s^{-1}. It is often measured at 2 m over a low grass sward at a standard meteorological station. This reference height is input to the FURSC function (Listing 7 Line 26 and Appendix B). Observations made at other heights must be converted (cf., van Keulen & Wolf (1986) p. 210). Windspeed during the day is generally higher than at night and the effective windspeed is taken as 1.33 times the average value (Listing 7 Line 52); this factor is somewhat arbitrarily chosen and may need adjustment in specific situations. Thermal air instability inside the canopy occurs spontaneously during the day when there is almost no wind and enhances gas exchange. This is accounted for by limiting the windspeed to a lower value of 0.2 m s^{-1} (Line 51).

6.1.7 *Carbon dioxide*

The CO_2 concentration was about 340 vppm (cm^3 m^{-3}) at sea level in 1986, and its value rises steadily (Goudriaan, 1987). The CO_2 concentration fluctuates very little during the year and usually does not change significantly inside the canopy. Its value at sea level is specified with a parameter. The volumetric CO_2 concentration decreases by 12% per 1000 m elevation (Listing 7 Line 59).

6.2 Derived and generated weather data

6.2.1 *Clear sky and overcast radiation*

The maximum amount of daily total global radiation can be computed accurately for any day and latitude (Figure 68) using the SUASTR and SUASTC subroutines of module T12 (Appendix B). These were constructed on the basis of van Keulen et al. (1982) and Spitters et al. (1986). The starting point is the solar constant (about 1400 J m^{-2} s^{-1}), i.e., the intensity of solar radiation measured outside the atmosphere and perpendicularly to the solar rays. Radiation at sea level on a perfectly clear day is about 25% lower than the solar constant due to absorption and reflection by water vapour and dust in the atmosphere.

Clear sky radiation at sea level can be used as a yardstick for monitored radiation (Subsection 6.1.2) and their ratio is the relative amount of radiation received. Observations taken at sea level are usually between 0.15 and 0.75 times the value for extraterrestrial radiation at the same location and the same date, but are less on very heavily overcast days and up to 0.9 times extraterrestrial radiation under extremely clear skies.

6.2.2 *Daylength*

Astronomical daylength is input for the photosynthesis computation in the FUPHOT function and SUPHOL subroutine and in calculating evaporation in the SUEVTR subroutine. Daylength provides the basis for splitting each 24-hour period into day and night fractions (L1Q, Listing 4 Line 144). Its value is accurately obtained from a set of mathematical equations in the SUASTR subroutine, using latitude and date as inputs.

To compute daylength for photoperiod-sensitive species, it must be realized that, even when the sun is still below the horizon the light level is high enough to trigger the photoperiodicity mechanism. Implied in the SUASTC subroutine is assumption that daylength for photoperiodism is the time per day that the sun is at inclinations higher than –4 degrees. Photoperiodic daylength is longer than astronomical daylength by about 0.5 h at the equator and by about 0.8 h or more in temperate zones, depending on the date in the year. The light level to which photoperiodism is sensitive is quite low and not well quantified. Vergara & Chang (1985) determined it to be 1.5-15 mW m^{-2} for rice crops; Salisbury (1981) determined the level to be higher. As a compromise, a value of 50 mW m^{-2} is used here, which corresponds with a sun angle of about −4 degrees. Because calculating the photoperiodic daylength makes no sense when the sun is continuously at higher inclinations, the SUASTR subroutine is limited to –66.5 + 4 and 66.5 – 4 degrees of latitude.

Solar height at noon (SUNH, degrees) can be computed by adding to the SUASTC subroutine:

SUNH = ASIN(COSLD + SINLD) / RAD

6.2.3 *Net radiation*

Net radiation (all wavelenghts included) is the balance of incoming short wave radiation (wavelength 400-1400 nm) minus its reflection and outgoing thermal radiation (>3000 nm), plus incoming thermal radiation (about 12.000 nm). Its calculation is part of the computation of the energy balance for evapotranspiration (SUEVTR subroutine, Appendix B, cf., van Keulen & Wolf (1986) p. 67). Reflection of short wave radiation is about 0.2-0.3 for crops. Reflection from a soil surface (i.e., its albedo) is 0.1-0.4 (see Table 28 Subsection 5.2.4). Reflection increases strongly at low inclinations of the sun (Menenti, 1984), but this is unimportant as the light level is then low. Net thermal radiation is computed according to the Brunt equation from surface temperature (as an indicator of the outgoing long wave radiation) and from cloudiness and air humidity (as indicators of the intensity of incoming long wave radiation from the sky). Values for net radiation can be obtained as output from the program by adding its name to the list of output variables of SUEVTR.

Figure 68 in Subsection 6.2.1 provides an example of the range of values of

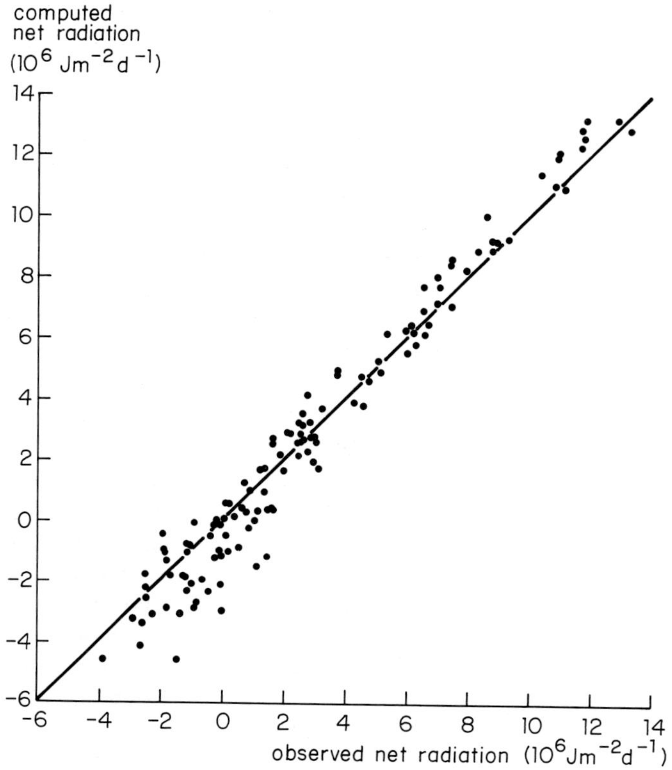

Figure 68. The measured and observed values of daily total net radiation on each third day for a full year in Wageningen, the Netherlands; the line indicates a 1:1 ratio (Data source: Department of Physics and Meteorology of the Agricultural university, Wageningen).

net radiation during a full year in the Netherlands. In this example observations and computations agree closely; significant differences only occur in winter.

6.2.4 Potential evapotranspiration

Simulation of the transpiration rate of a well-watered canopy was described in Section 4.1. Computed rates of 0-5 mm d^{-1} for a grass sward during a growing season in the Netherlands (see Figure 45 in Subsection 4.1.2) compared fairly well with measured rates and observations by the Department of Physics and Meteorology of the Agricultural University in Wageningen.

Similar simulations were carried out for a healthy rice crop with a closed canopy in the Philippines in the dry and wet seasons when the evapotranspiration rates were 4.5-9 and 3-5.5 mm d^{-1}, respectively (unpublished observations by the Climate Unit of the International Rice Research Institute). Because the canopy was closed, almost all evapotranspiration was transpiration. In the rice crop the 5-day average of simulated and measured values corresponded well (Figure 45), but values per individual day differed as much as 15% on average. The discrepancies of day-to-day values are probably largely due to difficulties in measuring this variable over a 24-hour period with the small field lysimeters used, because measured values do not show a consistent relation to weather variables. Soil evaporation was also included in this measurement, but its value must have been negligible. It therefore seems that potential rates of canopy transpiration can be obtained more easily, and at least as well, by simulation, rather than by measurement.

Potential evapotranspiration is a useful variable when characterizing a climate. At meteorological stations it is sometimes determined as the rate of transpiration of a standard grass sward well supplied with water and nutrients. As many variables as possible are then fixed. However, it seems easier, and for many purposes at least as good, to compute such rates rather than to measure them (cf., van Keulen & Wolf, 1986 p. 74). This rate is referred to here as the standard simulation of evaporation of grass (EVG, mm d^{-1}). It can be computed as canopy transpiration, separating the effects of radiation (EVGR) and drying power (EVGD), but with more variables constant: the reflection coefficient is 0.24, no night-time transpiration, leaf resistance is 150 s m^{-1} in the upper 2.5 m^2 m^{-2} of leaves, boundary layer resistance is 12 s m^{-1} and canopy resistance RSTP equals 132. / (1. + 0.54 · WDSAD) (GELGAM, 1984):

```
EVG = EVGR + EVGD * 2.5
EVGR,EVGD = SUEVTR (RDTC, RDTM, 0.24, DLA / 24., TPAD,...
            VPA, 150., 12., RSTP)
```

EVG can vary considerably during the year and between sites, as illustrated in Figure 69 (also Figure 65 Subsection 6.1.1).

Evaporation from a free water surface can be similarly approximated by setting the reflection coefficient at 0.07, permitting night-time evaporation, using the average day temperature, putting leaf resistance equal to 0.0, and assuming the boundary layer and turbulence resistances to be the same as for the grass sward:

```
EVW = EVWR + EVWD
EVWR, EVWD = SUEVTR (RDTC, RDTM, 0.07, 1.00, TPAV,...
             VPA, 0.0, 12.0, RSTP)
```

The calculations of EVG and EVW yield only approximations and these can be different from the approximations by Doorenbos & Kassam (1979). One source of differences is that the temperature of the evaporating surface (e.g., an evaporation pan) is one or two degrees below the average air temperature

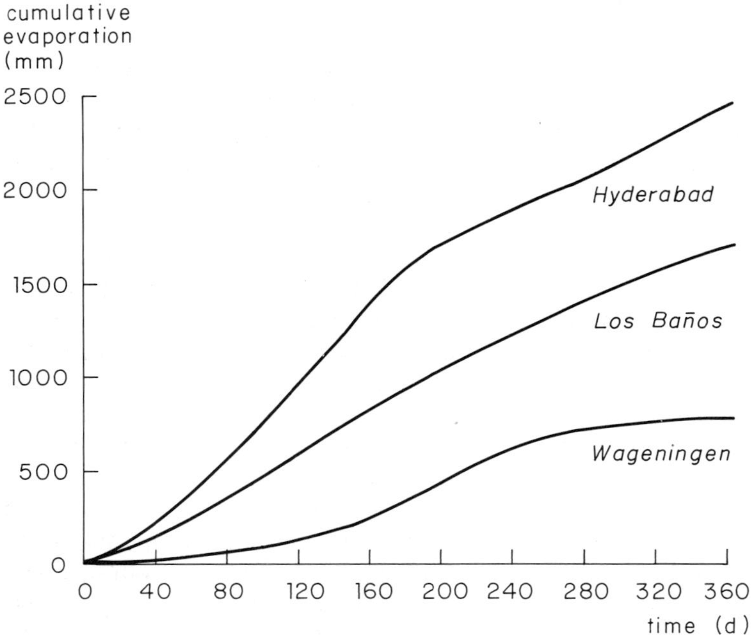

Figure 69. The simulated cumulative transpiration of a well-watered, standard grass sward in Wageningen, The Netherlands, Los Baños, the Philippines and Hyderabad, India. (For weather data, see Figure 65).

in a relatively dry environment (Tamisin, IRRI, personal communication), leading to rates 10-15% lower than those computed with the preceding lines.

Still another evapotranspiration rate (EVP, in mm d^{-1}), based exclusively on meteorological and physical inputs, is often used for climatic characterization and site comparison. It is determined with a correlation Penman method, described by Doorenbos & Kassam (1979, p. 17). It may be helpful to calculate EVP to perform more extensive comparisons with evaporation data locally obtained. For an average site the correlation can be translated into:

```
EVP = 1.0 * (0.75 * (RDTN / 2.47E6) + (1.0 - 0.75) * ...
   0.27 * (1. + WDSAV * 0.864) * (FUVP(TPAV) - VPA) * 10.)
```

The equation makes use of variables already in the program; RDTN is to be calculated as in SUEVTR (Appendix B) with FRD = 1.0 and RF = 0.0. The average site is at sea level, 25 °C, with conditions not very dry or windy. The original paper should be consulted for constants for meteorologically different sites. According to the authors, this value of EVP equals evaporation by a Class A pan multiplied by a factor of about 0.7 (0.5 to 0.8, depending on weather conditions).

6.2.5 *Generating weather data*

It is not often that weather data for more than a few full years is available from a single meteorological station. This is insufficient to test the stability of crop yields over a 10-25 year period using simulation. The next best alternative to a large set of historical weather data, is a large set of weather data generated from observations taken over a few years. These can be generated by using information contained in the historical data: the relations between values of variables on successive days and between the values of all variables on individual days. Richardson investigated these relations for precipitation and temperature in the USA, and his computer programs were modified and extended by Geng et al. (1985a,b). Several programs to generate daily weather variables were moulded into an easy-to-use program (Supit, 1986). With this program, daily values for solar radiation, precipitation, maximum and minimum temperature and windspeed are generated for any number of years from as little as two years of historical data. Humidity is not included.

When generating new weather data, twelve monthly precipitation totals plus the number of wet days for each month appear to be just as good as taking 365 daily values (Geng et al., 1986). This requires at least 10 times less data than taking daily values and means that sufficient rainfall data to support crop simulation can be collected from remote areas without frequent measurements. This not only spreads resources, but is particularly helpful because rain varies more than any other meteorological variable over short distances and requires a denser recording network than other weather characteristics.

7 Listings of modules

This chapter contains listings of modules discussed in the previous chapters. Listings 1-11 are written in the simulation language Continuous System Modelling Program (CSMP, IBM, 1975; Subsection 1.4.1). Abbeviations are explained in Listing 12.
Copies of the listings and the crop data on a floppy disk can be obtained from the authors through PUDOC (P.O. Box 4, Wageningen, 6700 AA, The Netherlands) by sending one high capacity 5.25 inch floppy disk with the note 'MACROS modules and data'. For a copy of a PCSMP version for IBM PC/AT compatibles plus a short manual, send two high capacity floppy disks with the note 'PCSMP'.

Listing 1. A CSMP program to compute growth efficiency characteristics.

```
TITLE PROGRAM TO COMPUTE CRG, CPG AND FC (JUNE 1986)
STORAGE COMP(6)
CRG=FCARB*(1.211+0.064) + FPROT*(1.793+0.094+(0.852+0.045)*LEG)+...
    FFAT *(3.030+0.159) + FLIGN*(2.119+0.112) +...
    FOA  *(0.906+0.048) + FMIN *(0.000+0.120)
**CARBOHYDRATE REQUIREMENT GROWTH IN G GLUCOSE PER G PRODUCT
CPG=FCARB*(0.123+0.093) + FPROT*(0.679+0.138+(1.250+0.066)*LEG)+...
    FFAT *(1.606+0.234) + FLIGN*(0.576+0.164) + ...
    FOA  *(-0.045+0.070)+ FMIN *(0.000+0.176)
**CO2 PRODUCTION DURING GROWTH IN G CO2 PER G PRODUCT
FC =FCARB*0.451 + FPROT*0.532 + FFAT*0.774 +...
    FLIGN*0.690 + FOA*0.375   + FMIN*0.000
**FRACTION CARBON IN G C PER G PRODUCT
EC =FCARB*17.3 + FPROT*22.7 + FFAT*37.7 + ...
    FLIGN*29.9 + FOA*13.9 + FMIN*0.
**ENERGY CONTENT PRODUCT, IN KJ PER G
ECAF=EC/(1.-FMIN)
**ENERGY CONTENT ASH FREE MATERIAL
ENEFF=EC/(CRG*15.6)
**ENERGY EFFICIENCY CONVERSION
ENEFFA=ECAF/(CRG*15.6)
**ENERGY EFFICIENCY CONVERSION, EXPRESSED ON ASH FREE BASIS
CAEFDM=FC/(CRG*0.400)
**CARBON EFFICIENCY, FRACTION, FOR TOTAL DRY MATTER
FCARB=COMP(1)
FPROT=COMP(2)
FFAT =COMP(3)
FLIGN=COMP(4)
FOA  =COMP(5)
FMIN =COMP(6)
TOTAL=COMP(1)+COMP(2)+COMP(3)+COMP(4)+COMP(5)+COMP(6)
```

```
BALANS=CRG*0.400 - CPG*0.273 - FC*1.000
**TOTAL MUST EQUAL 1.000, BALANS MUST REMAIN 0.000
TIMER DELT=1., FINTIM=1., PRDEL=1.
PRINT CRG,CPG,FC,TOTAL,BALANS,EC,ECAF,ENEFF,CAEFDM
**DATA
TITLE RICE LEAVES
PARAM LEG=0.
TABLE COMP(1-6)=0.53, 0.20, 0.04, 0.04, 0.04, 0.15
END
TITLE GROUNDNUT, SEED + POD
TABLE COMP(1-6)=0.14, 0.27, 0.39, 0.14, 0.03, 0.03
PARAM LEG=1.
END
STOP
ENDJOB
```

Listing 2. Module TIL to simulate development of tillers, florets and grains in rice. (To be inserted in module L1Q, Listing 4.)

```
TITLE TIL Tillers, florets and grains (JANUARY 1989)
**To replace line 36 in L1Q.CSM
*To be included in INITIAL
NTII  =WLVI/WTI
*To be included in DYNAMIC
NTI   =INTGRL(NTII,(GNTI-LNTI)*FADL)
GNTI  =DSTF*AMAX1(0.,(NTIP-NTI)/TCFT)
LNTI  =DSTD*AMAX1(0.,(NTI-NTIP)/TCDT)
NTIP  =CAGCR/CNTI
DSTF  =NOR(DST1-DS,DS-DST2)
DSTD  =NOR(DST1-DS,DS-(DST2+0.15))
CNTI  =AFGEN(CNTIT,DS)
NTIPL =NTI/NTII

NFL   =INTGRL(0.,GNFL*FADL)
GNFL  =DSFL*AMIN1(NFLMX-NFL,NFLP-NFL)/TCFF
NFLP  =CAGCR/CNFL
CNFL  =0.7*GGRMN
NFLMX =NFLMXT*NTI
DSFL  =NOR(DSF1-DS,DS-DSF2)

NGR   =INTGRL(0.,GNGR*FADL)
GNGR  =DSGR*AMAX1(0.,AMIN1(NGRP-NGR,NGRMX-NGR)/TCFG)
NGRP  =CAGCR/GGRMN
NGRMX =NFL
DSGR  =NOR(DSG1-DS,DS-DSG2)
GGRMN =WGRMX/GFP
GFP   =1./(1.33*DRR)
GGRMX =GGRMN*2.
WGR   =WSO/(AMAX1(NGR,1000.))

GSOM  =NGR*GGRMX*AFGEN(GGRT,TPAA)

PARAM DST1  =0.3, DSF1  =0.7, DSG1  =0.95
PARAM DST2  =0.75,DSF2  =0.95,DSG2  =1.15
PARAM TCFT  =15., TCFF  =7.,  TCFG  =3., TCDT  =10.
PARAM WTI   =1.0E-5, NFLMXT =100., WGRMX =23.5E-6
FUNCTION GGRT  = 10.,0.0, 15.,0.0, 18.,0.75, ...
                 23.,1.0, 27.,0.9, 40.,0.0
```

```
FUNCTION CNTIT = 0.0,5.E-6,  0.3,5.E-6,  0.75,25.E-6, ...
                 1.0,75.E-6, 2.1,75.E-6
FINISH WGR   =WGRMX, DS    =2.0, CELVN=3.0
```

Listing 3. Basic module for crop growth simulation (L1D).

```
TITLE L1D (JULY 1987)
FIXED IDATE,I,NL
STORAGE RDTMT(365),TPHT(365),TPLT(365),RAINT(365),...
   HUAAT(365),WDST(365),TKL(10),TYL(10)
INITIAL
   WRTI  =WLVI
   ALVI  =WLVI/(SLC*AFGEN(SLT,DSI))
   CPEW  =1.
   DREW  =1.
   PCEW  =1.
DYNAMIC
**WEIGHTS OF CROP COMPONENTS
**Explanation in sections 3.2, 2.2, 3.4
   WLV   =INTGRL(WLVI,GLV-LLV)
   WST   =INTGRL(WSTI,GST*(1.-FSTR))
   WIR   =INTGRL(0.,GST*(FSTR*(FCST/0.444))-LSTR)
   WSO   =INTGRL(WSOI,GSO)
   WEPSO =WSO*FEPSO
   WRT   =INTGRL(WRTI,GRT-LRT)
   WSS   =WLV+WST+WSO+WIR
   WCR   =WSS+WRT
   WLVD  =INTGRL(0.,LLV)
   WRTD  =INTGRL(0.,LRT)
**GROWTH RATES AND LOSS RATES
**Explanation in sections 2.4, 3.2, 2.2
   GLV   =CAGLV/CRGLV
   GST   =CAGST/CRGST
   GRT   =CAGRT/CRGRT
   GSO   =CAGSO/CRGSO
   LLV   =WLV*AFGEN(LLVT,DS)
   LRT   =WRT*AFGEN(LRTT,DS)
   LSTR  =INSW(AFGEN(CASTT,DS)-0.01,WIR*0.1,0.)
**CARBOHYDRATE AVAILABLE FOR GROWTH, EXPORT
**Explanation in sections 3.2, 2.4, 2.3, 2.2
   CAGCR =PCGW*0.682-RMCR*0.682+LSTR*1.111*0.947
   CAGSS =CAGCR*AFGEN(CASST,DS)*CPEW
   CAGRT =CAGCR-CAGSS
   CAGLV =CAGSS*AFGEN(CALVT,DS)
   CAGST =CAGSS*AFGEN(CASTT,DS)
   CAGSO =CAGSS-CAGLV-CAGST
   CELV  =PCGW-(RMLV+RMST+0.5*RMMA)
   CELVN =INTGRL(0.,INSW(CELV,1.,-CELVN/DELT))
**PHOTOSYNTHESIS, GROSS AND NET
**Explanation in sections 2.1, 3.3, 3.4
   PCGW  =PCGC*PCEW
   PCGC  =FUPHOT(PLMX,PLEA,ALV,RDTM,DATE,LAT)
```

```
   PLMX  =PLMXP*AFGEN(PLMTT,TPAD)*LIMIT(200.,600.,SLA)/SLC
   PLEA  =PLEI*AFGEN(PLETT,TPAD)
   PCGT  =INTGRL(0.,PCGW)
   RCRT  =INTGRL(0.,RMCR+RGCR)
   PCNT  =INTGRL(0.,PCGW-(RMCR+RGCR))

**RESPIRATION
**Explanation in sections 2.4, 2.3
   RMCT  =INTGRL(0.,RMCR)
   RMCR  =RMLV+RMST+RMSO+RMRT+RMMA
   RMLV  =WLV*RMCLV*TPEM*0.75
   RMST  =WST*0.010*TPEM+WIR*0.0
   RMRT  =WRT*0.015*TPEM
   RMSO  =AMIN1(1000.,WSO)*0.015*TPEM
   TPEM  =Q10**((TPAV-TPR)/10.)

   RMMA  =0.20*PCGW*0.5

   RGCR  =RGLV+RGST+RGSO+RGRT+RLSR
   RGLV  =GLV*CPGLV
   RGST  =GST*CPGST
   RGSO  =GSO*CPGSO
   RGRT  =GRT*CPGRT
   RLSR  =LSTR*1.111*0.053*1.467

**CARBON BALANCE CHECK
**Explanation in section 3.4
   CKCRD =FUCCHK(CKCIN,CKCFL,TIME)
   CKCIN =(WLV-WLVI)*FCLV+(WST-WSTI)*FCST+...
          (WSO-WSOI)*FCSO+(WRT-WRTI)*FCRT+WIR*0.444
   CKCFL =PCNT*0.2727-(WLVD*FCLV+WRTD*FCRT)

**LEAF AREA
**Explanation in section 3.3
   ALV   =INTGRL(ALVI,GLA-LLA+GSA)
   GLA   =GLV/SLN
   LLA   =LLV/SLA
   GSA   =0.5*GST/SSC
   SLN   =SLC*AFGEN(SLT,DS)
   SLA   =(WLV+0.5*WST*(SLC/SSC))/ALV

**PHENOLOGICAL DEVELOPMENT OF THE CROP
**Explanation in section 3.1
   DS    =INTGRL(DSI,INSW(DS-1.,DRV,DRR))
   DRV   =DRCV*DRED*DREW*AFGEN(DRVTT,TPAV)
   DRED  =AFGEN(DRDT,DLP)
   DRR   =DRCR*AFGEN(DRRTT,TPAV)

**WEATHER DATA AND TIME
**Explanation in chapter 6 and section 3.4
   RDTM  =RDTMT(IDATE)*RDUCF
   RDTC,DLA,DLP=SUASTR(DATE,LAT)
   TPAV  =(TPLT(IDATE)+TPHT(IDATE))/2.
   TPAD  =(TPHT(IDATE)+TPAV)/2.

   DATE  =AMOD(DATEB+TIME+364.,365.)+1.
   IDATE =DATE
```

```
**RUN CONTROL AND OUTPUT
METHOD RECT
TIMER DELT=1.,TIME=0.,FINTIM=1000.,PRDEL=10.,OUTDEL=10.
FINISH  DS    =2., CELVN =3.

PRINT DATE, WLV,WST,WIR,WSO,WRT ,GLV,GST,GSO,GRT ,...
     SLA,PLMX,ALV,DS, TPAV,RDTM ,PCGT,RCRT,RMCT
PRTPLOT WLV, WLVT, WLVST,  WLVSO
PAGE GROUP
   WLVT  =WLV+WLVD
   WLVST =WLVT+WST+WIR
   WLVSO =WLVST+WSO
   HI    =WSO/WSS
   RSH   =RMLV+RMST+RMSO+RMMA+RGLV+RGST+RGSO+RLSR
   WSTR  =WST+WIR
```

Listing 4. Crop growth module with quarter-day time periods (L1Q).

```
TITLE L1Q (AUGUST 1988)
FIXED IDATE, MINO, MAXO
STORAGE RDTMT(365),TPHT(365),TPLT(365),RAINT(365),HUAAT(365),WDST(365)

INITIAL
   WRTI  =WLVI
   WARI  =WLVI*0.05
   ALVI  =WLVI/(SLC*AFGEN(SLT,DSI))

DYNAMIC
**WEIGHT CROP COMPONENTS
**Explanation in sections 3.2, 2.4, 2.2, 3.4
   WLV   =INTGRL(WLVI,(GLV-LLV)*FADL)
   WAR   =INTGRL(WARI,(GAR-CUGCR)*FADL)
   WARR  =WAR/(WLV+1.E-10)
   WST   =INTGRL(WSTI,GST*FADL)
   WSR   =INTGRL(0.,(GSR-LSR)*FADL)
   WSO   =INTGRL(WSOI,GSO*FADL)
   WEPSO =WSO*FEPSO
   WRT   =INTGRL(WRTI,(GRT-LRT)*FADL)
   WSS   =WLV+WST+WSO+WAR+WSR
   WCR   =WSS+WRT
   WIR   =INTGRL(0.,((GAR-CUGCR)*0.900+(GSR-LSR))*FADL)
   WLVD  =INTGRL(0.,(0.5*LLV)*FADL)
   WRTD  =INTGRL(0.,LRT*FADL)

**GROWTH RATES AND LOSS RATES
**Explanation in sections 2.4, 2.2, 3.2, 3.4
   GAR   =PCGD*0.682-RMCR*0.682 +...
          LSR*1.111*0.947+0.5*LLV*(FCLV/0.400)*0.947
   GLV   =CAGLV/CRGLV
   GST   =CAGST/CRGST*...
         (CRGST*(1.0-FSTR)/(FSTR*(1.111/0.947-CRGST)+CRGST))
   GRT   =CAGRT/CRGRT
   GSO   =AMIN1(CAGSO/CRGSO,GSOM)
   GSOM  =(WSO+10.)*GSORM
   GSR   =INSW((WST+WSR)*(FSTR+0.10)-WSR,0.,GSRP)
   GSRP  =(CAGSS-GLV*CRGLV-GST*CRGST-GSO*CRGSO)*0.947/1.111

   LLV   =WLV*0.15*MCLV
```

```
   LRT   =WRT*0.15*MCRT
   LSR   =WSR*0.20*MCSR
   MCLV  =INSW(GSOAVM*0.8-GSOAV-10.,0.,1.)
   MCRT  =MCLV
   MCSR  =INSW(GSOAVM*1.0-GSOAV-10.,0.,1.)
   GSOAVM=INTGRL(0.,AMAX1(0.,GSOAV-GSOAVM)*FADL)
   GSOAV =INTGRL(0.,((GSO-GSOAV)/2.)*FADL)

**CARBOHYDRATE AVAILABLE AND CONSUMED FOR GROWTH, EXPORT
**Explanation in sections 3.2, 2.4, 2.2, 3.4
   CUGCR =CUGLV+CUGST+CUGSO+CUGRT+CUGSR
   CUGLV =GLV*CRGLV
   CUGST =GST*CRGST
   CUGSO =GSO*CRGSO
   CUGRT =GRT*CRGRT
   CUGSR =GSR*1.111/0.947

   CAGCR =LIMIT(0.,(WAR-0.05*WLV)/(DELT*FADL),(WAR-0.05*WLV)*1.5)
   CAGSS =CAGCR*AFGEN(CASST,DS)
   CAGRT =CAGCR-CAGSS
   CAGLV =CAGSS*AFGEN(CALVT,DS)
   CAGST =CAGSS*AFGEN(CASTT,DS)
   CAGSO =CAGSS-CAGLV-CAGST

   CELV  =INTGRL(10.,(PCGD-(RMLV+RMST+0.5*RMMA))*FADL-...
          INSW(DTIME-0.01,CELV/DELT,0.))
   CELVN =INTGRL(0.,INSW(CELV,1.,-CELVN/DELT))

**PHOTOSYNTHESIS, GROSS AND NET
**Explanation in sections 2.1, 3.3, 3.4
   PCGD  =PCGC/(DLA/24.)
   PCGC  =FUPHOT(PLMX,PLEA,ALV,RDTM,DATE,LAT)
   PLMXT =PLMXP*LIMIT(200.,600.,SLA)/SLC
   PLMX  =PLMXT*AFGEN(PLMTT,TPAA)*(1.-ELV/8000.)*...
          AFGEN(PLMHT,0.75*VPD)*INSW(WARR-0.3,1.,0.3)
   PLEA  =PLEI*AFGEN(PLETT,TPAA)

   PCGT  =INTGRL(0.,PCGD*FADL)
   RCRT  =INTGRL(0.,(RMCR+RGCR)*FADL)
   PCNT  =INTGRL(0.,(PCGD -(RMCR+RGCR))*FADL)

**RESPIRATION
**Explanation in sections 2.4, 2.3
   RMCT  =INTGRL(0.,RMCR*FADL)
   RMCR  =RMLV+RMST+RMSO+RMRT +RMMA
   RMLV  =INSW(0.5-NIGHT,RMLVN,RMLVD)
   RMLVN =WLV*RMCLV*TPEM
   RMLVD =WLV*RMCLV*TPEM*0.5
   RMST  =WST*0.010*TPEM+WSR*0.0+WAR*0.0
   RMRT  =WRT*0.015*TPEM
   RMSO  =AMIN1(1000.,WSO)*0.015*TPEM
   TPEM  =Q10**((TPAA-TPR)/10.)

   RMMA  =0.20*PCGDV*0.5
   PCGDV =INTGRL(0.,(PCGD-PCGDV)/1.*FADL)

   RGCR  =RGLV+RGST+RGSO+RGRT+RGSR+RLSR+RLLV
   RGLV  =GLV*CPGLV
```

```
   RGST  =GST*CPGST
   RGSO  =GSO*CPGSO
   RGRT  =GRT*CPGRT
   RGSR  =GSR/0.947*1.111*0.053*1.467
   RLSR  =LSR*1.111*0.053*1.467
   RLLV  =(LLV*0.5)*(FCLV/0.400)*0.053*1.467

**CARBON BALANCE CHECK
**Explanation in section 3.4
   CKCRD =FUCCHK(CKCIN,CKCFL,TIME)
   CKCIN =(WLV-WLVI)*FCLV+(WST-WSTI)*FCST+...
          (WSO-WSOI)*FCSO+(WRT-WRTI)*FCRT+WIR *0.444
   CKCFL =PCNT*0.2727-(WLVD*FCLV+WRTD*FCRT)

**AREA OF LEAVES
**Explanation in section 3.3
   ALV   =INTGRL(ALVI,(GLA-LLA+GSA)*FADL)
   GLA   =GLV/SLN
   LLA   =LLV/SLA
   GSA   =0.5*GST/SSC
   SLN   =SLC*AFGEN(SLT,DS)
   SLA   =(WLV+0.5*WST*(SLC/SSC))/ALV

**PHENOLOGICAL DEVELOPMENT
**Explanation in section 3.1
   DS    =INTGRL(DSI,INSW(DS-1.,DRV,DRR)*FADL)
   DRV   =DRCV*DRED*AFGEN(DRVTT,TPAA)
   DRED  =AFGEN(DRDT,DLP)
   DRR   =DRCR*AFGEN(DRRTT,TPAA)

**WEATHER DATA, TIME AND DATE
**Explanation in chapter 6 and sections 1.4, 3.4
   RDTM  =INSW(0.5-NIGHT,0.,RDTMT(IDATE)*RDUCF)
   RDTC,DLA,DLP=SUASTR(DATE,LAT)
   TPAA  =FUTP(IDATE,DTIME,TPHT,TPLT,0.15,0.45,0.90,0.60)
   VPD   =AMAX1(0.,FUVP(TPAA)-HUAAT(IDATE))

   DATE  =AMOD(DATEB+TIME+364.,365.)+1.
   IDATE =DATE
   DTIME =TIME-AINT(TIME)
   NIGHT =INSW(AND(DTIME-0.1,0.6-DTIME)-0.1, 1.,0.)

**RUN CONTROL AND OUTPUT
METHOD RECT
TIMER DELT=0.25,TIME=0.,FINTIM=1000.,PRDEL=10.,OUTDEL=10.
   FADL  =INSW(0.5-NIGHT,2.-DLA/12.,DLA/12.)
FINISH DS   =2., CELVN= 3.

PRINT DATE, WLV,WST,WSO,WRT,WAR,WSR,WARR, GLV,GST,GSO,...
   GRT,GSRP,GSR,SLA,PLMX,ALV,DS, TPAA,PCGT,RCRT,RMCT,LLV
PRTPLOT WLV, WLVT, WLVST, WLVSO
PAGE GROUP
   WLVT  =WLV+WLVD+WAR
   WLVST =WLVT+WST+WSR
   WLVSO =WLVST+WSO
   WSTR  =WST+WSR
   HI    =WSO/WSS
PRTPLOT PCGD, RSH, PCNSH, WARR, TPAA
```

```
PAGE GROUP=3
   RSH   =RMLV+RMST+RMSO+RMMA +RGLV+RGST+RGSO+RGSR+RLSR+RLLV
   PCNSH =PCGD-RSH
```

Listing 5. Crop data for rice (variety IR36).

```
TITLE OSIR36.DAT ORYZA SATIVA, RICE, CV IR36
**PHOTOSYNTHESIS AND RESPIRATION; TABLES 3,4,5,8,11,23
PARAMETER PLMXP =47., PLEI  =0.50
FUNCTION  PLMTT = 0.0,0.0,  10.,0.0, 25.,1.00, 30.,1.00,...
                  42.,0.0,  45.,0.0
FUNCTION  PLMHT = 0.0,1.00, 1.0,1.0, 2.0,0.99, 3.0,0.86,...
                  4.0,0.71
FUNCTION  PLETT = 0.0,1.0,  15.,1.0, 25.,0.90, 35.,0.60,...
                  45.,0.2,  50.,0.01
PARAMETER CRGLV=1.326, CRGST=1.326, CRGSO=1.462, CRGRT=1.326
PARAMETER CPGLV=0.408, CPGST=0.365, CPGSO=0.357, CPGRT=0.365
PARAMETER FCLV =0.419, FCST =0.431, FCSO =0.487, FCRT =0.431
PARAMETER RMCLV=0.02 , TPR  =25.,   Q10  =2.
**CONSISTENCY CHECK: 12/30*CRGLV=1.0*FCLV+12/44*CPGLV
**BIOMASS PARTITIONING AND AGING; TABLES 7,17
FUNCTION CALVT = 0.0,0.51, 0.5,0.51, 0.6,0.47, 0.7,0.32,...
                 0.8,0.26, 1.0,0.00, 1.1,0.00, 2.5,0.00
FUNCTION CASTT = 0.0,0.49, 0.5,0.49, 0.6,0.53, 0.7,0.68,...
                 0.8,0.74, 1.0,1.00, 1.1,0.27, 1.2,0.00,...
                 2.1,0.0
FUNCTION CASST = 0.0,0.86, 0.5,0.86, 0.6,0.86, 0.7,0.95,...
                 0.8,0.94, 1.0,0.89, 1.1,1.00, 2.5,1.00
PARAMETER FSTR  =0.25, FEPSO =0.8, GSORM =0.5
FUNCTION LLVT = 0.0,0.0, 1.0,0.0, 1.3,0.007, 1.8,0.012,...
                2.5,0.012
FUNCTION LRTT = 0.0,0.0, 1.0,0.0, 1.3,0.011, 1.8,0.010,...
                2.5,0.010

**PHENOLOGICAL DEVELOPMENT; TABLES 12,13,14,15,16,19,20,21
PARAMETER DRCV  =0.013, DRCR  =0.028
FUNCTION DRVTT = 10.,0.10, 19.,0.80, 25.,1.00, 27.,1.10,...
                 32.,1.20, 40.,1.00, 45.,1.00
FUNCTION DRRTT = 10.,0.45, 19.,0.75, 25.,0.90, 28.,1.00,...
                 30.,1.10, 40.,1.10, 45.,1.10
FUNCTION DRDT  = 0.0,1.0,  24.,1.0
FUNCTION DRWT  = 0.0,1.0,   1.,1.0
PARAMETER SLC  = 370., SSC   =1000., WDLV  =0.015
FUNCTION SLT   = 0.0,0.82, 0.6,1.0, 2.1,1.0
FUNCTION PLHTT = 0.0,0.0 , 1.0,1.0, 2.1,1.0

**WATER RELATIONS AND ROOT GROWTH; TABLES 22,24,25
PARAMETER WSSC  =0.5, WFSC  =1.0, FIEC  =0.65
PARAMETER ZRTMC =0.7, GZRTC =0.03

**INITIALIZATION
PARAMETER DATEB =197.
PARAMETER WLVI  =6.8,  WSTI  =6.8, WSOI  =0.0
PARAMETER DSI   =0.18, ZRTI  =0.20
FINISH DS     =2.0, CELVN =3.0, TPAV  =3.0
```

Listing 6. A program to simulate soil water balance dynamics for homogeneous unsaturated soils (SWD).

```
TITLE SWD SOIL WATER DYNAMICS (MARCH 1988)
STORAGE MS(10),MFLP(10),TKL(10),K(10),KAV(11),DZ(11),FLX(11)
FIXED I,NL

   WCL   =INTGRL(WCLI,DWCLDT,10)
**WATER CONTENTS OF 10 LAYERS

PROCEDURE DWCLDT=PRDW(FLX)
**CALCULATION OF RATE OF CHANGE OF WATER CONTENT
   DO 1 I=1,NL
     DWCLDT(I)=(FLX(I)-FLX(I+1))/(TKL(I)*CONV)
1  CONTINUE
ENDPROCEDURE

PROCEDURE FLX=PRFLX(K,DZ,MFLP,MFLPTP,MFLPBT)
**CALCULATION OF FLUXES BETWEEN LAYERS (DOWNWARDS POSITIVE)
   KAV(1)    =KSAT*EXP(-KMSA*MSTP)
   FLX(1)    =(MFLPTP-MFLP(1))/DZ(1)+KAV(1)
   DO 2 I=2,NL
     KAV(I)    =SQRT(K(I)*K(I-1))
     FLX(I)    =(MFLP(I-1)-MFLP(I))/DZ(I)+KAV(I)
2  CONTINUE
   KAV(NL+1)    =KSAT*EXP(-KMSA*MSBT)
   FLX(NL+1)    =(MFLP(NL)-MFLPBT)/DZ(NL+1)+KAV(NL+1)
ENDPROCEDURE

PROCEDURE MFLP,K,MFLPTP,MFLPBT=PRMFLP(WCL)
**CALCULATION MATRIC SUCTION AND HYDRAULIC CONDUCTIVITY
   MFLPTP=-(KSAT/KMSA)*(1.-EXP(-KMSA*MSTP))
   DO 3 I=1,NL
     MS(I)     =EXP(SQRT(-1.*ALOG(WCL(I)/WCST)/MSWCA))-1.
     K(I)      =KSAT*EXP(-KMSA*MS(I))
     IF (K(I).LE.1.E-10) K(I)=0.
     MFLP(I)   =-(KSAT/KMSA)*(1.-EXP(-KMSA*MS(I)))
3  CONTINUE
   MFLPBT=-(KSAT/KMSA)*(1.-EXP(-KMSA*MSBT))
ENDPROCEDURE

PROCEDURE DZ=PRDZ(TKL)
**CALCULATE DZ
   DZ(1)     =0.5*CONV*TKL(1)
   DO 4 I=2,NL
     DZ(I)     =0.5*CONV*(TKL(I)+TKL(I-1))
4  CONTINUE
   DZ(NL+1)     =0.5*CONV*TKL(NL)
ENDPROCEDURE

**RUN CONTROL, OUTPUT
METHOD RECT
TIMER FINTIM=20., PRDEL=2.5, DELT=0.01
PRINT WCL(1-10),MFLP(1)

PARAM MSTP=1000., MSBT=0., CONV=100.
TABLE WCLI(1-10)=10*0.05, TKL(1-10)=10*0.10
PARAM MSWCA=0.0164, KSAT=5.0, KMSA=0.0231, WCST=0.503
```

```
PARAM NL=10
END

TITLE CLAY (TYPE 17)
TABLE WCLI(1-10)=10*0.05
PARAM MSWCA=0.0088, KSAT=3.5, KMSA=0.0174, WCST=0.453
END
TITLE SAND (TYPE 4)
TABLE WCLI(1-10)=10*0.05
PARAM MSWCA=0.0255, KSAT=50.0, KMSA=0.0500, WCST=0.364
TIMER DELT=0.002
END
STOP
ENDJOB
```

Listing 7. Module for canopy transpiration at Production Level 2 (L2C).

```
TITLE L2C (JULY 1987)
*To be included in INITIAL:
   TPSI  =(TPLT(IDATE)+TPHT(IDATE))/2.
   IDATE =DATEB

*To be included in DYNAMIC:
**EFFECTS OF WATER SHORTAGE
**Explanations in sections 4.2, 4.3
   CPEW  =AMIN1(1.,0.5+TRW/(TRC+1.E-10))
   DREW  =AFGEN(DRWT,TRW/(TRC+1.E-10))
   PCEW  =TRW/(TRC+1.E-10)
   PLEH  =AFGEN(PLMHT,VPDC)

**POTENTIAL TRANSPIRATION AND DIFFUSION RESISTANCES CANOPY
**Explanation in sections 4.1, 4.4
   TRC   =TRCPR*(1.-EXP(-0.5*ALV))+TRCPD*AMIN1(2.5,ALV)
   TRCPR,TRCPD=SUEVTR(RDTC,RDTM,0.25,DLA/24.,TPAD,VPA,...
          RSLL,RSBL,RSTL)

   RSLL  =LIMIT(RSLLM,2000.,(CO2E-CO2I)/(PLNA+1.E-10)*...
          (68.4*24.0/1.6)-RSBL-RSTL)
   CO2I  =CO2E*FIEC
   RSLLM =(CO2E-CO2I)/(PLMX*0.9+1.E-10)*(68.4/1.6)-10.
   PLNA  =(PCGC/(DLA/24.)-RMLV*0.33)/(AMIN1(2.5,ALV+1.E-10))
   RSBL  =0.5*172.*SQRT(WDLV/(WDSAD*0.6))
   RSTL  =FURSC(WDSAD,AMIN1(2.5,ALV),PLHT,2.)

   TRRM  =TRC/(ZRT+1.E-10)

**ROOTED DEPTH AND CROP HEIGHT
**Explanation in section 4.2
   ZRT   =INTGRL(ZRTI,GZRT*AND(ZRTM-ZRT,1.0-DS))
   ZRTM  =AMIN1(ZRTMC,ZRTMS,TKLT)
   GZRT  =GZRTC*WSERT*TERT

   PLHT  =AFGEN(PLHTT,DS)

**POTENTIAL EVAPORATION SOIL
**Explanation in section 5.1
   EVSC  =EVSPR*EXP(-0.5*ALV)+EVSPD
   EVSPR,EVSPD=SUEVTR(RDTC,RDTM,RFS,1.00,TPAV,VPA,...
```

```
            0.00,RSBS,RSTS)
    RFS   =RFSD*(1.-0.5*WCL1/WCST1)
    RSBS  =172.*SQRT(WDCL/WDSS)
    WDSS  =FUWRED(WDLV,ALV,PLHT,WDSAV)
    RSTS  =FURSC(WDSAV,1.,0.1*PLHT,0.63*PLHT)
**EXTRA WEATHER DATA
**Explanation in section 6.1, 5.1
    WDSAV =AMAX1(0.2,WDST(IDATE))
    WDSAD =1.33*WDSAV
    VPA   =AMIN1(FUVP(TPAD),HUAAT(IDATE))
    RAIN  =RAINT(IDATE)
    VPDC  =(FUVP(TPAD)-VPA)*AMIN1(1.,30./RSTL)
    DSLR  =INTGRL(1.,...
            INSW(RAINT(IDATE+1)-0.5,1.,1.00001-DSLR)/DELT)
    CO2E  =340.*0.88**(ELV/1000.)
    TPS   =INTGRL(TPSI,(TPAV-TPS)/5.)
    WUPC  =TRC *1.E4/(PCGC+1.E-10)
    WUPT  =TRWT*1.E4/(PCGT+1.E-10)
```

Listing 8. Module to simulate water movement in free draining soils (Production Level 2, permanently unsaturated soils: L2SU).

```
TITLE L2SU water balance SAHEL (JULY 1987)
*To be included in INITIAL:
    WL1I  =WCLI1*TKL1*1.E4
    WL2I  =WCLI2*TKL2*1.E4
    WL3I  =WCLI3*TKL3*1.E4
    TKLT  =TKL1+TKL2+TKL3
*To be included in DYNAMIC:
**ACTUAL TRANSPIRATION (WATER UPTAKE)
**Explanation in section 5.2
    TRW   =TRWL1+TRWL2+TRWL3
    TRWL1 =TRRM*WSE1*ZRT1
    TRWL2 =TRRM*WSE2*ZRT2
    TRWL3 =TRRM*WSE3*ZRT3
    TRWT  =INTGRL(0.,TRW)
    WSE1  =FUWS(TRC,ALV,WCL1,WSSC,WFSC,WCWP1,WCFC1,WCST1)
    WSE2  =FUWS(TRC,ALV,WCL2,WSSC,WFSC,WCWP2,WCFC2,WCST2)
    WSE3  =FUWS(TRC,ALV,WCL3,WSSC,WFSC,WCWP3,WCFC3,WCST3)
    ZRT1  =LIMIT(0.,TKL1,ZRT)
    ZRT2  =LIMIT(0.,TKL2,ZRT-TKL1)
    ZRT3  =LIMIT(0.,TKL3,ZRT-TKL1-TKL2)
**GROWTH ROOTED DEPTH
**Explanation in subsection 4.2.3
    WSERT =INSW(ZRT-TKL1,WSE1,INSW(ZRT-TKL1-TKL2,WSE2,WSE3))
    TERT  =AFGEN(PLMTT,TPS)
**AVAILABLE AND TOTAL SOIL WATER
**Explanation in sections 5.2, 5.4
    WCL1  =WL1/(TKL1*1.E4)
```

```
   WCL2  =WL2/(TKL2*1.E4)
   WCL3  =WL3/(TKL3*1.E4)

   WL1   =INTGRL(WL1I,(WLFL1-WLFL2-EVSW1-TRWL1)*10.0)
   WL2   =INTGRL(WL2I,(WLFL2-WLFL3-EVSW2-TRWL2)*10.0)
   WL3   =INTGRL(WL3I,(WLFL3-WLFL4-EVSW3-TRWL3)*10.0)
   WCUM  =(WL1+WL2+WL3)/10.0
   WLFL1 =RAIN*(1.0-FRNOF)
   WLFL2 =AMAX1(0.,WLFL1-(WCFC1*TKL1*1000.-WL1*0.10)/DELT)
   WLFL3 =AMAX1(0.,WLFL2-(WCFC2*TKL2*1000.-WL2*0.10)/DELT)
   WLFL4 =AMAX1(0.,WLFL3-(WCFC3*TKL3*1000.-WL3*0.10)/DELT)

**EVAPORATION
**Explanation in section 5.2
   EVSW  =INSW(DSLR-1.1,EVSH,EVSD)
   EVSH  =AMIN1(EVSC,...
          (WL1*0.0001-WCAD1*TKL1)*1000./DELT+WLFL1)
   EVSD  =AMIN1(EVSC,...
          0.6*EVSC*(SQRT(DSLR)-SQRT(DSLR-1.))+WLFL1)
   EVSW1 =EVSW*(FEVL1/FEVLT)
   EVSW2 =EVSW*(FEVL2/FEVLT)
   EVSW3 =EVSW*(FEVL3/FEVLT)
   FEVL1 =AMAX1(WL1-WCAD1*TKL1*1.E4,0.)*...
          EXP(-EES*(0.25*TKL1))
   FEVL2 =AMAX1(WL2-WCAD2*TKL2*1.E4,0.)*...
          EXP(-EES*(TKL1+(0.25*TKL2)))
   FEVL3 =AMAX1(WL3-WCAD3*TKL3*1.E4,0.)*...
          EXP(-EES*(TKL1+TKL2+(0.25*TKL3)))
   FEVLT =FEVL1+FEVL2+FEVL3

**WATER BALANCE CHECK
**Explanation in section 5.4
   CKWRD =FUWCHK(CKWFL,CKWIN,TIME)
   CKWFL =INTGRL(0.,(WLFL1-EVSW-TRW-WLFL4)*10.)
   CKWIN =WL1-WL1I+WL2-WL2I+WL3-WL3I

**OUTPUT
   PRTPLOT TRC,TRW,EVSC,EVCW
   PAGE GROUP
   PRTPLOT WCL1,WCL2,WCL3,RAIN
   PAGE GROUP=3
   PRTPLOT ZRT,WSE1,WUPC
```

Listing 9. Module to simulate water movement in soils with impeded drainage (Production Level 2, temporarily saturated soils: L2SS).

```
TITLE L2SS water balance SAWAH includes subroutines (AUGUST 1988)
*before INITIAL
FIXED ITYL
STORAGE TRWL(10),WCLEQI(10),WCLMQI(10),KMSA1T(20),KMSA2T(20)
STORAGE KMSMXT(20),KSTT(20),MSWCAT(20),WCSTT(20)
/      COMMON /SLDPTH/ ZL(10)
/      COMMON /VOLWAT/ WCAD(10),WCFC(10),WCST(10),WCWP(10)
/      COMMON /HYDCON/ KMSMX(10),KMSA1(10),KMSA2(10),KST(10)
/      COMMON /PFCURV/ MSWCA(10)

*To be included in INITIAL:
**INITIALIZATION OF LAYER SOIL WATER CONTENTS
```

```
**Explanation in Section 5.1, 5.3, Subsection 5.4.5
PROCEDURE WCLQTI,WCUMI=PRWCLI(WCLEQI,WCLMQI,WCLISC)
   WCUMI  =0.
   DO 1 I=1,NL
      WCLQTI(I)=INSW(WCLISC,WCLEQI(I),WCLMQI(I))
      WCUMI    =WCUMI+WCLQTI(I)*TKL(I)*1000.
 1 CONTINUE
ENDPROCEDURE
   WCLISC=AND(-WCLIS,ZWI-TKLT)-0.5
   ZWI   =AFGEN(ZWTB,DATEB)

PROCEDURE WCLEQI,TKLT=PRWCLE(NL,TKL,TYL,ZWI)
   DO 2 I=1,NL
      ITYL    =TYL(I)
      KST(I)  =KSTT  (ITYL)
      KMSMX(I)=KMSMXT(ITYL)
      KMSA1(I)=KMSA1T(ITYL)
      KMSA2(I)=KMSA2T(ITYL)
      MSWCA(I)=MSWCAT(ITYL)
      WCST(I) =WCSTT (ITYL)
      WCFC(I) =FUWCMS(I,100.0)
      WCWP(I) =FUWCMS(I,1.6E4)
      WCAD(I) =FUWCMS(I,1.0E7)
 2 CONTINUE
   WCLCH,WLOCH,WCLEQI,EVSW,RUNOF,DRSL,WCUMCH,ZECH,TKLT...
      =SUSAWA(1,WCLQT,WLOQT,NL,TRWL,EVSC,RAIN,ZWI,TKL,...
      TYL,1.0,DTMIN,DTMX1,DTFX,WLOMX,ZEQT,CSA,CSB,CSC2)
ENDPROCEDURE

   ZEQTI =0.02*(1.-WCL1/WCST1)
   WCL1  =WCLQTI(1)
   WCST1 =WCST(1)

*To be included in DYNAMIC:
**SOIL WATER, PONDED WATER, DEPTH EVAPORATION FRONT
**Explanation Sections 5.3, 5.4.4, 5.4.5
   WCLQT =INTGRL(WCLQTI,WCLCH,10)
   WLOQT =INTGRL(WLOQTI,WLOCH)
   WCUM  =INTGRL(WCUMI ,WCUMCH*1000.)
   ZEQT  =INTGRL(ZEQTI ,ZECH)

   WCLCH,WLOCH,WCLEQI,EVSW,RUNOF,DRSL,WCUMCH,ZECH,TKLT=...
      SUSAWA(2,WCLQT,WLOQT,NL,TRWL,EVSC,RAIN,ZW,TKL,TYL,...
      1.0,DTMIN,DTMX1,DTFX,WLOMX,ZEQT,CSA,CSB,CSC2)
   ZW    =AFGEN(ZWTB,DATE)
   WCL1  =WCLQT(1)

**ACTUAL TRANSPIRATION AND EFFECT WATER STRESS
**Explanation Subsections 4.2.2, 5.3.7
   TRWT  =INTGRL(0.0,TRW)
PROCEDURE TRW,TRWL,WSERT=PRTRAN(TRC,ALV,WCLQT,ZRT,TRRM)
   TRW   =0.0
   ZLL   =0.0
   WSERT =0.0
   DO 3 I=1,NL
      WSE=FUWS(TRC,ALV,WCLQT(I),WSSC,WFSC,WCWP(I),WCFC(I),WCST(I))
      ZRTL   =AMIN1(TKL(I),AMAX1(ZRT-ZLL,0.))
      WLA    =AMAX1(0.,(WCLQT(I)-WCWP(I))*TKL(I)*1000.)
```

```
      TRWL(I)=AMIN1(WSE*ZRTL*TRRM,WLA/DELT)
      TRW    =TRW+TRWL(I)
      WSEE   =INSW(AND(WCLQT(I)+0.05-WCST(I),ZRT-0.2)-0.5,WSE,0.)
      IF(ZRT.LT.(ZLL+TKL(I)).AND.(ZRT.GE.ZLL)) WSERT =WSEE
      ZLL    =ZLL+TKL(I)
 3  CONTINUE
ENDPROCEDURE
   TERT  =AFGEN(PLMTT,TPS)

**WATER BALANCE CHECK
**Explanation Section 5.4.4
   CKWIN =INTGRL(0.,(WCUMCH+WLOCH)*1000.)
   CKWFL =INTGRL(0.,RAIN-RUNOF-EVSW-TRW-DRSL)
   CKWRD =FUWCHK(CKWFL,CKWIN,TIME)

**OUTPUT
PRTPLOT EVSW,TRW,DRSL,RAIN,RUNOF
PAGE GROUP=3
PRTPLOT WCLQT(1),WCLQT(3),WCLQT(5),WCLQT(7),WCUM
PAGE GROUP=4
PRTPLOT ZRT,ZEQT,ZW,WLOQT
```

Listing 10. Data characterizing a loamy soil for L2SU and L2SS.

```
TITLE STANDARD DATA LOAMY SOIL, DEEP GROUNDWATER TABLE (APRIL 1988)
**DATA FOR MODULE L2SU
PARAMETER TKL1  =0.2,  TKL2  =0.3,  TKL3  =0.5
PARAMETER WCFC1 =0.36, WCWP1 =0.11, WCAD1 =0.01, WCST1 =0.50
PARAMETER WCFC2 =0.36, WCWP2 =0.11, WCAD2 =0.01, WCST2 =0.50
PARAMETER WCFC3 =0.36, WCWP3 =0.11, WCAD3 =0.01, WCST3 =0.50
INCON     WCLI1 =0.36, WCLI2 =0.36, WCLI3 =0.36

**DATA FOR MODULE L2SS
PARAMETER NL   =10
TABLE TKL(1-10)=10*0.10, TYL(1-10)=10*13.,  WCLMQI(1-10)=10*0.36
PARAMETER WCLIS =-1.,   WLOMX =0.02
FUNCTION  ZWTB  =0.,3.0, 366.,3.0
INCON     WLOQTI=0.0
PARAMETER DTMIN =0.001, DTMX1 =0.1, DTFX  =0.03

**SURFACE AND OTHER SOIL CHARACTERISTICS
PARAMETER FRNOF =0.0, RFSD  =0.2, WDCL  =0.05, ZRTMS =0.9
PARAMETER EES   =20., CSC2  =0.1, CSA   =0.15, CSB   =10.

**CHARACTERISTICS SOIL TYPES 1-20
TABLE KMSA1T(1-20)=...
.1960,.1385,.0821,.0500,.0269,.0562,.0378,.0395,.0750,.0490,...
.0240,.0200,.0231,.0353,.0237,.0248,.0274,.0480,.0380,.1045
TABLE KMSA2T(1-20)=...
 .08,  .63, 3.30,10.90,15.00, 5.26, 2.10,16.40,  .24,22.60,...
26.50,47.30,14.40,33.60, 3.60, 1.69, 2.77,28.20, 4.86, 6.82
TABLE KMSMXT(1-20)=...
 80.0, 90.0,125.0,175.0,165.0,100.0,135.0,200.0,150.0,130.0,...
300.0,300.0,300.0,200.0,300.0,300.0,300.0, 50.0, 80.0, 50.0
TABLE KSTT(1-20)  =...
1120.00,300.00,110.00, 50.00,  1.00,  2.30,   .36, 26.50,...
  16.50, 14.50, 12.00,  6.50,  5.00, 23.50,  1.50,   .98,...
   3.50,  1.30,   .22,  5.30
```

```
TABLE MSWCAT(1-20)=...
.0853,.0450,.0366,.0255,.0135,.0153,.0243,.0299,.0251,.0156,...
.0186,.0165,.0164,.0101,.0108,.0051,.0085,.0059,.0043,.0108
TABLE WCSTT(1-20) =...
.3950,.3650,.3500,.3640,.4700,.3940,.3010,.4390,.4650,.4550,...
.5040,.5090,.5030,.4320,.4750,.4450,.4530,.5070,.5400,.8630
```

Listing 11. Daily total global radiation in MJ m^{-2} d^{-1}. (The CSMP table is formatted to quickly locate any calendar date or Julian date.)

```
TITLE METEO LOS BANOS (IRRI), PHILIPPINES, 1984
PARAMETER LAT   =14.17, ELV   =10., RUDCF =1.E6
TABLE RDTMT(1-365) = ...
     9.85,12.98, 8.88, 4.57,10.52,13.59, 9.63,14.80,12.52,11.66, ...
     9.91,17.65,17.33,13.19,15.76,13.51,12.05,11.95,14.80,11.84, ...
    18.33,16.08, 5.00,10.02, 7.39, 8.17,10.77, 7.46,16.01,11.31, ...
    14.02, ...
          14.58,12.27,16.87,20.89,10.27,13.91,11.95, 9.77,12.87, ...
    16.72,19.50,16.51,17.97,21.96,17.33,20.25,20.93,13.98,16.58, ...
    20.00,17.37,15.44,11.31,21.35,23.03,13.73,21.29,15.30, ...
                                                          17.58, ...
    21.18,22.04,18.47,13.91, 8.56,14.09,20.54,21.75,22.14,14.69, ...
    20.75,15.94,20.64,22.71,21.18,24.35,25.56,18.36,23.92,25.35, ...
    25.42,24.53,16.12,23.28,15.37,20.22,12.77,14.44,22.32,21.07, ...
    23.57,25.38,17.58,21.71,25.46,20.25,26.06,25.49,23.21,23.96, ...
    23.78,20.29,24.28,24.17,24.28,23.46,21.75,15.73,17.51,20.18, ...
    20.18,20.18,23.89,12.63,22.60,20.57,19.50,16.05,16.79,16.15, ...
    11.41,19.40,10.70,18.54,11.88,17.11,16.87,20.68,17.79,14.05, ...
    21.14,25.46,22.89,24.53,17.61,15.08,19.65,21.75,18.72,17.15, ...
    14.41,16.19,12.77,20.97,21.82,21.03,21.93,23.67,24.35,16.15, ...
    19.57, ...
          17.40,22.60,19.90,19.72,13.62,18.40,13.23,18.29,21.86, ...
    16.33,13.16,18.93,21.86,19.79,11.20, 9.17,15.05, 6.49,10.41, ...
    11.45, 5.25, 8.99, 6.85,13.37, 8.88, 7.49, 8.99,22.36,15.58, ...
     5.25, ...
           7.56,12.52,19.93,18.90,10.45,11.98,17.65,17.08,15.48, ...
    18.26,19.00,23.60,23.17,17.40,20.68,24.28,19.25,16.90,21.64, ...
    21.78,16.83,22.75,20.43,14.44,17.11,17.29,20.22,23.42,22.36, ...
    13.66,14.84, ...
                15.12,12.70,21.03,17.83, 9.70, 6.60,13.09, 7.53, ...
    16.19,18.50,12.13, 9.02, 5.39, 3.25, 9.13,16.37,10.84,14.26, ...
    17.08,19.97,23.07, 6.67, 5.67,11.56,10.73,11.02, 9.74, 9.10, ...
     8.99,12.13,16.15, ...
                      18.11,18.22,11.16,22.89,14.90,12.48,24.71, ...
    21.35,23.10,20.07,19.04,23.10,24.60,19.08,20.22,20.75,18.08, ...
    21.29,16.87, 9.31,10.84,13.55,17.93,11.27, 9.99,10.06,16.19, ...
    16.65,22.42,20.04, ...
                      15.83,20.89,12.73, 8.31,13.37,11.88,10.81, ...
     8.92, 7.96, 7.24,20.25,19.04, 9.74,18.93,15.73,21.64,21.96, ...
    11.98, 3.64, 6.82, 1.65, 6.07,12.13,10.20,11.52,20.75,10.13, ...
    14.48, 7.92,11.41, 3.68, ...
                            12.05,16.26,20.25,12.87, 5.28,18.47, ...
    17.11,18.86,16.55,11.24,12.44,20.11,17.26,16.01,19.04,17.54, ...
    16.65,16.58,18.40, 8.20,16.44,14.02,11.84, 8.06,17.86, 1.08, ...
     7.49, 7.78,13.51, 6.60, ...
                             5.28,15.16,14.16,10.27,12.27,13.69, ...
     9.85, 9.88,19.00,18.54,15.30,11.16, 7.10,10.09,13.55,11.06, ...
    11.52,12.34,13.94,11.45,13.84,10.56, 9.56, 7.74, 9.53,11.70, ...
```

```
      14.41,14.69, 8.17,15.26, 8.42

      TABLE RAINT(1-365)=.........................................
      TABLE TPLT(1-365) =.........................................
      TABLE TPHT(1-365) =.........................................
      TABLE HUAAT(1-365)=.........................................
      TABLE WDST(1-365) =.........................................
      **Each table should contain 365 elements for one year simulation
```

Listing 12. Abbreviations in the modules of Listings 1-11. (For an explanation of CSMP functions and labels, see IBM (1975) or Basstanie & van Laar (1982).)

Abbreviation	Explanation	Dimension
AFGEN	CSMP function	-
ALV(I)	area leaves (initial)	ha ha^{-1}
AMAX1	CSMP function	-
AMIN1	CSMP function	-
AMOD	CSMP function	-
AND	CSMP and FORTRAN function	-
CAG(CR,LV,RT,SO,SS,ST)	carbohydrates (glucose) available for growth of total crop (CR), leaves (LV), roots (RT), storage organs (SO), shoot plus storage organs (SS) and stems (ST)	kg $ha^{-1}d^{-1}$
CALVT	relation of fraction CAGLV/CAGSS to DS	-
CASST	relation of fraction CAGSS/CAGCR to DS	-
CASTT	relation of fraction CAGST/CAGSS to DS	-
CELV	carbohydrate export (glucose, 24 h total) from leaves plus stems, excluding remobilization	kg $ha^{-1}d^{-1}$
CELVN	number of days that CELV is negative	d
CKCFL	sum of integrated carbon fluxes into and out of the crop	kg ha^{-1}
CKCIN	carbon in the crop accumulated since simulation started	kg ha^{-1}
CKCRD	difference between carbon added to the crop since initialization and the net total of integrated carbon fluxes, relative to their sum	-
CKWFL	sum of integrated water fluxes into and out of soil compartments	mm
CKWIN	change in total soil water content since initialization	mm
CKWRD	difference between water added to the soil since initiation and the sum of integrated water fluxes, relative to this sum	-
CNFL	carbohydrates needed to initiate and maintain 1 floret	kg $ha^{-1}d^{-1}$
CNTI	carbohydrates needed to initiate and maintain 1 tiller	kg $ha^{-1}d^{-1}$
CNTIT	relation of CNTI to DS	-
COMMON	FORTRAN label	-
CO2E	CO_2 concentration ambient air	vppm
CO2I	CO_2 concentration in stomatal cavity	vppm
CPEW	effect of water stress on carbohydrate partitioning	-
CPG(LV,RT,SO,ST)	weight of CO_2 produced during formation (=growth) of of dry matter of leaves (LV), roots (RT), storage organs (SO) and stem (ST)	kg kg^{-1}
CRG(LV,RT,SO,ST)	weight of carbohydrates required for growth of leaves (LV), roots (RT), storage organs (SO), stems (ST)	kg kg^{-1}
CSA	soil evaporation constant (*A* in Eq. 17)	cm^2 d^{-1}

CSB	soil evaporation constant (*B* in Eq. 17)	-
CSC2	soil evaporation constant (c_2 in Eq. 12)	$cm^2\ d^{-1}$
CUG(CR,LV,RT,SO,SR,ST)	weight of carbohydrates used for growth of the whole crop (CR), leaves (LV) roots (RT), storage organs (SO), shielded reserves (SR) and stems (ST)	$kg\ ha^{-1}d^{-1}$
DATE(B)	Julian date (at beginning of simulation)	-
DELT	CSMP time period for integration	d
DLA	daylength, astronomical	h
DLP	daylength effective for photoperiodism	h
DR(R,V)	development rate crop in vegetative and reproductive phase	d^{-1}
DRC(R,V)	development rate constant in the vegetative (V) and reproductive (R) phase	d^{-1}
DRDT	relation of DRED to daylength	-
DRED	effect of daylength in DRV	-
DREW	effect of water stress in DRV	-
DRRTT	relation of DRR to temperature	-
DRSL	water drained from deepest soil layer (equals WLFL4)	$mm\ d^{-1}$
DRVTT	relation of DRV to temperature	-
DRWT	relation of DREW to level of water stress	-
DS(I)	phenologial development stage crop (initial)	-
DSF1,2	DS when floret formation starts, ends (module TIL)	-
DSFL	variable with value 1.0 during floret formation, else 0.0	-
DSG1,2	DS when grain formation starts, ends (module TIL)	-
DSGR	variable with value 1.0 during grain formation, else 0.0	-
DST1,2	DS when tiller formation starts, ends (module TIL)	-
DSTI	variable with value 1.0 during tiller formation, else 0.0	-
DSLR	number of days since last rain	-
DTFX	fixed timestep for SAWAH	d
DTIME	time in current day	d
DTMIN	minimum time period for integration in SAWAH	d
DTMX1	maximum time period for integration in SAWAH	d
DWCLDT	rate of change of soil water content (program SWD)	d^{-1}
DYNAM(IC)	CSMP label	-
DZ	distance between compartment centres (program SWD)	m
EES	extinction coefficient for evaporation in soil	m^{-1}
ELV	elevation of growth site above sea level	m
END	CSMP label	-
ENDJOB	CSMP label	-
ENDPRO(CEDURE)	CSMP label	-
EVSC	potential soil evaporation rate for current weather conditions and crop	$mm\ d^{-1}$
EVSD	evaporation rate soil on dry days (i.e. almost no rain)	$mm\ d^{-1}$
EVSH	evaporation rate soil on humid days	$mm\ d^{-1}$
EVSPD	potential evaporation soil due to drying power air	$mm\ d^{-1}$
EVSPR	potential evaporation soil due to radiation	$mm\ d^{-1}$
EVSW	evaporation rate from the soil (actual value; *e* in Eq.12)	$mm\ d^{-1}$
EVSW1-3	EVSW for individual soil compartments	$mm\ d^{-1}$
FADL	fraction to adapt time period to account for daylength	-
FC(LV,RT,SO,ST)	fraction weight carbon of total dry weight in leaves (LV), roots (RT), storage organs (SO) and stems (ST)	$kg\ kg^{-1}$
FEPSO	fraction economic product in storage organs (dry weights)	$kg\ kg^{-1}$
FEVL1-3/FEVLT	fraction of EVSW from soil compartments 1-3	-
F1,2(1-5)	leaf area fraction in 0-30 and 30-60 degree leaf angle classes for layers 1 to 5	-

FIEC	ratio of CO2I vs CO2E	-
FINISH	CSMP function	-
FINTIM	CSMP function (finish time simulation)	d
FIXED	CSMP function	-
FLX	water flux density (program SWD; q in Eq. 4)	$cm\ d^{-1}$
FRNOF	fraction of precipitation that runs off field	-
FSTR	fraction stem weight at flowering that is remobilizable	$kg\ kg^{-1}$
FUCCHK	user defined function for carbon balance check	-
FUNCTION	CSMP or FORTRAN function	-
FUPHOT	user defined function for canopy photosynthesis	-
FURSC	user defined function for canopy resistance	-
FUTP	user defined function for temperature	-
FUVP	user defined function for vapour pressure	-
FUWCHK	user defined function for water balance check	-
FUWRED	user defined function for windspeed reduction	-
FUWS	user defined function for water stress	-
GAR	growth rate of available reserves (glucose)	$kg\ ha^{-1}d^{-1}$
G(CR,LV,RT,SO,SR,ST)	growth rate (dry matter) of the whole crop (CR), leaves (LV), roots (RT), storage organs (SO), shielded reserves (SR, starch) and stems (ST)	$kg\ ha^{-1}d^{-1}$
GFP	grain filling period	d
GGRMN	minimal growth rate of one grain	$kg\ d^{-1}$
GGRMX	maximal growth rate of one grain	$kg\ d^{-1}$
GGRT	relation of temperature to growth rate of grains	-
GLA	growth rate leaf area	$ha\ ha^{-1}d^{-1}$
GN(FL,GR,TI)	growth of number of florets, grains, tillers	$ha^{-1}d^{-1}$
GSA	growth rate photosynthetically active stem area	$ha\ ha^{-1}d^{-1}$
GSOAV	running average of GSO	$kg\ ha^{-1}d^{-1}$
GSOAVM	maximum value of GSOAV	$kg\ ha^{-1}d^{-1}$
GSOM	maximum growth rate storage organs	$kg\ ha^{-1}d^{-1}$
GSORM	maximum relative growth rate storage organs	$kg\ kg^{-1}d^{-1}$
GSRP	potential rate of GSR	$kg\ ha^{-1}d^{-1}$
GZRT	growth rate rooting depth	$m\ d^{-1}$
GZRTC	maximum value of GZRT	$m\ d^{-1}$
HI	harvest index (based on above ground dry matter)	$kg\ kg^{-1}$
HUAAT	table of values of VPA during year	kPa
I	index in DO-loops and dimensioned variables	-
IDATE	integer value of DATE	d
INIT(IAL)	CSMP label	-
INSW	CSMP function	-
INTGRL	CSMP function	-
ITYL	integer value soil Type number	-
K	hydraulic conductivity (program SWD; k in Eq.3)	$cm\ d^{-1}$
KAV	average hydraulic conductivity of adjacent layers (SWD)	$cm\ d^{-1}$
KMSA	parameter in exponent of unsaturated conductivity (SWD)	cm^{-1}
KMSA1	soil characteristic (α in Eq.3)	-
KMSA1T	table of characteristic of soil types	-
KMSA2	soil characteristic (a in Eq.3)	-
KMSA2T	table of characteristic of soil types	-
KMSMX	soil characteristic ($\lvert h \rvert_{max}$ in Eq.3)	-
KMSMXT	table of characteristic of soil types	-
KSAT	saturated hydraulic conductivity (program SWD; k_s in Eq.3)	$cm\ d^{-1}$
KST	saturated hydraulic conductivity (k_s in Eq.3)	$cm\ d^{-1}$
KSTT	table of characteristic of soil types	-

LAT	latitude (south of equator negative values)	degree
LIMIT	CSMP funcion	-
LLA	rate of loss of leaf area	$ha\ ha^{-1}d^{-1}$
LLV	rate of loss of leaf weight (dry matter)	$kg\ ha^{-1}d^{-1}$
LLVT	relation of relative loss rate of leaves to DS	-
LNTI	loss of number of tillers	$ha^{-1}d^{-1}$
LRT	rate of loss of root weight (dry matter)	$kg\ ha^{-1}d^{-1}$
LRTT	relation of relative loss rate due to aging to DS	-
LSR	rate of loss of shielded reserves to WAR	$kg\ ha^{-1}d^{-1}$
LSTR	loss rate of stem reserves (starch)	$kg\ ha^{-1}d^{-1}$
MCLV	trigger for mobilization of carbohydrates from leaves	-
MCRT	trigger for mobilization of carbohydrates from roots	-
MCSR	trigger for mobilization of carbohydrates from shielded reserves	-
METHOD	CSMP label	-
MFLP	matric flux potential (program SWD; Φ in Eq.5)	$cm^2\ d^{-1}$
MFLPBT	MFLP at bottom of the profile (program SWD)	$cm^2\ d^{-1}$
MFLPTP	MFLP at top of the profile (program SWD)	$cm^2\ d^{-1}$
MS	matric suction (program SWD; $\|h\|$ in Eq.2)	cm
MSBT	MS at the bottom of the profile (program SWD)	cm
MSTP	MS at the top of the profile (program SWD)	cm
MSWCA	soil characteristic (gamma in Eq.2)	-
MSWCAT	table of characteristic of soil types	-
NFL(MX,P,MXT)	number of florets (maximum, potential, maximum per tiller) (module TIL)	ha^{-1}
NGR(MX,P)	number of grains (maximum, potential) (module TIL)	ha^{-1}
NIGHT	variable to indicate day part: night (1) or day (0)	-
NL	number of soil compartments simulated in L2SS	-
NOR	CSMP and FORTRAN function	-
NTI	number of tillers, including number of main stems (NTII)	ha^{-1}
NTII	initial number of 'tillers', i.e. the number main stems	ha^{-1}
NTIP	potential number of tillers (limited by carbohydrates)	ha^{-1}
OUTDEL	CSMP function (output interval)	d
PAGE	CSMP label	-
PARAM(ETER)	CSMP label	-
PCEW	effect of water stress on PCGC	-
PCGC	photosynthesis canopy, gross, in current weather and physiological state (level 1), as CO_2 per daytime period	$kg\ ha^{-1}d^{-1}$
PCGD	PCGC expressed per 24 h (equal to PCGC for 1 d time steps)	$kg\ ha^{-1}d^{-1}$
PCGDV	running average of PCGD	$kg\ ha^{-1}d^{-1}$
PCGT	PCGC totaled since start of simulation	$kg\ ha^{-1}$
PCGW	photosynthesis canopy, gross, reduced by water shortage (level 2), as CO_2	$kg\ ha^{-1}d^{-1}$
PCNSH	net photosynthesis above ground part crop	$kg\ ha^{-1}d^{-1}$
PCNT	net canopy photosynthesis totaled since start simulation	$kg\ ha^{-1}$
PLEA	PLEI at actual temperature	$kg\ CO_2 ha^{-1}h^{-1}/ /(J\ m^{-2}\ s^{-1})$
PLEH	direct effect air humidity on PLMX	-
PLEI	initial efficiency use absorbed light by individual leaves,as PLEA	
PLETT	relation of PLEI to temperature	-
PLHT	plant height	m
PLHTT	relation of PLHT to DS	-
PLMHT	relation of PLMXP to air humidity	-

PLMTT	relation of PLMXP to temperature	-
PLMX	maximum rate of photosynthesis of single leaves (CO_2) in current conditions	kg $ha^{-1}h^{-1}$
PLMXP	PLMX for standard SLC and optimal conditions	kg $ha^{-1}h^{-1}$
PLMXT	PLMX adjusted for leaf thickness	kg $ha^{-1}h^{-1}$
PLNA	daytime average of leaf net photosynthesis per unit area	kg $ha^{-1}d^{-1}$
PRDEL	CSMP function (print interval)	d
PRINT	CSMP function	-
PROCED(URE)	CSMP label	-
PRTPLOT	CSMP function	-
Q10	Q10 of maintenance respiration sensitivity to temperature	-
RAIN	precipitation	mm d^{-1}
RAINT	table of daily precipitation values during a year	mm d^{-1}
RCRT	respiration crop, totaled (for CO_2)	kg ha^{-1}
RDTC	radiation daily total global above atmosphere (400-1400 nm)	J $m^{-2}d^{-1}$
RDTM	radiation, daily total global, measured (400-1400 nm)	J $m^{-2}d^{-1}$
RDTMT	table of measured daily total global radiation during year	-
RDUCF	radiation units conversion factor	variable
RFS	reflection coefficient soil for RDTM	-
RFSD	RFS for dry soil	-
RG(CR,LV,RT,SO,SR,ST)	respiration (in CO_2) due to growth of the whole crop (CR), leaves (LV), roots (RT), storage organs (SO), shielded reserves (SR) and stems (ST)	kg $ha^{-1}d^{-1}$
RLLV	respiration caused by remobilization from dying leaves	kg $ha^{-1}d^{-1}$
RLSR	respiration caused by remobilization (loss) of shielded reserves	kg $ha^{-1}d^{-1}$
RMCLV	standard coefficient for leaf maintenance respiration (CO_2)	kg $kg^{-1}d^{-1}$
RM(CR,LV,RT,SO,ST)	maintenance respiration (CO_2) of whole crop (CR), leaves (LV), roots (RT), storage organs (SO), stems (ST)	kg $ha^{-1}d^{-1}$
RMCT	RMCR, totaled since initialization	kg ha^{-1}
RMLV(D,N)	RMLV in daytime (D) and nighttime (N)	kg $ha^{-1}d^{-1}$
RMMA	maintenance respiration due to metabolic activity	kg $ha^{-1}d^{-1}$
RSB(L,S)	boundary layer resistance for water vapour diffusion from average leaf (L) or soil (S)	s m^{-1}
RSH	respiration rate of shoot (growth plus maintenance resp)	kg $ha^{-1}d^{-1}$
RSLL	leaf resistance for water vapour diffusion in average leaf	s m^{-1}
RSLLM	minimum value of RSLL	s m^{-1}
RST(L,S)	resistance to diffusion for water vapour, CO_2 and heat due to turbulence in canopy from average leaf (L) or soil (S)	s m^{-1}
RUNOF	water flowing from surface to other fields	mm d^{-1}
SLA	specific leaf weight, actual value (eventually corrected for contribution stem area)	kg ha^{-1}
SLC	specific leaf weight constant	kg ha^{-1}
SLN	SLA for new leaves	kg ha^{-1}
SLT	relation of SLA to DS	-
SQRT	CSMP function	-
SSC	specific stem weight constant (SLC analogy)	kg ha^{-1}
STOP	CSMP label	-
STORAGE	CSMP label	-

SUASTR	user defined subroutine for astronomical variables	-
SUEVTR	user defined subroutine for evapotranspiration	-
SUPHOL	user defined subroutine for canopy photosynthesis	-
SUSAWA	user defined subroutine for soil water balance	-
TABLE	CSMP label	-
TCD	time constant for dying of tillers	d
TCF	time constant for formation of plant organs	d
TERT	effect of temperature on root growth rate	-
TIMER	CSMP label	-
TITLE	CSMP label	-
TKL(I)1-3	thickness soil compartment I, 1-3	m
TKLT	thickness of combined soil compartments	m
TPA(A,D,V)	actual air temperature at each DTIME (A), in daytime (D) and 24h average (V)	°C
TPEM	temperature effect on maintenance respiration	-
TPHT	table of maximum day temperatures during a year	°C
TPLT	table of minimum night temperatures during a year	°C
TPR	reference temperature for maintenance respiration	°C
TPS(I)	temperature of the soil (initial)	°C
TRC	transpiration rate canopy, potential value for current weather and crop (level 1)	mm d^{-1}
TRCP(D,R)	potential transpiration canopy due to drying power air (D) and absorbed radiation (R)	mm d^{-1}
TRRM	potential transpiration rate per unit rooted length	mm $d^{-1}m^{-1}$
TRW	transpiration rate canopy, actual value with water stress (level 2)	mm d^{-1}
TRWL1-3,(1-NL)	TRW from individual compartments 1-3 or 1-NL	mm d^{-1}
TRWT	TRW totaled since start of simulation	mm
TYL(1-NL)	number indicating soil type of compartment	-
VPA	humidity of the air, early morning value	kPa
VPD	vapour pressure difference	kPa
VPDC	vapour pressure deficit in canopy, daytime average	kPa
WAR(I)	available carbohydrate (glucose) in leaves (initial)	kg ha^{-1}
WARR	WAR relative to WLV	-
WC(AD,FC,ST,WP)	volumetric water content of soil when air dry (AD), at wilting point (WP), field capacity (FC) and saturation (ST, equals relative total pore space); these variables are indexed 1-NL in L2SS, and numbered 1-3 in L2SU	m^3 m^{-3}
WCL(I)1-3	relative soil water content per layer in L2SS (initial) (θ in Eq.2)	m^3 m^{-3}
WCLCH(1-NL)	rate of change of WCLQT	m^3 $m^{-3}d^{-1}$
WCLEQI(1-NL)	initial value of WCLQT in equilibrium situation	m^3 m^{-3}
WCLIS	switch parameter for soil water initialization	-
WCLISC	switch water soil layers initial (see Section 5.4.4)	-
WCLMQI(1-NL)	initial value of WCLQT from observations	m^3 m^{-3}
WCLQT(I)(1-NL)	same as WCL(I)1-3 in module with impeded drainage	m^3 m^{-3}
WCR	weight crop, including roots	kg ha^{-1}
WCST	volumetric water content at saturation (θ in Eq.2)	m^3 m^{-3}
WCSTT	table of water content at saturation for soil types	-
WCUM(I)	total water in soil profile (initial)	mm
WCUMCH	rate of change of WCUM	mm d^{-1}
WDCL	width of soil cloth (WDLV analogue)	m
WDLV	width of leaves	m
WDS(AD,AV,S)	wind speed, daytime average (AD), 24 h average (AV), and near the soil surface (S)	m s^{-1}

WDST	table of daily values observed wind speeds during a year	$m\ s^{-1}$
WEPSO	weight economic part of storage organs	$kg\ ha^{-1}$
WFSC	flooding stress sensitivity coefficient	-
WGR(MX)	average weight of grains, filled plus unfilled (maximum)	kg
WIR	weight increment reserves (starch) since start simulation	$kg\ ha^{-1}$
WLA	water available to the crop in a layer	$m^3\ ha^{-1}$
WLFL1-4	fluxes of water into layers 1-3 and out of layer 3	$mm\ d^{-1}$
WL1-3(I)	volumetric soil water content per compartment (initial)	$m^3\ ha^{-1}$
WL0QT(I)	water standing above soil surface (initial)	m
WL0CH	rate of change of WL0QT	$m\ d^{-1}$
WL0MX	maximum level of water on the surface (bund height)	m
WLV(I)	weight leaves (initial)	$kg\ ha^{-1}$
WLVD	weight dead leaves	$kg\ ha^{-1}$
WLVSO	total above ground dry weight	$kg\ ha^{-1}$
WLVST	sum of WLV, WLVD and WST	$kg\ ha^{-1}$
WLVT	sum of WLV and WLVD	$kg\ ha^{-1}$
WRT(I)	weight roots (initial)	$kg\ ha^{-1}$
WRTD	weight dead roots	$kg\ ha^{-1}$
WSE(1-3)	effect of water stress on water uptake in layers 1-3	-
WSERT	effect of water stress on root water uptake	-
WSO(I)	weight storage organs (initial)	$kg\ ha^{-1}$
WSR(I)	weight of shielded reserves (starch) in stem (initial)	$kg\ ha^{-1}$
WSS	weight shoot plus storage organs	$kg\ ha^{-1}$
WSSC	water stress sensitivity coefficient	-
WST(I)	weight stems (initial) minus WSR or WIR contained in it	$kg\ ha^{-1}$
WSTR	stem weight (WST+WSR or WST+WIR, depending upon module)	$kg\ ha^{-1}$
WUPC	water use efficiency, current, relative to net photosynthesis leaves (water transpired per kg CO_2 fixed, net, daytime)	$kg\ kg^{-1}$
WUPT	WUPC of total net photosynthesis and transpiration	$kg\ kg^{-1}$
ZEQT(I)	depth evaporation front in upper soil layer (z_E in Eq.12)	m
ZECH	rate of change of ZEQT	$m\ d^{-1}$
ZL(1-NL)	depth of upper boundary of each soil compartment	m
ZLL	depth upper boundary compartment (L2SS)	m
ZREF	reference height windspeed observations	m
ZRT(I)	rooting depth (initial)	m
ZRT1-3	ZRT differentiated per soil compartment	m
ZRTL	rooting depth in individual layers (L2SS)	-
ZRTM	maximum for ZRT	m
ZRTM(C,S)	maximum rooting depth for crop (C) and soil (S)	m
ZW (I)	depth of free water table (initial)	m
ZWTB	table with observed ZW versus time	-

8 References

Aggarwal, P.K. & Penning de Vries, F.W.T., 1989. Potential production of wheat in southeast Asia. Agricultural Systems 30(1):49-69.

Alberda, Th., 1969. The effect of low temperature on dry matter production, chlorophyll concentration and photosynthesis of maize plants of different ages. Acta Botanica Neerlandica 18(1):39-49.

Alberda, Th. et al., 1977. Crop photosynthesis: methods and compilation of data obtained with mobile field equipment. Agricultural Research Reports, 865. Pudoc, Wageningen.

Alberda, Th., 1984. Production and water use of several food and fodder crops under irrigation in the desert of southwestern Peru. Agricultural Research Reports, 928. Pudoc, Wageningen.

Amthor, J.S., 1984. The role of maintenance respiration in plant growth. Plant, cell and environment 7:561-569.

Angus, J.F. & Zandstra, H.G., 1980. Climatic factors and the modeling of rice growth and yield. In: Agroclimatology of the rice crop. International Rice Research Institute, Los Baños, pp. 189-199.

Angus, J.F., Mackenzie, D.H., Morton, R. & Schafer, C.A., 1981. Phasic development in field crops. II. Thermal and photoperiodic responses of spring wheat. Field Crops Research 4:269-283.

Azcon-Bieto, J., Lambers, H. & Day, D.A., 1983. Effect of photosynthesis and carbohydrate status on respiratory rates and the involvement of the alternative pathway in leaf respiration. Plant Physiology 72:598-603.

Bakema, A.H. & Jansen, D.M., 1987. Using a simulation model to evaluate weather effects. In: Impact of weather parameters on the growth and yield of rice. International Rice Research Institute, Los Baños, pp. 283-290.

Baker, C.H. & Curry, R.B., 1976. Structure of agricultural simulations: a philosophic view. Agricultural Systems 1: 201-218.

Bassett, D.M., Stockton, J.R. & Dickens, W.L., 1970. Root growth of cotton as measured by P^{32} uptake. Agronomy Journal 62:200-203.

Basstanie, L.J.M. & Laar, H.H. van, 1982. Introduction of CSMP by an elementary simulation program. In: F.W.T. Penning de Vries and H.H. van Laar (Eds.): Simulation of plant growth and crop production. Simulation Monographs. Pudoc, Wageningen, pp. 50-65.

Bell, C.J., 1982. A model of stomatal control. Photosynthetica 16(4):486-495.

Benschop, M., 1986. TUCROS, een simulatiemodel voor de tulpecultivar "Apeldoorn". Simulation Report CABO-TT No. 6. Centre for Agrobiological Research, Wageningen.

Berge, H.F.M. ten, 1986. Heat and water transfer at the bare soil surface.

Aspects affecting thermal imagery. Ph.D. Thesis, Agricultural University, Wageningen.
Berge, H.F.M. ten, Metselaar, K. & Stroosnijder, L., 1987. Measurement of matric flux potential: a simple procedure for the hydraulic characterization of soils. Netherlands Journal of Agricultural Science, 35:371-384.
Berge, H.F.M. ten & Jansen, D.M., 1989. Water balance module SAWAH: users guide and outline. Simulation Report CABO-TT. Centre for Agrobiological Research, Wageningen (forthcoming).
Berry, J.A. & Downton, W.J.S., 1982. Environmental regulation of photosynthesis. In: Govindjee (Ed.): Photosynthesis, Vol II. Academic Press, New York, pp. 263-343.
Beuerlein, J.E. & Pendleton, J.W., 1971. Photosynthetic rates and light saturation curves of individual soybean leaves under field conditions. Crop Science 11:217-219.
Bhagsari, A.S. & Brown, R.H., 1976. Photosynthesis in peanut (*Arachis*) genotypes. Journal Peanut Science 3:1-5.
Bhagsari, A.S., Brown, R.H. & Schepers, J.S., 1976. Effect of moisture stress on photosynthesis and some related physiological characteristics in peanut. Crop Science 16: 712-715.
Bhagsari, A.S. & Brown, R.H., 1976. Relationship of net photosynthesis to carbon dioxide concentration and leaf characteristics in selected peanut (*Arachis*) genotypes. Journal Peanut Science 3:10-14.
Bierhuizen, J.F. & Slatyer, R.O., 1964. Photosynthesis of cotton leaves under a range of environmental conditions in relation to internal and external diffusive resistances. Australian Journal of Biological Sciences 17:348-359.
Bidwell, R.G.S., 1983. Carbon nutrition of plants: photosynthesis and respiration. In: Steward, F.C. & Bidwell, R.G.S. (Eds.), 1983. Plant Physiology, Vol VII: Energy and Carbon Metabolism. Academic Press, New York, pp. 287-458.
Bjorkman, O., 1981. Responses to different quantum flux densities. Encyclopedia Plant Physiology, New Series, Vol 12a. Springer Verlag, Berlin, pp. 57-107.
Bodlaender, K.B.A., 1986. Effects of drought on water use, photosynthesis and transpiration of potatoes. 1. Drought resistance and water use. In: Potato research of tomorrow. Pudoc, Wageningen, pp. 36-43.
Bouma, J., 1985. Morphology of lowland soils and flow of water and gases. In: Proceedings of the International Workshop on Physical Aspects of Soil Management in Rice-based Cropping Systems (PASMIRCS), International Rice Research Institute, Los Baños, 1985, pp. 199-215.
Breman, H., Cisse, A.M., Djiteye, M.A. & Elberse, W.Th., 1982. Le potentiel botanique des paturages. In: F.W.T. Penning de Vries & M.A. Djiteye (Eds.): La productivite des paturages saheliens. Agricultural Research Reports 918. Pudoc, Wageningen.
Brockington, N.R., 1979. Computer modelling in Agriculture. Oxford Uni-

versity Press, Oxford.
Brown, K.F. & Biscoe, P.V., 1985. Fibrous root growth and water use of sugar beet. Journal of Agricultural Science, 105:679-691.
Brunt, D., 1932. Notes on radiation in the atmosphere. I. Quarterly Journal of the Royal Meteorological Society 58:389-420.
Bull, T.A., 1969. Photosynthetic efficiencies and photorespiration in Calvin cycle and C_4 dicarboxylic acid plants. Crop Science, 9:726-729.
Campbell, G.S., 1977. An introduction to environmental biophysics. Springer Verlag, New York.
Charles-Edwards, D.A., 1981. The mathematics of photosynthesis and productivity. Academic Press, London.
Cock, J.H. & Yoshida, S., 1973. Changing sink and source relations in rice (*Oryza sativa L.*) using carbon dioxide enrichment in the field. Soil Science and Plant Nutrition 19(3):229-234.
Cock, J.H., Franklin, D., Sandoval, G., & Juri, P., 1979. The ideal cassava plant for maximum yield. Crop Science 19:271-279.
Cook, M.G. & Evans, L.T. 1983. Some physiological aspects of the domestication and improvement of rice (*Oryza spp.*). Field Crops Research 6:219-238.
Constable, G.A. & Rawson, H.M., 1980. Effect of leaf position, expansion and age on photosynthesis, transpiration and water use efficiency of cotton. Australian Journal of Plant Physiology 7:89-100.
Dagley, S. & Nicholson, D.E., 1970. An introduction to metabolic pathways. Blackwell Scientific Publications, Oxford.
Dantuma, G., 1973. Rates of photosynthesis in leaves of wheat and barley varieties. Netherlands Journal of Agricultural Science 21:181-187.
Davidoff, B. & Selim, H.M., 1986. Goodness of fit for eight water infiltration models. Soil Science Society of America Journal 50:759-764.
Day, W., Legg, B.J., French, B.K., Johnston, A.E., Lawlor, D.W. & Jeffers, W. de C., 1978. A drough experiment using mobile shelters: the effect of drought on barley yield, water use and nutrient uptake. Journal Agricultural Science, Cambridge 91:599-623.
Dayan, E., Keulen, H. van & Dovrat, A., 1981. Tiller dynamics and growth of Rhodes grass after defoliation: a model named TILDYN. Agro-ecosystems 7:101-112.
Dent, J.B. & Blackie, M.J., 1979. Systems Simulation in Agriculture. Applied Science Publishers, London.
Dingkuhn, M., 1985. Physiologische Faktoren der Trockenheitsresistenz bei Reis. Ph.D. Thesis, University of Hamburg, West Germany.
Doorenbos, J., Kassam, A.H., Bentvelder, C. & Uittenboogaard, G., 1978. Yield response to water. U.N. Economic Commission to West Asia, Rome.
Doorenbos, J. & Kassam, A.H., 1979. Yield response to water. FAO irrigation and drainage paper, No.33. FAO, Rome.
Dornhoff, G.M. & Shibles, R.M., 1970. Varietal differences in net photosyn-

thesis of soybean leaves. Crop Science 10:42-45.
Driessen, P.M., 1986a. The waterbalance of soil. In: H. van Keulen & J. Wolf (Eds.): Modelling of agricultural production: weather, soils and crops. Simulation Monographs. Pudoc, Wageningen, pp. 76-116.
Driessen, P.M., 1986b. Soil physical data. In: H. van Keulen & J. Wolf (Eds.): Modelling of agricultural production: weather, soils and crops. Simulation Monographs. Pudoc, Wageningen, pp. 212-223.
Eastin, J.D., 1983. Sorghum. In: Potential productivity of field crops under different environments. International Rice Research Institute, Los Baños, pp.181-204.
Elderen, E. van, 1987. Scheduling farm operations, a simulation model. Simulation Monographs. Pudoc, Wageningen.
Ehleringer, J., 1976. Leaf absorptance and photosynthesis as affected by pubescence in the genus Encelia. Annual Report Director Department Plant Biology, Carnegie Institution, pp. 413-418.
Ehleringer, J. & Bjorkman, O., 1976. Carbon dioxide and temperature dependence of the quantum yield for CO_2 uptake in C_3 and C_4 plants. Annual Report of the Director Department of Plant Biology, Carnegie Institution, pp. 418-421.
El-Sharkawy, M.A. & Hesketh, J.D., 1964a. Effect of temperature and H_2O deficit on leaf photosynthetic rate of different species. Crop Science 4:514-518.
El-Sharkawy, M.A. & Hesketh, J.D., 1964b. Effect of stomatal differences among species on leaf photosynthesis. Crop Science 4:619-621.
El-Sharkawy, M.A., Hesketh J.D. & Muramoto, H. 1965. Leaf photosynthesis rates and other growth characteristics among 26 species of gossypium. Crop Science 5:173-175.
El-Sharkawy, M.A., Cock, J.H. & de Cadena, G., 1984a. Influence of differences in leaf anatomy on net photosynthetic rates of some cultivars of cassava. Photosynthesis Research 5:235-242.
El-Sharkawy, M.A., Cock, J.H. & Held, K.A.A., 1984b. Water use efficiency of cassava II. Differing sensitivity of stomata to air humidity in cassava and other warm-climate species. Crop Science 24:503-507.
Encyclopedia of Plant Physiology, New Series, Vol. 1-19, 1975-1986. Springer Verlag, Heidelberg.
Eswaran, H., 1985. Interpreting physical aspects of wetland soil management from soil taxonomy. In: Proceedings of the International Workshop on Physical Aspects of Soil Management in Rice-based Cropping Systems (PASMIRCS). International Rice Research Institute, Los Baños, pp. 17-30.
Evans, L.T. & Wardlaw, I.F., 1976. Aspects of the comparative physiology of grain yield in cereals. Advances in Agronomy 28:301-359.
Farquhar, G.D. & von Caemmerer, S., 1982. Modelling of photosynthetic response to environmental conditions. Encyclopedia Plant Physiology, New

Series 12b. Springer Verlag, Berlin, pp. 549-587.
Feddes, R.A., Kowalik, P.J., & Zaradny, H., 1978. Simulation of field water use and crop yield. Simulation Monographs. Pudoc, Wageningen.
Forrester, J.W., 1961. Industrial dynamics. MIT Press, Boston.
Forward, D.F., 1983. Respiration, a holistic approach to metabolism. In: Steward & Bidwell, pp. 458-510.
Frère, M., 1987. The FAO agroclimatological data base: its use in studying rice-weather relationships. In: The impact of weather parameters on the growth and yield of rice. International Rice Research Institute, Los Baños, pp. 41-46.
Fukai, S., Kuroda, E. & Yamagishi, T., 1985. Leaf gas exchange of upland and lowland rice cultivars. Photosynthesis Research 7:127-135.
Gardner, W.R., 1958. Some steady state solutions of the unsaturated moisture flow equation with application to evaporation from a water table. Soil Science 85:228-232.
Gates, D.M., Strain, B.R. & Weber, J.A., 1983. Ecophysiological effects of changing atmospheric CO_2 concentration. Encyclopedia Plant Physiology, New Series, Vol 12D. Springer Verlag, Berlin, pp. 503-526.
GELGAM, 1984. Herziening van de berekening van de gewasverdamping in het hydrologisch model GELGAM. Rapport van de ad hoc groep Verdamping. Dienst Waterbeheer, Provincie Gelderland, Arnhem.
Geng S., Penning de Vries, F.W.T. & Supit, I. 1985a. Analysis and Simulation of weather variables: rain and wind in Wageningen. Simulation Reports CABO-TT No 4. Centre for Agrobiological Research, Wageningen.
Geng S., Penning de Vries, F.W.T. & Supit, I., 1985b. Analysis and Simulation of weather variables: temperature and solar radiation. Simulation Reports CABO-TT No 5. Centre for Agrobiological Research, Wageningen.
Geng S., Penning de Vries, F.W.T. & Supit, I., 1986. A simple method for generating daily rainfall data. Agricultural and Forest Meteorology 36:363-376.
Gifford, R.M. & Evans, L.T., 1981. Photosynthesis, carbon partitioning and yield. Annual Review Plant Physiology 32: 485-509.
Gifford, R.M. & Jenkins, C.L.D., 1982. Prospects of applying knowledge of photosynthesis toward improving crop production. In: Govindjee (Ed.): Photosynthesis, Vol II. Academic Press, London, pp. 419-457.
Giraldez, J.V. & Sposito, G., 1985. Infiltration in swelling soils. Water Resources Research 21:33-44.
Goudriaan, J., 1977. Crop micrometeorology: a simulation study. Simulation Monographs. Pudoc, Wageningen.
Goudriaan, J., 1982a. Some techniques in dynamic simulation. In: F.W.T. Penning de Vries & H.H. van Laar (Eds.): Simulation of plant growth and crop production. Simulation Monographs. Pudoc, Wageningen, pp. 98-113.
Goudriaan, J., 1982b. Potential production processes. In: F.W.T. Penning de

Vries and H.H. van Laar (Eds.): Simulation of plant growth and crop production. Simulation Monographs. Pudoc, Wageningen, pp. 98-113.
Goudriaan, J., 1986. A simple and fast numerical method for the computation of daily totals of crop photosynthesis. Agricultural and Forest Meteorology 38:249-254.
Goudriaan, J., 1987. The biosphere as a driving force in the global carbon cycle. Netherlands Journal of Agricultural Science 35:177-187.
Goudriaan, J., 1988. The bare bones of leaf angle distribution in radiation models for canopy photosynthesis and radiation exchange. Agricultural and Forest Meteorology 43: 155-169.
Goudriaan, J., 1989. Modelling of ageing, development, delays and dispersion. In: Rabbinge et al. (Eds.): Simulation and systems management in crop protection. Simulation Monographs, Pudoc, Wageningen.
Goudriaan, J. & Laar, H.H. van, 1978a. Calculation of daily totals of the gross CO_2 assimilation of leaf canopies. Netherlands Journal of Agricultural Science 26:373-382.
Goudriaan, J. & Laar, H.H. van, 1978b. Relations between leaf resistance, CO_2-concentrations and CO_2-assimilation in maize, beans, lalang grass and sunflower. Photosynthetica 12(3):241-249.
Goudriaan, J. & Ruiter, H.E. de, 1983. Plant growth in response to CO_2 enrichment, at two levels of nitrogen and phosphorus supply. 1. Dry matter, leaf area and development. Netherlands Journal of Agricultural Science 31: 157-169.
Goudriaan, J., Laar, H.H. van, Keulen, H. van & Louwerse, W., 1984. Simulation of the effect of increased atmospheric CO_2 on assimilation and transpiration of a closed crop canopy. Wissenschafliche Zeitschrift Humbolt Universitaet Berlin, Math.-Nat. R. 33(4):352-356.
Graham, D. & Chapman, E.A., 1979. Interactions between photosynthesis and respiration in higher plants. Encyclopedia Plant Physiology 6:150-162.
Grashoff, C., Klein Hulze, J.A., & Smid, G.H., 1987. Opbrengstinstabiliteit en oogstzekerheid bij veldbonen en erwten. CABO Reports. Centre for Agrobiological Research, Wageningen.
Gregory, P.J., McGowan, M., Biscoe, P.V. & Hunter, B., 1978. Water relations of winter wheat. 1. The growth of the root system. Journal of Agricultural Science, Cambridge 91:91-102.
Gregory, P.J. & Reddy, M.S., 1982. Root growth in an intercrop of pearl millet/groundnut. Field Crops Research 5:241-252.
Gulmon, S.L. & Chu, C.C., 1981. The effects of light and nitrogen on photosynthesis, leaf characteristics and dry matter allocation in the chaparral shrub *Diplacus aurantiacus*. Oecologia 49:207-212.
Gupta, S.C., Larson, W.E. & Allmaras, R.R., 1984. Predicting soil temperature and soil heat flux under different tillage-surface residue conditions. Soil Science Society of America Journal 48(2):223-232.
Gupta, P.C. & O'Toole, J.C., 1986. Upland rice, a global perspective. In-

ternational Rice Research Institute, Los Baños.
Hahn, S.K. & Hozyo, Y., 1983. Sweet potato and yam. In: Potential productivity of field crops under different environments. International Rice Research Institute, Los Baños, pp. 319-340.
Halevy, A.H., 1985. Handbook of Flowering, Volumes I to V. CRC Press Inc, Boca Raton, Florida.
Hall, A.E., 1982. Mathematical models of plant water loss and plant water relations. In: O.L. Lange et al. (Eds.): Encyclopedia of Plant Physiology Vol.12B. Springer Verlag, Berlin, pp. 231-261.
Hall, A.E., & Loomis, R.S. 1972. Photosynthesis and respiration by healthy and beet yellow virus-infected sugar beets (*Beta vulgaris L.*). Crop Science 12:566-572.
Hansen, S. & Aslyng, H.C., 1984. Nitrogen balance in crop production, simulation model NITCROS. Hydrotechnical Laboratory, Royal Veterinary and Agricultural University, Copenhagen.
Hanway, J.J. & Weber, C.R., 1971. N,P and K percentages in soybean (*Glycine max (L.) Merrill*) plant parts. Agronomy Journal 63:286-290.
Haverman, P.P., 1986. A characterization of the growth of the cowpea crop in the Sahelian zone. MSc thesis Department of Theoretical Production Ecology, Agricultural University, Wageningen.
Heemst, H.D.J. van, 1986. The distribution of dry matter during growth of a potato crop. Potato Research 29:55-66.
Hesketh, J.D. & Jones, J.W. (Eds.): 1980. Predicting photosynthesis for ecosystem models, I, II. CRC Press, Boca Raton, Florida 33431.
Hodges, T., Kanemasu, E.T. & Teare, I.D., 1979. Modeling dry matter accumulation and yield of grain sorghum. Canadian Journal of Plant Science 59:803-818.
Hofstra, G. & Hesketh, J.D., 1969. Effects of temperature on the gas exchange of leaves in the light and dark. Planta (Berl.) 85:228-237.
Hoogmoed, W.B. & Bouma, J., 1980. A simulation model for predicting infiltration into cracked clay soils. Soil Science Society of America Journal 44:458-461.
Horie, T., 1987. Dynamic models for evaluating and predicting the growth and yield of rice and sunflower crops from weather. Food and Fertilizer Technology Center, Technical Bulletin 99. Taipei, Taiwan.
Horie, T., Wit, C.T. de, Goudriaan, J. & Bensink, J., 1979. A formal template for the development of cucumber in its vegetative stage. I, II and III. Proceedings Nederlandse Academie Wetenschappen Serie. C. 82(4):433-479.
Horie, T., Nakagawa, H. & Kira, T., 1986. Studies on modelling and prediction of rice development processes. Japanese Journal of Crop Science 55(1):214-215.
Hozyo, Y., Shimotsubo, K., & Kato, S. 1982. The interrelationship between photosynthetic activity and tuberous root growth in sweet potato grafts. 5th International Symposium on Tropical Root and Tuber Crops. International

Rice Research Institute, Los Baños, pp. 105-113.
Huda, A.K.S., Sivakumar, M.V.K., Virmani, S.M., Seetharama, N., Singh Sardar & Sekaran, J.G., 1984. Modeling the effect of environmental factors on sorghum growth and development. In: Virmani (Ed.): Agrometeorology of sorghum and millet in the semi-arid tropics. International Crop Research Institute for the Semiarid Tropics, Patancheru, India, pp. 277-287.
Huffaker, R.C. & Miller, B.L., 1978. Reutilization of Ribulose Biphosphate Carboxylase. In: Siegelman and Hind (Eds.): Photosynthetic carbon assimilation, Plenum Publishing Corporation, pp. 139-152.
Huke, R.E., 1985. FARIDPUR: a computer-assisted instruction model for rainfed lowland rice. IRRI Research Paper Series 104, International Rice Research Institute, Los Baños, 14 pp.
IBM, 1975. Continuous system modeling program III. General system information manual (GH19-7000) and users manual (SH19-7001-2). IBM Data Processing Divison, White Plains, New York.
IBM/Microsoft, 1984. Fortran Compiler Version 2, IBM Personal Computer, Computer Language Series. Portsmouth, England.
ICAR (Indian Council for Agricultural Research), 1977. Crop planning in drought and flood prone areas. New Delhi.
Idso, S.B., Reginato, R.J., Jackson, R.D. & Pinter, P.J., 1981. Foliage and air temperatures: evidence for a dynamic 'equivalence point'. Agricultural Meteorology 24:223-226.
IIASA, 1980. Beware of pitfalls; a short guide to avoiding common errors in systems analysis. Executive Report, International Institute for Application of Systems Analysis (IIASA). IIASA, A-2361, Laxemburg, Austria.
Ingram, K.T. & McCloud, D.E., 1984. Simulation of potato crop growth and development. Crop Science 24:21-27.
IRRI, 1977. Annual Report for 1976. The International Rice Research Institute, Los Baños.
IRRI, 1984. Terminology for rice growing environments. International Rice Research Institute, Los Baños.
IRRI, 1985. Soil physics and rice. In: Proceedings of the International Workshop on Physical Aspects of Soil Management in Rice-based Cropping Systems (PASMIRCS). International Rice Research Institute, Los Baños.
Jackson, M.B. & Drew, M.C., 1984. Effects of flooding on growth and metabolism of herbaceous plants. In: T.T. Kozlowski (Ed.): Flooding and plant growth. Academic Press, London, pp. 47-128.
Jackson, B.S., Arkin, G.F. & Douglas, W.A., 1983. An evaluation of selected time steps in a crop growth simulation model. Extended Abstracts, 16th Conference of Agriculture and Forest Meteorology. American Meteorological Society, Boston, Mass., pp. 110-112.
Jansen, D.M. & Gosseye, P., 1986. Simulation of growth of millet (*Pennisetum americanum*) as influenced by water stress. Simulation Reports CABO-TT No. 10. Centre for Agrobiological Research, Wageningen.

Jarvis, P.G., 1981. Stomatal conductance, gaseous exchange and transpiration. In: Plants and their atmospheric environment. Blackwell Scientific Publications, Oxford, pp. 175-214.
Johnson, I.R. & Thornley, J.H.M., 1983. Vegetative crop growth model incorporating leaf area expansion and senescence, and applied to grass. Plant, Cell and Environment 6:721-729.
Jolliffe, P.A. & Tregunna, E.B., 1968. Effect of temperature, CO_2 concentration, and light intensity on oxygen inhibition of photosynthesis in wheat leaves. Plant Physiology 43:902-906.
Jones, L.H., 1971. Adaptive responses to temperature in dwarf french beans, *Phaseolus vulgaris*. Annals of Botany 35:581-596.
Kaigama, B.K., Teare, I.D., Stone, L.R. & Powers, W.L., 1977. Root and top growth of irrigated and nonirrigated grain sorghum. Crop Science, 17:555-559.
Keulen, H. van, 1975. Simulation of water use and herbage growth in arid regions. Simulation Monographs. Pudoc, Wageningen.
Keulen, H. van, 1976. A calculation method for potential rice production. Contributions of the Central Research Institute for Agriculture No. 21. Bogor, Indonesia, 26 pp.
Keulen, H. van, 1977. Simulation of water use and herbage growth in arid regions. Simulation Monographs. Pudoc, Wageningen.
Keulen, H. van, 1982. Crop production under semi-arid conditions, as determined by nitrogen and moisture availability. In: F.W.T. Penning de Vries & H.H. van Laar, (Eds.), Simulation of plant growth and crop production. Simulation Monographs. Pudoc, Wageningen, pp. 234-249.
Keulen, H. van & De Milliano, W.A.J., 1984. Potential wheat yields in Zambia: a simulation approach. Agricultural Systems 14:171-192.
Keulen, H. van, Laar, H.H. van, Louwerse, W. & Goudriaan, J., 1980. Physiological aspects of increased CO_2 concentration. Experientia 36:786-792.
Keulen, H. van, Penning de Vries, F.W.T. & Drees, E.M., 1982. A summary model for crop growth. In: F.W.T. Penning de Vries and H.H. van Laar (Eds.): Simulation of plant growth and crop production. Simulation Monographs. Pudoc, Wageningen, pp. 87-97.
Keulen, H. van & Heemst, H.D.J. van, 1986. Meteorological data. In: H. van Keulen & J. Wolf (Eds.): Modelling of agricultural production: weather, soils and crops. Simulation Monographs. Pudoc, Wageningen, pp. 208-211.
Keulen, H. van & Wolf, J., 1986 (Eds.): Modelling of agricultural production: weather, soils and crops. Simulation Monographs. Pudoc, Wageningen, 464 pp.
Keulen, H. van, & Seligman, N.G., 1987. Simulation of water use, nitrogen nutrition and growth of a spring wheat crop. Simulation Monographs. Pudoc, Wageningen.
Khan, M.A. & Tsunoda, S., 1970. Evolutionary trends in leaf photosynthesis and related leaf characteristics among cultivated wheat species and its wild

relatives. Japanese Journal Plant Breeding 20(3):133-140.
Kingwell, R.S. & Pannell, D.J., 1987. A bioeconomic model of dryland farming. Simulation Monographs. Pudoc, Wageningen.
Klute, A., 1952. A numerical method for solving the flow equation for water in unsaturated materials. Soil Science 73:105-117.
Koerner, Ch. & May, R., 1979. Stomatal behaviour in alpine plant communities between 600 and 2600 m above sea level. In: J. Grace, E.D. Ford & P.G. Jarris (Eds.): Plants and their atmospheric environment. Blackwell Scientific Publisher, Oxford, U.K., pp. 205-218.
Koller, D. & Hadas, A., 1982. Environmental aspects of the germination of seeds. In: O.L. Lange et al. (Eds.): Encyclopedia of Plant Physiology Vol. 12B: 401-431. Springer Verlag, Berlin.
Koorevaar, P., Menelik, G. & Dirksen, C., 1983. Elements of Soil Physics. Developments in Soil Science 13. Elsevier Scientific Publisher, Amsterdam, 228 pp.
Kropff, M.J., 1987. Physiological effects of sulphur dioxide. I. The effect of SO_2 on photosynthesis and stomatal regulation of *Vicia Faba*. Plant, Cell and Environment 10:753-760.
Kwon, Y.W., 1984. Fundamental physiological constraints in breeding for higher yielding cold tolerant rice variety. International Crop Science Symposium on Potential Productivity and Yield Constraints of Rice in East Asia. Kyushu University, Fukuoka, Japan.
Lambers, J.Th., 1979. Energy metabolism in higher plants in different environments. Ph.D. thesis, University of Groningen.
Lambert, J.R., Baker, D.N. & Phene, C.J., 1976. Dynamic simulation of processes in the soil under growing row crops: RHIZOS. In: Computers applied to management of large scale agricultural enterprises. Proceedings US-USSR seminar, Moscow Riga Kishinev.
Landivar, J.A., 1979. The application of cotton simulation model GOSSYM in genetic feasibility studies. MSc thesis, Mississippi State University, Mississippi.
Lantinga, E.A., 1985. Productivity of grasslands under continuous and rotational grazing. Ph.D. thesis, Agricultural University, Wageningen.
Leffelaar, P.A. & Ferrari, T.J., 1989. Some elements of dynamic simulation. In: Rabbinge et al.(Eds.): Simulation and systems management in crop protection. Simulation Monographs. Pudoc, Wageningen.
Lin, Z.F. & Ehleringer, J., 1982. Changes in spectral properties of leaves as related to chlorophyll content and age of Papaya. Photosynthetica 16(4) :520-525.
Linneman, H., DeHoogh, J., Keyzer, M.A. & Heemst, H.D.J. van, 1979. MOIRA, model of international relations in agriculture. North Holland Publishing Company, Amsterdam.
Lof, H., 1976. Water use efficiency and competition between arid zone annuals, especially the grasses Phalaris minor and Hordeum murinum. Agri-

cultural Research Reports 853. Pudoc, Wageningen.
Loomis, R.S., Rabbinge, R. & Ng, E., 1979. Explanatory models in crop physiology. Annual Review Plant Physiology 30: 339-367.
Louwerse, N. & Zweerde, W. van de, 1977. Photosynthesis transpiration and leaf morphology of *Phaseolus vulgaris* and *Zea mays* grown at different irradiances in artificial and sunlight. Photosynthetica 11(1):11-21.
Louwerse, W., 1980. Effects of CO_2 concentration and irradiance on the stomatal behaviour of maize, barley and sunflower plants in the field. Plant, cell and environment 3:391-398.
Louwerse, W & Eikhoudt, J.W., 1975. A mobile laboratory for measuring photosynthesis, respiration and transpiration of field crops. Photosynthetica 9: 31-34
Louwerse, W. & Oorschot, J.L.P. van, 1969. An assembly for routine measurements of photosynthesis, respiration and transpiration of intact plants under controlled conditions. Photosynthetica 3:305-315.
Ludwig, L.J., Saeki, T. & Evans, L.T., 1965. Photosynthesis in artificial communities of cotton plants in relation to leaf area. Australian Journal of Biological Sciences 18:1103-1118.
MacKerron, D.K.L., 1984. Potato spout emergence in soil as a function of temperature and time of planting. In: F.A. Winiger, and A. Stockli (Eds): Conference papers 9th Biennial Conference of the European Association for Potato Research. Interlaken, Switzerland, pp. 364-365.
Makino, A., Mae, T. & Ohira, K., 1984. Changes in photosynthetic capacity in rice leaves from emergence through senescence. Analysis from Ribulose –1,5– biphosphate carboxylase and leaf conductance. Plant and Cell Physiology 25(3):511-521.
Marletto, V. & Keulen, H. van, 1984. Winter wheat experiments in the Netherlands and Italy analyzed by the SUCROS model. Simulation Reports CABO-TT No. 3. Centre for Agrobiological Research, Wageningen.
McCree, K.J., 1974. Equations for the rate of dark respiration of white clover and grain sorghum, as functions of dry weight, photosynthetic rate and temperature. Crop Science 14:509-514.
McCree, K.J., 1982. Photosynthetically active radiation. Encyclopedia Plant Physiology 12a. Springer Verlag, Berlin, pp. 41-55.
McCree, K.J., 1983. Carbon balance as a function of plant size in sorghum plants. Crop Science 23: 1173-1177.
McCree, K.J. & Kresovich, S., 1978. Growth and maintenance requirements of white clover as a function of daylength. Crop Science 18:22-25.
McDermitt, D.K. & Loomis, R.S., 1981. Elemental composition of biomass and its relation to energy content, growth efficiency, and growth yield. Annals of Botany 48:275-290.
McKeague, J.A., Eilers, R.G., Thomasson, A.J., Reeve, M.J., Bouma, J., Grossman, R.B., Favrot, J.C., Renger, M. & Strebel, O., 1984. Tentative assessment of soil survey approaches to the characterization and interpreta-

tion of air-water properties of soils. Geoderma 34:69-100.
McKinion, J.M., 1980. Dynamic simulation: a positive feedback mechanism for experimental research in the biological sciences. Agricultural Systems 5:239-250.
Menenti, M., 1984. Physical aspects and determination of evaporation in deserts applying remote sensing techniques. Ph.D. thesis, Institute for Land and Water Management, Wageningen.
Miedema, P., 1982. The effects of low temperature on *Zea mays*. Advances in Agronomy 35:93-128.
Minchin, F.R., Summerfield, R.J., Hadley, P., Roberts, E.H. & Rawsthorne, S., 1981. Carbon and nitrogen nutrition of nodulated roots of grain legumes. Plant, Cell and Environment 4:5-26.
Monteith, J., 1973. Principles of environmental physics. Edward Arnold, London.
Moorby, J., 1981. Transport systems in plants. Longman, London and New York.
Moorman, F.R. & Breemen, N. van, 1978. Rice: soil, water, land. International Rice Research Institute, Los Baños, 185 pp.
Morris, R.A., 1987. Characterizing rainfall distributions at IRRI cropping systems research sites in the Philipines. In: Impact of weather parameters on the growth and yield of rice. International Rice Research Institute, Los Baños, pp. 189-214.
Muramoto, H., Hesketh, J.D. & El-Sharkawy, M.A., 1965. Relationships among the rate of leaf area development, photosynthesis rate and rate of dry matter production among American cultivated cottons and other species. Crop Science 5:163-166.
Nevins, D.J. & Loomis, R.S., 1970. A method for determining net photosynthesis and transpiration of plant leaves. Crop Science 10:3-6.
Ng, N. & Loomis, R.S., 1984. Simulation of growth and yield of the potato crop. Simulation Monographs. Pudoc, Wageningen.
Oldeman, L.R. & Frere, M., 1982. A study of the agroclimatology of the humid tropics of Southeast Asia. FAO, Rome.
Oldeman, L.R., Seshu, D.V. & Cady, F.B., 1987. Response of rice to weather variables. In: Impact of weather parameters on the growth and yield of rice. International Rice Research Institute, Los Baños, pp. 5-40.
O'Toole, J.C. & Cruz, R.T., 1980. Response of leaf water potential, stomatal resistance and leaf rolling to water stress. Plant Physiol. 65: 428-432.
Pallas, J.E. & Samish, Y.B., 1984. Photosynthetic response of peanut. Crop Science 14:478-482.
Parlange, J.-Y., 1971. Theory of water movement in soils: 2. One-dimensional infiltration. Soil Science 111:170-174.
Passioura, J.B., 1977. Grain yield, harvest index, and water use of wheat. The Journal of the Australian Institute of Agricultural Science 43 (3-4):117-120.
Penman, H.L., 1956. Evaporation: an introductory survey. Netherlands Jour-

nal of Agricultural Science 4:9-29.
Penning de Vries, F.W.T., 1975. The cost of maintenance processes in plant cells. Annals of Botany 39:77-92.
Penning de Vries, F.W.T., 1977. Evaluation of simulation models in agriculture and biology: conclusions of a workshop. Agricultural Systems 2:99-107.
Penning de Vries, F.W.T., 1982. Le potentiel physiologique des pâturages et des cultures agricoles; La production potentielle des pâturages naturelles. In: F.W.T. Penning de Vries and M.A. Djiteye (Eds.): La productivité des pâturages sahéliens. Agricultural Research Reports 918. Pudoc, Wageningen, pp. 87-98, 165-181.
Penning de Vries, F.W.T., 1983. Modeling of growth and production. Encyclopedia of Plant Physiology, New Series, Vol. 12D. Springer Verlag, Berlin, pp. 117-150.
Penning de Vries, F.W.T., 1987. Some opportunities for using crop models in rice. In: Impact of weather parameters on the growth and yield of rice. International Rice Research Institute, Los Baños, pp. 291-300.
Penning de Vries, F.W.T., Brunsting, A.B. & Laar, H.H. van, 1974. Products, requirements and efficiency of biological synthesis, a quantitative approach. Journal Theoretical Biology 45:339-377.
Penning de Vries, F.W.T. & Laar, H.H. van, 1976. Substrate utilization in germinating seeds. In: Landsberg, J.J. & C.V. Cutting (Eds.): Environmental effects on crop physiology. Academic Press, London, New York.
Penning de Vries, F.W.T., Witlage, J.M. & Kremer, D., 1979. Rates of respiration and of increase in structural dry matter in young wheat, ryegrass and maize plants in relation to temperature, to water stress and to their sugar content. Annals of Botany 44:595-609.
Penning de Vries, F.W.T. & Laar, H.H. van, (Eds.) 1982: Simulation of plant growth and crop production. Simulation Monographs. Pudoc, Wageningen.
Penning de Vries, F.W.T., Laar, H.H. van & Chardon, M.C.M., 1983. Bioenergetics of growth of seeds, fruits and storage organs. In: I. Yoshida (Ed.): Potential productivity of field crops under different environments. International Rice Research Institute, Los Baños, pp. 38-59.
Penning de Vries, F.W.T., Rabbinge, R., Jansen, D.M. & Bakema, A., 1988. Transfer of systems analysis and simulation in agriculture to developing countries. Agricultural Administration and Extension 29: 85-96.
Press, W. et al., 1986. The art of scientific computing. ISBN O-521-30811-9. In: Numerical recipes. Cambridge University Press, U.S.A. 700 pp.
Raats, P.A.C., 1970. Steady infiltration from line sources and furrows. Soil Science Society of America: Proceedings 34:709-714.
Rabbinge, R., 1982. Pests, diseases and crop production. In: F.W.T. Penning de Vries and H.H. van Laar (Eds.): Simulation of plant growth and crop production. Simulation Monographs, Pudoc, Wageningen, pp. 253-265.
Rabbinge, R., Ward, S.A., & Laar, H.H. van, 1989. Simulation and systems management in crop protection. Simulation Monographs. Pudoc, Wagen-

ingen 1989.
Rawls, W.J., Brakensiek, D.L., & Saxton, K.E., 1982. Estimation of soil water properties. 1982 Transactions of the ASEA. Paper No. 81-2510.
Rawson, H.M. & Constable, G.A., 1980. Carbon production of sunflower cultivars in field and controlled environments. I. Photosynthesis and transpiration of leaves, stems and heads. Australian Journal of Plant Physiology 7: 555-573.
Rawson, H.M., Begg, J.E. & Woodward, R.G., 1977. The effect of atmospheric humidity on photosynthesis, transpiration and water use efficiency of leaves of several plant species. Planta 134:5-10.
Ray, T.B. & Black, C.C., 1979. The C_4 pathway and its regulation. In: M. Gibbs and E. Latzko (Eds.), Encyclopedia of Plant Physiology, New Series, Vol 6. Springer Verlag, Berlin, pp. 77-101.
Rasche, K., 1979. Movements of stomata. In: Encyclopedia of Plant Physiology, New Series, Vol 7. Springer Verlag, Berlin, pp. 383-441.
Robertson, G.W., 1968. A biometeorological time scale for a cereal crop involving day and night temperature and photoperiod. International Journal of Biometeorology 12: 191-223.
Rose, C.W., 1966. Agricultural physics. Pergamon Press, New York.
Ryle, G.J.A., Brockington, N.R., Powell, C.E., & Cross, B., 1973. The measurement and prediction of organ growth in uniculm barley. Annals of Botany 37:233-246.
Rijtema, 1969. Soil moisture forecasting. Bulletin 513. Institute of Land and Water Management Research (ICW), Wageningen, 17 pp.
Salisbury, F.B., 1981. Responses to Photoperiod. Encyclopedia Plant Physiology, New Series, Vol. 12A. Springer Verlag, Berlin, pp. 135-168.
Saxena, N.P., Natarajan, M. & Reddy, M.S. Chickpea, pigeonpea and groundnut. In: Potential productivity of field crops under different environments. International Rice Research Institute, Los Baños, pp. 281-305.
Saxton, K.E., Romberger, J.S. & Rawls, W.J., 1986. Estimating generalized soil water characteristics by texture. Soil Science Society of America Journal 50:1031-1036.
Schapendonk, A.H.C.M. & Bouma, P., 1984. First growth of cucumber in relation to assimilate supply and sink capacity. Scientia Horticulture 23:21-33.
Schuh, W.M. & Bauder, J.W., 1986. Effect of soil properties on hydraulic conductivity-moisture relationships. Soil Science Society of America Journal 50:848-855.
Schulze, E.-D. & Hall, A.E., 1982. Stomatal responses, water loss and CO_2 assimilation rates of plants in contrasting environments. In: O.L. Lange, P.S. Nobel, C.B. Osmond & H. Ziegler (Eds.): Encyclopedia of Plant Physiology New Series, Volume 12B. Springer Verlag, Berlin, pp. 181-230.
Shaykewich, C.F. & Stroosnijder, L., 1977. The concept of matric flux potential applied to simulation of evaporation from soil. Netherlands Journal of

Agricultural Science 25:63-82.
Sharma, P.K., De Datta, S.K. & Redulla, C.A., 1987. Root growth and yield response of rainfed lowland rice to planting method. Experimental Agriculture 23:305-313.
Sibma, L., 1987. Ontwikkeling en groei van mais (*Zea mays L.*) onder Nederlandse omstandigheden. Pudoc, Wageningen.
Sinclair, T.R., Hoffer, R.M. & Schreiber, M.M., 1971. Reflectance and internal structure of leaves from several crops during a growing season. Agronomy Journal 63:864-868.
Sinclair, T.R. & Wit, C.T. de, 1976. Analysis of carbon and nitrogen limitations to soybean yield. Agronomy Journal 68:319-324.
Sivakumar, M.V.K., Seetharama, N., Singh, S. & Bidinger, F.R., 1979. Water relations, growth and dry matter accumulation of sorghum under postrainy season conditions. Agronomy Journal 71:843-847.
Slatyer, R.O., 1967. Plant water Relationships. Academic Press, London.
Smith, R.E. & Parlange, J.-Y., 1978. A parameter efficient hydrologic infiltration model. Water resources research 14:533-538.
Spitters, C.J.T., 1984. A simple simulation model for crop-weed competition. 7th International Symposium on Weed Biology, Ecology and Systematics. COLUMA-EWRS, Paris, pp. 355-366.
Spitters, C.J.T., 1986. Separating the diffuse and direct component of global radiation and its implications for modeling canopy photosynthesis. Part II. Agricultural and Forest Meteorology 38:231-242.
Spitters, C.J.T., Toussaint, H.A.J.M. & Goudriaan, J., 1986. Separating the diffuse and direct component of global radiation and its implications for modeling canopy photosynthesis. Part I. Agricultural and Forest Meteorology 38:217-229.
Spitters, C.J.T., Keulen, H. van & Kraalingen, D.W.G. van, 1989. A simple and universal crop growth simulation: SUCROS87. In: R. Rabbinge, S.A. Ward & H.H. van Laar (Eds): Simulation and systems management in crop protection. Simulation Monograph 32. Pudoc, Wageningen.
Stansel, J.W., 1975. The rice plant – its development and yield. Six decades of rice research in Texas. Research Monograph 4. Texas Agricultural Experiment Station, pp. 9-21.
Steinhorst, R.K., Hunt, H.W., Innis, G.S. & Haydock, K.P., 1978. Sensitivity analysis of the ELM model. In: G.S. Innis (Ed.): Grassland simulation model. Ecological Studies 26, Springer Verlag, New York.
Steward, F.C. & Bidwell, R.G.S. (Eds.), 1983. Plant Physiology, Vol VII: Energy and Carbon Metabolism. Academic Press, New York.
Steyeart, L.T., le Duc, S.K., & McQuigg, J.D., 1978. Atmospheric pressure and wheat yield modeling. Agricultural Meteorology 18:23-24.
Stone, L.R., Teare, I.D., Nickell, C.D. & Mayaki, W.C., 1976. Soybean root development and soil water depletion. Agronomy Journal 68:677-680.
Stroosnijder, L., 1982. Simulation of the soil water balance. In: F.W.T. Pen-

Stroosnijder, L., 1982. Simulation of the soil water balance. In: F.W.T. Penning de Vries and H.H. van Laar (Eds.): Simulation of plant growth and crop production. Simulation Monographs. Pudoc, Wageningen, pp. 175-193.

Summerfield, R.J., Minchin, F.R., Roberts E.H. and Hadley, P., 1983. In: Potential productivity of field crops under different environments. International Rice Research Institute, Los Baños, pp. 249-280.

Summerfield, R.J. & Roberts, E.H. 1987. Effects of illuminance on flowering in long and short day grain legumes: a reappraisal and unifying model. In: J.G. Atherton (Eds.): Manipulation of flowering. Butterworth, London, pp. 203-223.

Supit, I., 1986. Manual for generation of daily weather data. Simulation Reports CABO-TT No. 7. Centre for Agrobiological Research, Wageningen, 46 pp.

Tanner, C.B. & Sinclair, T.R., 1983. Efficient use of water in crop production: research or re-search. In: H.M. Taylor, W.R. Jordan & T.R. Sinclair (Eds.): Limitations to efficient water use in crop production, ASA-CSSA-SSSA, Madison, pp. 1-27.

Taylor, H.M. & Klepper, B., 1973. Rooting density and water extraction patterns for corn. Agronomy Journal 65:965-968.

Taylor, H.M. & Klepper, B., 1974. Water relations and cotton. 1. Root growth and water use as related to top growth and soil water content. Agronomy Journal 66:584-588.

Taylor, H.M. & Klepper, B., 1978. The role of rooting characteristics in the supply of water to plants. Advances in Agronomy 30:99-128.

Teubner, F., 1985. Messung der Photosynthese und Transpiration an Weizen, Kartoffel und Sonnenblume. Diplomarbeit University Bayreuth, West Germany.

TNO, 1977. Precipitation and measurements of precipitation. Proceedings and Informations No. 23. Verslagen en Mededelingen van de Commissie Hydrologies Onderzoek, Central Organization for Applied Scientific Research, the Hague.

USDA Soil Survey Staff, 1975. Soil taxonaomy: A basic system of soil classification for making and interpreting soil surveys. Agricultural Handbook 436. U.S. Government Printers Office, Washington, DC.

Vanderlip, R.L. & Arkin, G.F., 1977. Simulating accumulation and distribution of dry matter in grain sorghum. Agronomy Journal 69: 917-923.

Vauclin, M., Haverkamp, R. & Vachaud, G. 1979. Resolution numerique d'une equation de diffusion non-lineaire. Presse universitaire de Grenoble, France, 183 pp.

Veen, B.W. & Boone, F.R., 1981. The influence of mechanical resistance and phosphate supply on morphology and function of corn roots. Plant and Soil 63:77-81.

Veihmeyer, F.J. & Hendrickson, A.H., 1949. Methods of measuring field ca-

pacity and wilting percentages of soils. Soil Science 68:75-94.
Veldkamp, H.J., 1986. Physiological causes of yield variation in cassava (*Manihot esculenta* Crantz). Agricultural University Wageningen Papers, 85-6, Agricultural University, Wageningen, Netherlands.
Vergara, B.S. & Chang, T.T., 1985. The flowering response of the rice plant to photoperiod. International Rice Research Institute, Los Baños.
Versteeg, M.N., 1985. Factors influencing the productivity of irrigated crops in southern Peru, in relation to prediction by simulation models. Ph.D. thesis, Agricultural University, Wageningen.
Vertregt, N. & Penning de Vries, F.W.T., 1987. A rapid method for determining the efficiency of biosynthesis of plant biomass. Journal of Theoretical Biology 128:109-119.
Visser, A.J.C. de, 1984. Interactions between energy and nitrogen metabolism in *Pisum sativum*. Ph. D. thesis. University of Groningen, Groningen.
Vos, J., Drees, E.M. & Penning de Vries, F.W.T., 1982. Modelling of postfloral growth of wheat. In: F.W.T. Penning de Vries and H.H. van Laar (Eds.): Simulation of plant growth and crop production. Simulation Monographs. Pudoc, Wageningen, pp. 144-151.
Vos, J. & Groenwold, J., 1983. Estimation of root densities by observation tubes and endoscope. Plant and Soil 74: 295-300.
Vos, J. & Groenwold, J., 1986. Root growth of potato crops on a marine-clay soil. Plant and Soil 94:17-33.
Warrington, I.J. & Kanemasu, E.T., 1983. Corn growth response to temperature and photoperiod. I. Seedling emergence, tassel initiation and anthesis. Agronomy Journal 75:749-754.
Werk, K.S., Ehleringer, J., Forseth, I.N. & Cook, C.S., 1983. Photosynthetic characteristics of Sonoran desert winter annuals. Oecologia (Berl.) 59(1) :101-105.
Whisler, F.D., 1983. Sensitivity tests of the environmental variables in IRRIMOD. IRRI Research Paper Series 94. International Rice Research Institute, Los Baños.
Whisler, F.D., Acock, B., Baker, D.N., Fye, R.E., Hodges, H.F., Lambert, J.R., Lemon, H.E., Mckinion, J.M. & Reddy, V.R., 1986. Crop simulation models in agronomic systems. Advances in Agronomy 40: 141-208.
Wilson, D., 1982. Response to selection for dark respiration rate of mature leaves of *Lolium perenne* and its effects on growth of young plants and simulated swards. Annals of Botany 49:303-312.
Wit, C.T. de, 1958. Transpiration and crop yields. Agricultural Research Reports 64.6. Pudoc, Wageningen, 88 pp.
Wit, C.T. de, 1965. Photosynthesis of leaf canopies. Agricultural Research Reports 663. Pudoc, Wageningen.
Wit, C.T. de, 1967. Photosynthesis: its relationship to overpopulation. In: Harvesting the sun. Academic Press, New York.
Wit, C.T. de, 1968. Plant production. Miscellaneous Papers 3. Agricultural

University, Wageningen, pp. 25-50.
Wit, C.T. de, 1970. Dynamic concepts in biology. In: I. Setlik (Ed.): Prediction and measurement of photosynthetic productivity. Proceedings International Biological Program/Plant Production Technical Meeting Trebon. Pudoc, Wageningen, pp. 17-23.
Wit, C.T. de, Brouwer, R. & Penning de Vries, F.W.T., 1970. The simulation of photosynthetic systems. In: I. Setlik (Ed.): Prediction and measurement of photosynthetic productivity. Proceedings International Biological Program/Plant Production Technical Meeting Trebon. Pudoc, Wageningen, pp. 47-70.
Wit, C.T. de & Keulen, H. van, 1972. Simulation of transport processes in soils. Simulation Monographs. Pudoc, Wageningen.
Wit, C.T. de et al., 1978. Simulation of assimilation, respiration and transpiration of crops. Simulation Monographs. Pudoc, Wageningen.
Wit, C.T. de & Goudriaan, J., 1978. Simulation of ecological processes. Simulation Monographs, Pudoc, Wageningen.
Wit, C.T. de, Laar, H.H. van & Keulen, H. van, 1979. Physiological potential of crop production. In: J. Sneep and A.J.T. Hendiksen (Eds.): Plant breeding perspectives. Pudoc, Wageningen, pp. 47-82.
Wit, C.T. de & Penning de Vries, F.W.T., 1982. La synthese et la simulation des systemes de production primaire. In: Penning de Vries, F.W.T. & M.A.Djiteye (Eds): La productivite des paturages saheliens. Agricultural Research Reports 918. Pudoc, Wageningen.
Wit, C.T. de & Penning de Vries, F.W.T., 1983. Crop growth models without hormones. Netherlands Journal of Agricultural Science 31:313-323.
Wit, C.T. de, Keulen, H. van, Seligman, N.G. & Sfarim, I., 1988. Application of interactive multiple goal linear programming to agricultural development. Agricultural Systems 26:211-230.
Wittenback, V.A., Lin, W. & Hebert, R.R., 1982. Vacuolar localization of proteases and degradation of chloroplasts in mesophyl protoplasts from senescing primary wheat leaves. Plant Physiology 69:98-102.
Woldendorp, J.W., 1978. The rhizosphere as part of the plant-soil system. In: A.H.J. Frijsen & J.W. Woldendorp (Eds.): Structure and functioning of plant populations. North Holland Publishing Company, Amsterdam, pp. 237-267.
Wösten, J.H.M., Bannink, M.H. & Beuving, J., 1986. Waterretentie- en doorlatendheidskarakteristieken van boven-en ondergronden in Nederland: de Staringreeks. Institute for Land and Water Management, rapport 18. Wageningen.
Yoshida, S., 1981. Fundamentals of rice crop science. International Rice Research Institute, Los Baños.
Yoshida, S. & Hasegawa, S., 1982. The rice root system: its development and function. In: Drought resistance and crops with emphasis on rice. International Rice Research Institute, Los Baños, pp. 97-114.

Zadoks, J.C., Rijsdijk, F.H. & Rabbinge, R., 1984. EPIPRE, a systems approach to supervised control of pests and diseases in wheat in the Netherlands. In: G.R. Conway (Ed.): Pest and pathogen control: Strategy, tactical and policy models. International series on applied systems analysis. Wiley and Sons, pp. 34-351.

9 Index

Appendix A. Error messages from SUBROUTINES and FUNCTIONS

Message number	Name subroutine	Condition	Possible cause
1.1	FUPHOT	ALV < 0 or ALV > 25	Wrong initialization of leaf area or weight; too much leaf death
1.2	FUPHOT	PLMX < 0 or PLMX > 100	Wrong initialization of PLMXP; reduction of PLMXP excessive
1.3	FUPHOT	PLEA < 0 or PLEA > 0.7	Wrong initialization of PLEI; reduction of PLEI excessive
2.1	SUASTC	DATE < 0. or > 365.	Wrong calculation of Julian date
2.2	SUASTC	AOB < -1.0 or > 1.0	LAT < -66.5 or > 66.5 degrees
2.3	SUASTC	RDTM < 0. or > RDTC	Wrong conversion of measured radiation into J $m^{-2}d^{-1}$
3.1	SUEVTR	FRD < 0. or > 1.	Wrong calculation of fraction
3.2	SUEVTR	VPAS < 0. or > 12.55	Wrong calculation of TPAD
3.3	SUEVTR	VPA < 0. or > VPAS	Wrong conversion of measured air humidity into kPa
4.1	SUMSKM	MS < 0. or > 1.E8	Wrong calculation of MS
5.1	SUPHOL	ALVL(I2) < 0 or > 25	Wrong calculation of green area of a layer
5.2	SUPHOL	ALVDL(I2) < 0 or > 25	Wrong calculation of dead area of a layer
5.3	SUPHOL	PLMXL(I2) < 0 or > 100	Wrong calculation of PLMX(I)
5.4	SUPHOL	PLEAL(I2) < 0 or > 0.7	Wrong calculation of PLEA(I)
6.1	SUSLIN	TYL(I) < 1. or > 20.	Wrong number in TABLE TYL
6.2	SUSLIN	WCLQTI(I) < WCAD(I) or > WCST(I)	Wrong initialization water contents; check measurements, units
7.1	SUWCMS	WCLQT < WCAD(I) or > WCST(I)	Wrong calculation of water content in a soil compartment
7.2	SUWCMS	MS < 0. or > 1.E8	Wrong calculation of MS

Appendix B. Listing of module T12 with SUBROUTINES and FUNCTIONS used in Listings 1-11

```
TITLE T12.CSM, MAY 88
END
STOP
      FUNCTION FUCCHK(CKCIN,CKCFL,TIME)
C check on crop carbon balance. used in L1D, L1Q. 03/87
      FUCCHK=2.0*(CKCIN-CKCFL)/(CKCIN+CKCFL+1.E-10)
      IF(ABS(FUCCHK).GT.0.01) WRITE (6,10) FUCCHK, CKCIN, CKCFL, TIME
 10   FORMAT(/'* * *error in carbon balance, please check* * *',/,'CKCRD
     $=',F6.3,' CKCIN=',F8.2,' CKCFL=',F8.2,' AT  TIME=',F6.1)
      RETURN
      END

      FUNCTION FUPHOT(PLMX,PLEA,ALV,RDTM,DATE,LAT)
C computes canopy photosynthesis. used in L1D, L1Q. 8/87
      IMPLICIT REAL (A-Z)
      INTEGER IT,I
      DATA KDIF/0.7155/,PI/3.1415926/,SCV/0.200/,GAUSR/0.3872893/
      CALL SUERRM(1.1,ALV, 0., 25.,6.)
      CALL SUERRM(1.2,PLMX,0.,100.,6.)
      CALL SUERRM(1.3,PLEA,0., 0.7,6.)
      CALL SUASTC(DATE,LAT,RDTM,RDTC,FRDIF,COSLD,SINLD,DSINBE,SOLC,DLA)
      GDFG  =0.
      IF(PLMX*PLEA*ALV.LE.0.0) GOTO 50
      ALVL  =AMIN1(10.,ALV)
      REFH  =(1.-SQRT(1.0-SCV))/(1.+SQRT(1.-SCV))
      DO 40 IT=1,3
        HOUR  =12.0+DLA*0.5*(0.5+(IT-2)*GAUSR)
        SINB  =AMAX1(0.,SINLD+COSLD*COS(2.*PI*(HOUR+12.)/24.))
        REFS  =REFH*2./(1.+1.6*SINB)
        PAR   =0.5*RDTM*SINB*(1.0+0.4*SINB)/DSINBE
        PARDIF=AMIN1(PAR,SINB*FRDIF*(RDTM/RDTC)*0.5*SOLC)
        PARDIR=PAR-PARDIF
        KDIRBL=(0.5/SINB)*KDIF/(0.8*SQRT(1.-SCV))
        KDIRT =KDIRBL*SQRT(1.-SCV)
        FGROS =0.
        DO 30 I=1,3
          ALVC  =0.5*ALVL+GAUSR*(I-2)*ALVL
          VISDF =(1.-REFS)*PARDIF*KDIF*EXP(-KDIF*ALVC)
          VIST  =(1.-REFS)*PARDIR*KDIRT*EXP(-KDIRT*ALVC)
          VISD  =(1.-SCV)*PARDIR*KDIRBL*EXP(-KDIRBL*ALVC)
          VISSHD=VISDF+VIST-VISD
          FGRSH =PLMX*(1.-EXP(-VISSHD*PLEA/PLMX))
          VISPP =(1.-SCV)*PARDIR/SINB
          IF (VISPP.LE.0.) GO TO 10
          FGRSUN=PLMX*(1.-(PLMX-FGRSH)*(1.-EXP(- VISPP*PLEA/PLMX))/
    $             (PLEA*VISPP))
          GO TO 20
10        FGRSUN=FGRSH
20        CONTINUE
          FSSLA =EXP(-KDIRBL*ALVC)
```

```
           FGL    =FSSLA*FGRSUN+(1.-FSSLA)*FGRSH
           IF(I.EQ.2) FGL    =FGL*1.6
           FGROS =FGROS+FGL
 30      CONTINUE
         FGROS =FGROS*ALVL/3.6
         IF(IT.EQ.2) FGROS =FGROS*1.6
         GDFG   =GDFG+FGROS
 40    CONTINUE
 50    FUPHOT=GDFG*DLA/3.6
       RETURN
       END

       FUNCTION FURSC(WDS,ALV,PLHT,ZREF)
C calculates canopy resistance upper layers. in L2C. 4/87
       ZR     =AMAX1(ZREF,PLHT+1.)
       D      =AMAX1(0.1,0.63*PLHT)
       ZNOT   =AMAX1(0.05,0.1*PLHT)
       ALVX   =AMAX1(1.,ALV)
       WDSX   =AMAX1(0.2,WDS)
       FURSC  =0.74*(ALOG((ZR-D)/ZNOT))**2/(0.16*WDSX)*ALVX
       RETURN
       END

       FUNCTION FUTP(IDATE,DTIME,TPHT,TPLT,FA,FB,FC,FD)
C approximates daily course of air temperature. in L1Q. 9/85
       DIMENSION TPHT(365),TPLT(365)
       IF(IDATE.EQ.366) IDATE=365
       FUTP   =FA*TPHT(MAX0(1,IDATE-1))+(1.-FA)*TPLT(IDATE)
       IF(DTIME.GT.0.2) FUTP   =FB*TPHT(IDATE)+(1.-FB)*TPLT(IDATE)
       IF(DTIME.GT.0.4) FUTP   =FC*TPHT(IDATE)+(1.-FC)*TPLT(IDATE)
       IF(DTIME.GT.0.6) FUTP   =FD*TPHT(IDATE)+
      $                          (1.-FD)*TPLT(MIN0(365,IDATE+1))
       RETURN
       END

       FUNCTION FUVP(TP)
C vapour pressure (kPa) relation to temperature. in L1Q, L2C. 9/85
       FUVP   =0.100*6.11*EXP(17.47*TP/(TP+239.))
       RETURN
       END

       FUNCTION FUWCHK(CKWFL,CKWIN,TIME)
C check on soil water balance. used in L2SU, L2SS. 3/87
       FUWCHK=2.0*(CKWIN-CKWFL)/(CKWIN+CKWFL+1.E-10)
       IF(ABS(FUWCHK).GT.0.01.AND.ABS(CKWIN).GT.0.2)
      $WRITE (6,10) FUWCHK,CKWIN,CKWFL,TIME
 10    FORMAT(/'* * *error in water balance, please check* * *',/,' CKWRD
      $=',F6.3,' CKWIN=',F8.2,' CKWFL=',F8.2,' AT TIME=',F6.1)
       RETURN
       END

       FUNCTION FUWCMS(I,MS)
C converts moisture suction into water contents. in L2SS. 9/87
       REAL MS
       CALL SUWCMS(I,2,WCL,MS)
       FUWCMS=WCL
       RETURN
       END
```

```
      FUNCTION FUWRED(WDLV,ALV,PLHT,WDS)
C calculates windspeed near soil surface. in L2C. 9/87
      IMPLICIT REAL (A-Z)
      PLHTX =AMAX1(0.05,PLHT)
      ALVX  =AMAX1(0.01,ALV)
      MIXL  =SQRT(1.2732*AMAX1(0.005,WDLV)/(ALVX/PLHTX))
      A     =SQRT(0.2*ALVX*PLHTX/(2.*MIXL*0.5))
      FUWRED=AMAX1(0.2,WDS)*EXP(-A*(1.0-0.05/PLHTX))
      RETURN
      END

      FUNCTION FUWS(TRC,ALV,WCL,WSSC,WFSC,WCWP,WCFC,WCST)
C computes reduction of water uptake, used in L2SU, L2SS. 5/87
      DATA A,B,ALVMAX/0.76,0.15,2./
      IF(WCL .LE. WCFC) THEN
        SDPF  =1./(A+B*ALVMAX*TRC/(ALV+1.E-10))-(1.-WSSC)*0.4
        IF(WSSC.LT.0.6) THEN
          SDPF  =SDPF+0.025*AMIN1(0.,ALVMAX*TRC/(ALV+1.E-10)-6.)/
     $              (1.+5.*WSSC+4.*WSSC*WSSC)
        ENDIF
        WCX   =WCWP+(WCFC-WCWP)*(1.00-AMIN1(1.,AMAX1(0.,SDPF)))
        FUWSX =(WCL-WCWP)/(WCX-WCWP+1.E-10)
      ELSE
        FUWSX =1.-(1.-WFSC)*(WCL-WCFC)/(WCST-WCFC+1.E-10)
      ENDIF
      FUWS  =AMIN1(1.,AMAX1(0.,FUWSX))
      RETURN
      END

      SUBROUTINE SUASTC(DATE,LAT,RDTM, RDTC,FRDIF,COSLD,
     $           SINLD,DSINBE,SOLC,DLA)
C astronomical standard computations. used in L1D, L1Q. 5/87
      IMPLICIT REAL (A-Z)
      DATA PI/3.1415926/,RAD/0.0174533/
      DEC   =-ASIN(SIN(23.45*RAD)*COS(2.*PI*(DATE+10.)/365.))
      COSLD =COS(DEC)*COS(LAT*RAD)
      SINLD =SIN(DEC)*SIN(LAT*RAD)
      AOB   =SINLD/COSLD
      CALL SUERRM(2.1,DATE,0.,365.,6.)
      CALL SUERRM(2.2,AOB,-1.0,1.0,6.)
      DLA   =12.*(1.+2.0*ASIN(AOB)/PI)
      DSINBE=3600.*(DLA*(SINLD+0.4*(SINLD*SINLD+COSLD*COSLD*0.5))+
     $        12.0*COSLD*(2.0+3.0*0.4*SINLD)*SQRT(1.-AOB*AOB)/PI)
      DSINB =3600.*(SINLD*DLA+24./PI*COSLD*SQRT(1.-AOB**2))
      SOLC  =1370.*(1.0+0.033*COS(2.*PI*DATE/365.))
      RDTC  =SOLC*DSINB
      CALL SUERRM(2.3,RDTM,0.,RDTC,6.)
      ATMTR =RDTM/RDTC
      IF(ATMTR.GT.0.75) FRDIF =0.23
      IF(ATMTR.LE.0.75.AND.ATMTR.GT.0.35) FRDIF =1.33-1.46*ATMTR
      IF(ATMTR.LE.0.35.AND.ATMTR.GT.0.07) FRDIF =1.-2.3*(ATMTR-0.07)**2
      IF(ATMTR.LE.0.07) FRDIF =1.00
      RETURN
      END

      SUBROUTINE SUASTR(DATE,LAT,RDTC,DLA,DLP)
C computes daylength, daily total radiation clear. in L1D, L1Q. 5/87
```

```
      IMPLICIT REAL (A-Z)
      DATA INSP/-4.0/,PI/3.1415926/,RAD/0.0174533/
      CALL SUASTC(DATE,LAT,RDTM,RDTC,FRDIF,COSLD,SINLD,DSINBE,SOLC,DLA)
      DLP   =12.*(PI+2.*ASIN((-SIN(INSP*RAD)+SINLD)/COSLD))/PI
      RETURN
      END

      SUBROUTINE SUCONV(SWICH2,TKL,ZL,ZLT,RAIN,ZW,WLO,WLOMX,
     $            RUNOF,TRWL,EVSW,EVSC,DRSL,NL)
C converts units between main program and subroutines. in L2SS. 4/88
      IMPLICIT REAL (A-Z)
      INTEGER I,NL,SWICH2
      DIMENSION TKL(10),ZL(10),TRWL(10)
      F1    =0.01
      F2    =10.0
      IF(SWICH2.EQ.1) F1=1./F1
      IF(SWICH2.EQ.1) F2=1./F2
      DO 10 I=1,NL
         TKL(I)   =TKL(I) *F1
         ZL(I)    =ZL(I)  *F1
         TRWL(I)  =TRWL(I)*F2
 10   CONTINUE
      ZLT   =ZLT  *F1
      ZW    =ZW   *F1
      WLO   =WLO  *F1
      WLOMX =WLOMX*F1
      RAIN  =RAIN *F2
      EVSW  =EVSW *F2
      EVSC  =EVSC *F2
      RUNOF =RUNOF*F2
      DRSL  =DRSL *F2
      RETURN
      END

      SUBROUTINE SUERRM(MNR,X,XMIN,XMAX,NUNIT)
C checks whether X is between limits. in FUNCTIONS, SUBROUTINES. 8/87
      IMPLICIT REAL (A-Z)
      INTEGER IUNIT
      IF((X.LT.XMIN*0.99).AND.(XMIN.NE.-99.)) GOTO 10
      IF((X.GT.XMAX*1.01).AND.(XMAX.NE.-99.)) GOTO 10
      RETURN
 10   IUNIT =IFIX(NUNIT)
      WRITE(IUNIT,20) MNR,X,XMIN,XMAX
      STOP
 20   FORMAT(//,' ***fatal error in variable or parameter value ***',
     $/,'    message number, value, minimum and maximum: ',/,10X,F4.1,
     $3(3X,E10.3))
      END

      SUBROUTINE SUEVTR(RDTC,RDTM,RF,FRD,TPAD,VPA,RSL,RSB,RST,EVPR,EVPD)
C potential evapotranspiration rates crop, soil. used in L2C. 7/87
      CALL SUERRM(3.1,FRD,0.,1.,6.)
      VPAS  =FUVP(TPAD)
      CALL SUERRM(3.2,VPAS,0.0,12.55,6.)
      CALL SUERRM(3.3,VPA, 0.0,VPAS, 6.)
      SLOPE =4158.6*10.*VPAS/(TPAD+239.)**2
      APSCH =0.67*(RSB+RST+RSL)/(RSB/0.93+RST)
      RLWI  =4.8972E-3*(TPAD+273.)**4*(0.618+0.0365*SQRT(10.*VPA))
```

```
      RLWO  =4.8972E-3*1.00*(TPAD+273.)**4
      RDTN  =RDTM*(1.-RF)-(RLWO-RLWI)*(RDTM/(0.75*RDTC))*FRD
      EVPR  =0.001*RDTN*SLOPE/((SLOPE+APSCH)*2390.)
      DRYP  =(VPAS-VPA)*10.*1200./(RSB+RST) *FRD
      EVPD  =86400.*0.001*DRYP/((SLOPE+APSCH)*2390.)
      RETURN
      END

      SUBROUTINE SUGRHD(TKL,NL,HGT,HGB)
C calculates  gravitational head at interfaces. in L2SS. 9/87
      DIMENSION TKL(10),HGT(10),HGB(10)
      HGT(1)   =0.
      HGB(1)   =-TKL(1)
      DO 10 I=2,NL
        HGT(I)   =HGT(I-1)-TKL(I-1)
        HGB(I)   =HGB(I-1)-TKL(I)
 10   CONTINUE
      RETURN
      END

      SUBROUTINE SUINTG(HPBP,SWICH4,SWICH5,HPP,FLX,TKL,WCL,
     $           MS,DTMIN,DTMX1,DTMX2,DTFX,INXSAT,DHH,
     $           JTOT,FLXSQ2,WLO,WLOMX,NL,DELT,DT)
C calculates SAWAH-timestep, integrates rates during day;in L2SS;08/88
      IMPLICIT REAL (A-Z)
      INTEGER NL,I,IX,ITEL1,ITEL2,SWICH4,SWICH5,INXSAT,JTOT
      COMMON /VOLWAT/ WCAD(10),DUMMY1(10),WCST(10),DUMMY2(10)
      DIMENSION WCL(10),HPP(11),INXSAT(10,10)
      DIMENSION WCLRCH(10),FLX(11),TKL(10),MS(10),DHH(10)
      DATA TINY/1.0E-10/
      HPTP  =WLO
      IF(HPTP.LE.0..OR.FLX(1).LE.0.) THEN
         DTSRF =DELT
      ELSE
         DTSRF =AMAX1(TINY,HPTP/(FLX(1)))
      ENDIF
      DT    =AMIN1(DELT,DTSRF,DTMX1,DTMX2)
      DO 10 I=1,NL
         WCLRCH(I)=(FLX(I)-FLX(I+1))/TKL(I)
         IF(WCLRCH(I).LT.-TINY)SATTIM=-(WCL(I)-WCAD(I)-TINY)/WCLRCH(I)
         IF(WCLRCH(I).GT. TINY)SATTIM= (WCST(I)-WCL(I)-TINY)/WCLRCH(I)
         IF(ABS(WCLRCH(I).LT.TINY)SATTIM=DELT
         DT    =AMIN1(SATTIM,DT)
10    CONTINUE
      IF(SWICH5.EQ.2) THEN
         DT    =AMIN1(DT,DTFX)
         IF(DHH(JTOT).LT.-TINY.OR.JTOT.EQ.0) THEN
            CONTINUE
         ELSE
            IX    =INXSAT(JTOT,1)-1
            IF(IX.LT.1) THEN
               CONTINUE
            ELSEIF(FLX(IX+1).GT.-0.1) THEN
               CONTINUE
            ELSEIF(FLXSQ2.LE.0.) THEN
               CONTINUE
            ELSE
               MSAL  =TKL(IX)/2.
```

```
                CALL SUWCMS(IX,0,WIX2,MSAL)
                MSACT =MS(IX)
                CALL SUWCMS(IX,0,WIX1,MSACT)
                IF(WIX2.LE.WIX1) THEN
                  CONTINUE
                ELSE
                  DLIM=-(WIX2-WIX1)*TKL(IX)/FLX(IX+1)
                  DT=AMIN1(DT,AMAX1(DLIM,TINY))
                ENDIF
             ENDIF
          ENDIF
          GOTO 100
       ELSE
          CONTINUE
       ENDIF
       GOTO 30
20     DT    =DT/2.
30     IF(DT.LT.DTMIN) GOTO 100
       IF(ABS(WCLRCH(1)).LT.TINY) GOTO 40
       IF(HPTP.GT.TINY) THEN
          MST1  =-(HPTP-DT*(FLX(1)))
       ELSE
          GOTO 40
       ENDIF
       WCT2  =WCL(1)+DT*FLX(1)/TKL(1)
       IF((WCT2-WCAD(1)).LE.-TINY) GOTO 20
       IF(WCT2.GE.WCST(1)) THEN
          MST2  =WCT2-WCST(1)
       ELSE
          CALL SUWCMS(1,SWICH4,WCT2,MST2)
       ENDIF
       DELZ  =0.5*TKL(1)
       DH    =-MST2+MST1-DELZ
       CH    =-DH*FLX(1)
       IF(CH.LT.-TINY) GOTO 20
40     DO 70 I=2,NL
          GOTO 60
50        DT    =DT/2.
60        ITEL1 =0
          ITEL2 =0
          IF(DT.LT.DTMIN) GOTO 100
          IF(ABS(WCLRCH(I-1)).LT.TINY.AND.ABS(WCLRCH(I)).LT.TINY) GOTO 70
          IF(ABS(WCLRCH(I-1)).LT.TINY) THEN
             IF(ABS(WCL(I-1)-WCST(I-1)).LT.TINY) THEN
                DELZ  =0.5*TKL(I)
                MST1  =0.
                ITEL2 =1
             ELSE
                DELZ  =0.5*(TKL(I-1)+TKL(I))
                MST1  =MS(I-1)
             ENDIF
          ELSE
             DELZ  =0.5*(TKL(I-1)+TKL(I))
             WCT1  =WCL(I-1)-DT*FLX(I)/TKL(I-1)
             IF((WCT1-WCAD(I-1)).LE.-TINY) GOTO 50
             IF(WCT1.GE.WCST(I-1)) THEN
                MST1  =WCT1-WCST(I-1)
             ELSE
```

```
         CALL SUWCMS(I-1,SWICH4,WCT1,MST1)
      ENDIF
   ENDIF
   IF(ABS(WCLRCH(I)).LT.TINY) THEN
      IF(ABS(WCL(I)-WCST(I)).LT.TINY) THEN
         DELZ  =0.5*TKL(I-1)
         MST2  =0.
      IF(HPP(I).GT.TINY) THEN
         IF(HPP(I).GE.TKL(I-1)) ITEL1 =1
            DELZ  =0.5*TKL(I-1)
            MST2  =0.
         ENDIF
      ELSE
         DELZ  =0.5*(TKL(I-1)+TKL(I))
         MST2  =MS(I)
      ENDIF
   ELSE
      DELZ  =0.5*(TKL(I-1)+TKL(I))
      IF(ITEL2.EQ.1) DELZ  =0.5*TKL(I)
      WCT2  =WCL(I)+DT*FLX(I)/TKL(I)
      IF((WCT2-WCAD(I)).LE.-TINY) GOTO 50
      IF(WCT2.GE.WCST(I)) THEN
         MST2  =WCT2-WCST(I)
      ELSE
         CALL SUWCMS(I,SWICH4,WCT2,MST2)
      ENDIF
   ENDIF
   DH     =-MST2+MST1-DELZ
   CH     =-DH*FLX(I)
   IF(ITEL1.EQ.1) CH     =+1.
   IF(CH.LT.-TINY) GOTO 50
70    CONTINUE
      GOTO 90
80    DT     =DT/2.
      IF(DT.LT.DTMIN) GOTO 100
90    IF(ABS(WCLRCH(NL)).LT.TINY) GOTO 100
      ITEL1 =0
      ITEL2 =0
      WCT1  =WCL(NL)-DT*FLX(NL+1)/TKL(NL)
      IF((WCT1-WCAD(NL)).LE.-TINY) GOTO 80
      IF(WCT1.GE.WCST(NL)) THEN
         MST1  =WCT1-WCST(NL)
      ELSE
         CALL SUWCMS(NL,SWICH4,WCT1,MST1)
      ENDIF
      DELZ  =0.5*TKL(NL)
      IF(HPBP.GT.TINY.AND.WCL(NL).LT.WCST(NL)) THEN
         IF(HPBP.GE.TKL(NL)) ITEL1  =1
         DELZ  =TKL(NL)
         MST2  =-HPBP
      ENDIF
      DH     =-MST2+MST1-DELZ
      CH     =-DH*FLX(NL+1)
      IF(ITEL1.EQ.1) CH     =+1
      IF(CH.LT.-TINY) GOTO 80
100   WL0    =AMAX1(0.,WL0-DT*FLX(1))
      DO 110 I=1,NL
         WCL(I)    =WCL(I)+DT*WCLRCH(I)
```

```
 110  CONTINUE
      RETURN
      END

      SUBROUTINE SUMFLP(SWICH3,I,MS,MFLP)
C calculates matrix flux potential. in L2SS. 8/87
      IMPLICIT REAL (A-Z)
      INTEGER I,IG,IX,SWICH3
      COMMON /HYDCON/DUMMY1(10),KMSA1(10),DUMMY2(10),KST(10)
      DIMENSION MSI(8),XGAUS(3),WGAUS(3)
      DATA MSI/0.,10.,50.,250.,750.,1500.,5000.,10000./
      DATA XGAUS/0.112702,0.5,0.887298/
      DATA WGAUS/0.277778,0.444444,0.277778/,TINY/1.E-10/
      MFLP  =0.
      IF(MS.GT.TINY) THEN
         IF(SWICH3.EQ.1) THEN
            MFLP  =(KST(I)/KMSA1(I))*(EXP(-KMSA1(I)*MS)-1.)
         ELSE
            DO 30 IX=2,7
               DMFLP =0.
               IF(MS.GT.MSI(IX-1)) THEN
                  MSX   =AMIN1(MSI(IX),MS)
                  IF(IX.EQ.2) THEN
                     DO 10 IG=1,3
                        X      =MSX*XGAUS(IG)
                        CALL SUMSKM(SWICH3,I,X,KMSX)
                        DMFLP =DMFLP+KMSX*WGAUS(IG)
 10                  CONTINUE
                     MFLP  =MFLP-DMFLP*(MSX-MSI(IX-1))
                  ELSE
                     DO 20 IG=1,3
                        X      =MSX*(MSI(IX-1)/MSX)**XGAUS(IG)
                        CALL SUMSKM(SWICH3,I,X,KMSX)
                        DMFLP =DMFLP+X*KMSX*WGAUS(IG)
 20                  CONTINUE
                     IF(DMFLP.LE.0.0) GOTO 40
                     MFLP  =MFLP-DMFLP*ALOG(MSX/MSI(IX-1))
                  ENDIF
               ENDIF
 30         CONTINUE
         ENDIF
      ENDIF
 40   CONTINUE
      RETURN
      END

      SUBROUTINE SUMSKM(SWICH3,I,MS,KMS)
C calculates hydraulic conductivity from suction. in L2SS. 8/87
      IMPLICIT REAL (A-Z)
      INTEGER SWICH3,I
      COMMON /HYDCON/ KMSMX(10),KMSA1(10),KMSA2(10),KST(10)
      DATA TINY,MSAD/1.E-10,1.E7/
      CALL SUERRM(4.1,MS,0.,1.E8,6.)
      IF(MS.LT.MSAD-TINY) THEN
        IF((SWICH3.EQ.2).AND.(MS.GT.KMSMX(I))) THEN
          KMS    =KMSA2(I)*(MS**(-1.4))
        ELSE
          KMS    =KST(I)*EXP(-KMSA1(I)*MS)
```

```
        ENDIF
      IF(KMS.LT.TINY) KMS    =0.
      ELSE
        KMS    =0.
      ENDIF
      RETURN
      END

      SUBROUTINE SUPHOL(IN,PLMX,PLEA,ALVL,ALVDL,F1,F2,RDTM,
     $           DATE,LAT,PCGC,PCGCL)
C computes canopy photosynthesis for 1-5 layers. for L1D,L1Q. 8/87
      IMPLICIT REAL (A-Z)
      INTEGER I1,I2,I3,I4,IN
      DIMENSION PLMX(5),PLEA(5),ALVL(5),ALVDL(5),PCGCL(5)
      DIMENSION F1(5),F2(5),F3(5),PHL(5,3)
      DATA PI/3.1415926/,SCV/0.200/,GAUSR/0.3872983/
      CALL SUASTC(DATE,LAT,RDTM,RDTC,FRDIF,COSLD,SINLD,DSINBE,SOLC,DLA)
      KBLTOP=0.97*F1(1)+0.85*F2(1)+0.65*(1.00-F1(1)-F2(1))
      KDFTOP=KBLTOP*SQRT(1.0-SCV)
      DO 20 I2=1,5
         PCGCL(I2)=0.0
         DO 10 I1=1,3
            PHL(I2,I1)=0.0
 10      CONTINUE
         CALL SUERRM(5.1,ALVL(I2), 0.,25.0,6.)
         CALL SUERRM(5.2,ALVDL(I2),0.,25.0,6.)
         CALL SUERRM(5.3,PLMX(I2), 0.,100.,6.)
         CALL SUERRM(5.4,PLEA(I2), 0., 0.7,6.)
 20   CONTINUE
      DO 60 I1=-1,1
         HOUR  =12.0+DLA*0.5*(0.5+I1*GAUSR)
         SINB  =AMAX1(0.,SINLD+COSLD*COS(2.*PI*(HOUR+12.)/24.))
         O15   =AMAX1(0.16,0.966*SINB)
         O45   =AMAX1(0.46,0.707*SINB)
         O75   =1.0-0.268*O15-0.732*O45
         OTOP  =F1(1)*O15+F2(1)*O45+(1.00-F1(1)-F2(1))*O75
         REFH  =(1.-SQRT(1.0-SCV))/(1.+SQRT(1.-SCV))
         REFV  =REFH*2.0*OTOP/(OTOP+KDFTOP*SINB)
         PAR   =0.5*RDTM*SINB*(1.0+0.4*SINB)/DSINBE
         PARDIF=AMIN1(PAR,SINB*FRDIF*(RDTM/RDTC)*0.5*SOLC)
         PARDIR=AMAX1(0.,PAR-PARDIF)
         RTDIF =(1.0-REFV)*PARDIF
         RTDIRT=(1.0-REFV)*PARDIR
         RTDIRD=(1.0-SCV )*PARDIR
         SUNPER=RTDIRD/SINB
         FSSL  =1.0
         DO 50 I2=1,IN
            IF ((ALVL(I2)*PLEA(I2)*PLMX(I2)).LE.0.) GOTO 50
            F3(I2)=1.00-F2(I2)-F1(I2)
            O     =F1(I2)*O15+F2(I2)*O45+F3(I2)*O75
            T2DS  =F1(I2)*0.034+F2(I2)*0.25+F3(I2)*0.47+
     $              SINB*SINB*(F1(I2)*0.90+F2(I2)*0.25-F3(I2)*0.42)
            RANGET=SQRT(12.0*AMAX1(0.,T2DS-O*O))
            KBL   =0.97*F1(I2)+0.85*F2(I2)+0.65*F3(I2)
            KDIF  =KBL*SQRT(1.0-SCV)
            KDIRBL=O/SINB
            KDIRT =KDIRBL*SQRT(1.-SCV)
            FGROS  =0.
```

```
          DO 40 I3=-1,1
             ALVC  =(0.5+GAUSR*I3)*(ALVL(I2)+ALVDL(I2))
             VISDF =RTDIF*KDIF*EXP(-KDIF*ALVC)
             VIST  =RTDIRT*KDIRT*EXP(-KDIRT*ALVC)
             VISD  =RTDIRD*KDIRBL*EXP(-KDIRBL*ALVC)
             VISSHD=VISDF+VIST-VISD
             FGRSH =PLMX(I2)*(1.-EXP(-VISSHD*PLEA(I2)/PLMX(I2)))
             FGRSUN=0.0
             DO 30 I4=-1,1
                SN     =0+RANGET*I4*GAUSR
                VISSUN=VISSHD+SN*SUNPER
                FGRS  =PLMX(I2)*(1.0-EXP(-VISSUN*PLEA(I2)/PLMX(I2)))
                IF (I4.EQ.0) FGRS  =FGRS*1.6
                FGRSUN=FGRSUN+FGRS
 30          CONTINUE
             FGRSUN=FGRSUN/3.6
             FSSLA =FSSL*EXP(-KDIRBL*ALVC)
             FGL   =FSSLA*FGRSUN+(1.-FSSLA)*FGRSH
             IF(I3.EQ.0) FGL   =FGL*1.6
             FGROS =FGROS+FGL
 40       CONTINUE
          FGROS =FGROS*ALVL(I2)/3.6
          IF(I1.EQ.0) FGROS =FGROS*1.6
          PHL(I2,I1+2)=FGROS*DLA/3.6
          RTDIF =RTDIF *EXP(-KDIF  *(ALVL(I2)+ALVDL(I2)))
          RTDIRT=RTDIRT*EXP(-KDIRT *(ALVL(I2)+ALVDL(I2)))
          RTDIRD=RTDIRD*EXP(-KDIRBL*(ALVL(I2)+ALVDL(I2)))
          FSSL  =FSSL  *EXP(-KDIRBL*(ALVL(I2)+ALVDL(I2)))
 50    CONTINUE
 60 CONTINUE
    DO 70 I2=1,IN
       PCGCL(I2)=PHL(I2,1)+PHL(I2,2)+PHL(I2,3)
 70 CONTINUE
    PCGC  =PCGCL(1)+PCGCL(2)+PCGCL(3)+PCGCL(4)+PCGCL(5)
    RETURN
    END

    SUBROUTINE SUSAWA(SWICH1,WCLQT,WLOQT,NL,TRWL,EVSC,
   $            RAIN,ZW,TKL,TYL,DELT,DTMIN,DTMX1,DTFX,
   $            WLOMX,ZEQT,CSA,CSB,CSC2,WCLCH,WLOCH,
   $            WCLEQI,EVSW,RUNOF,DRSL,WCUMCH,ZECH,ZLT)
C calculates the soil water balance for 24 h. in L2SS. 04/88
    IMPLICIT REAL (A-Z)
    INTEGER SWICH1,SWICH2,SWICH3,SWICH4,SWICH5
    INTEGER NL,I,INXSAT,JJTOT,JTOT,J
    COMMON /SLDPTH/ ZL(10)
    COMMON /VOLWAT/ WCAD(10),WCFC(10),WCST(10),WCWP(10)
    COMMON /HYDCON/ KMSMX(10),KMSA1(10),KMSA2(10),KST(10)
    COMMON /PFCURV/ MSWCA(10)
    DIMENSION INXSAT(10,10),JJTOT(10),WCL(10),TKL(10),MS(10)
    DIMENSION FLXSTT(11),FLXUNT(11),FLX(11),FLXINT(11)
    DIMENSION HGT(10),HGB(10),HPP(11),DHH(10),TRWL(10),TYL(10)
    DIMENSION WCLQT(10),WCLCH(10),WCLEQI(10)
    DATA TINY1,TINY2,SWICH3,SWICH4,SWICH5/1.E-3,1.E-4,1,1,2/
    DO 10 I=1,NL
       IF(SWICH1.EQ.1) ZL(I)=0.
       WCLCH(I) =0.
       WCL(I)   =WCLQT(I)
```

```
10    CONTINUE
      WLO   =WLOQT
      ZE    =ZEQT*100.
      EVSW  =0.
      DRSL  =0.
      RUNOF =0.
      CALL SUCONV(1,TKL,ZL,ZLT,RAIN,ZW,WLO,WLOMX,RUNOF,TRWL,EVSW,
     $          EVSC,DRSL,NL)
      IF(SWICH1.EQ.1) THEN
         CALL SUGRHD(TKL,NL,HGT,HGB)
         CALL SUSLIN(TYL,NL,TKL,ZW,ZLT,WCL)
         DO 20 I=1,NL
            WCLEQI(I)=WCL(I)
20       CONTINUE
      ELSE
         TIMTOT=0.
         DO 30 I=1,NL
            WCL(I)    =WCL(I)-DELT*TRWL(I)/TKL(I)
            FLXINT(I)=0.
30       CONTINUE
         FLXINT(NL+1)=0.
         DTMX2 =1.
         WLO   =WLO+RAIN*DELT
         HPBP  =ZLT-ZW
         IF(WLO.GE.EVSC*DELT) THEN
            EVSW  =EVSC
            EVSWX =0.
            WLO   =WLO-EVSC*DELT
         ELSEIF(WLO.GT.0.) THEN
            EVSW1 =WLO/DELT
            EVSW2 =AMIN1(EVSC-EVSW1,(WCL(1)-WCAD(1))*TKL(1)/DELT)
            EVSW  =EVSW1+EVSW2
            EVSWX =EVSW2
            WCL(1)=WCL(1)-EVSW2*DELT/TKL(1)
            WLO   =0.
         ELSE
            EVSW1 =EVSC
            EVSW2 =(WCL(1)-WCAD(1))*TKL(1)/DELT
            EVSW3 =CSC2/(ZE+TINY1)
            EVSW  =AMIN1(EVSW1,EVSW2,EVSW3)
            EVSWX =EVSW
            WCL(1)=WCL(1)-EVSW*DELT/TKL(1)
         ENDIF
40       JTOT  =0
         DO 50 J=1,NL
            JJTOT(J) =0
50       CONTINUE
         CALL SUSTCH(WCL,NL,INXSAT,JTOT,JJTOT)
         IF(JTOT.NE.0) THEN
          CALL SUSTHH(INXSAT,JTOT,JJTOT,NL,HGT,HGB,WLO,WLOMX,HPBP,
     $                HPP,DHH)
          CALL SUSTFL(NL,INXSAT,JTOT,JJTOT,TKL,DHH,FLXSTT,FLXSQ1,FLXSQ2)
         ELSE
            CONTINUE
         ENDIF
         IF(JTOT.EQ.1.AND.JJTOT(1).EQ.NL) THEN
           FLXUNT(1)=0.
         ELSE
```

```
          CALL SUUNST(SWICH3,SWICH4,WCL,TKL,NL,WLO,WLOMX,HPBP,
     $                MS,FLXUNT)
        ENDIF
        CALL SUSEFL(INXSAT,NL,JTOT,JJTOT,FLXSTT,FLXUNT,WLO,WLOMX,
     $              FLXSQ1,FLXSQ2,FLX)
        CALL SUINTG(HPBP,SWICH4,SWICH5,HPP,FLX,TKL,WCL,MS,DTMIN,DTMX1,
     $       DTMX2,DTFX,INXSAT,DHH,JTOT,FLXSQ2,WLO,WLOMX,NL,DELT,DT)
        DO 60 I=1,NL+1
           FLXINT(I)=FLXINT(I)+DT*FLX(I)
 60     CONTINUE
        TIMTOT=TIMTOT+DT
        DTMX2 =DELT-TIMTOT
        IF(DTMX2.GT.TINY2) GOTO 40
        IF(WLO.GT.WLOMX) THEN
           RUNOF =WLO-WLOMX
           WLO   =WLOMX
        ENDIF
        DRSL  =FLXINT(NL+1)/DELT
        CALL SUZECA(WCLQT(1),EVSC,RAIN,-EVSWX,FLXINT(2),
     $              WLO,DELT,ZE,CSA,CSB,CSC2)
      ENDIF
      CALL SUCONV(2,TKL,ZL,ZLT,RAIN,ZW,WLO,WLOMX,RUNOF,TRWL,EVSW,
     $            EVSC,DRSL,NL)
      ZECH  =ZE/100.-ZEQT
      WLOCH =WLO-WLOQT
      WCUMCH=0.
      DO 70 I=1,NL
         WCLCH(I)=WCL(I)-WCLQT(I)
         WCUMCH  =WCUMCH+WCLCH(I)*TKL(I)
 70   CONTINUE
      RETURN
      END

      SUBROUTINE SUSEFL(INXSAT,NL,JTOT,JJTOT,FLXSTT,FLXUNT,
     $           WLO,WLOMX,FLXSQ1,FLXSQ2,FLX)
C selects between saturated and unsaturated fluxes; in L2SS; 04/88
      IMPLICIT REAL (A-Z)
      INTEGER INXSAT,JJTOT,JTOT,J,JJ,I,IIN,IOUT,NL
      DIMENSION INXSAT(10,10),JJTOT(10),FLXSTT(11),FLXUNT(11),FLX(11)
      DATA TINY/0.001/
      HPTP=AMIN1(WLO,WLOMX)
      DO 10 I=1,NL+1
         FLX(I)   =FLXUNT(I)
 10   CONTINUE
      DO 70 J=1,JTOT
         IF(FLXSTT(J).GT.TINY) THEN
            IOUT  =INXSAT(J,JJTOT(J))+1
            IIN   =INXSAT(J,1)
            FLX(IOUT)=AMIN1(FLXSTT(J),FLXUNT(IOUT))
            IF(IOUT.EQ.NL+1) FLX(IOUT)=FLXSTT(J)
            IF (JJTOT(J).GT.1) THEN
               DO 20 JJ=2,JJTOT(J)
                  FLX(INXSAT(J,JJ))=FLX(IOUT)
 20            CONTINUE
            ENDIF
            FLX(IIN)=AMIN1(FLX(IOUT),FLXUNT(IIN))
            IF(IIN.EQ.1.AND.HPTP.GT.TINY) FLX(IIN)=FLX(IOUT)
         ELSEIF(FLXSTT(J).LT.-TINY) THEN
```

```
         IOUT  =INXSAT(J,1)
         IIN   =INXSAT(J,JJTOT(J))+1
         IF(FLXSQ2.LE.0.) THEN
            FLX(IOUT)=FLXSTT(J)
         ELSE
            IF(FLXUNT(IOUT).LT.-TINY) THEN
               FLX(IOUT)=FLXUNT(IOUT)
            ELSEIF(FLXUNT(IOUT).GT.TINY) THEN
               FLX(IOUT)=AMAX1(FLXSQ2,FLXUNT(IOUT))
            ELSE
               FLX(IOUT)=0.
            ENDIF
         ENDIF
         IF(IOUT.EQ.1) FLX(IOUT)=FLXSTT(J)
         IF(JJTOT(J).GT.1) THEN
            DO 30 JJ=2,JJTOT(J)
               FLX(INXSAT(J,JJ))=AMAX1(FLX(IOUT),FLXSTT(J))
30          CONTINUE
         ENDIF
         FLX(IIN)=AMAX1(FLX(IOUT),FLXSTT(J))
      ELSE
         IIN   =INXSAT(J,1)
         IOUT  =NL+1
         IF(FLXUNT(IIN).LE.-TINY) THEN
            FLX(IIN)=FLXUNT(IIN)
            IF(JJTOT(J).GT.1) THEN
               DO 40 JJ=2,JJTOT(J)
                  FLX(INXSAT(J,JJ))=AMAX1(FLXSQ1,FLX(IIN))
40             CONTINUE
            ENDIF
            FLX(IOUT)=AMAX1(FLXSQ1,FLX(IIN))
         ELSEIF(FLXUNT(IIN).GE.TINY) THEN
            FLX(IIN)=AMIN1(FLXUNT(IIN),FLXSQ2)
            IF(JJTOT(J).GT.1) THEN
               DO 50 JJ=2,JJTOT(J)
                  FLX(INXSAT(J,JJ))=FLX(IIN)
50             CONTINUE
            ENDIF
            FLX(IOUT)=FLX(IIN)
         ELSE
            FLX(IIN)=0.
            IF(JJTOT(J).GT.1) THEN
               DO 60 JJ=2,JJTOT(J)
                  FLX(INXSAT(J,JJ))=0.
60             CONTINUE
            ENDIF
            FLX(IOUT)=0.
         ENDIF
      ENDIF
70 CONTINUE
   DO 80 I=1,NL+1
      FLXUNT(I)=0.
      FLXSTT(I)=0.
80 CONTINUE
   RETURN
   END

   SUBROUTINE SUSLIN(TYL,NL,TKL,ZW,ZLT,WCL)
```

```
C derives soil parameters. in L2SS. 03/88
      IMPLICIT REAL (A-Z)
      INTEGER NL,I,IX
      COMMON /VOLWAT/ WCAD(10),WCFC(10),WCST(10),WCWP(10)
      COMMON /SLDPTH/ ZL(10)
      DIMENSION TYL(10),TKL(10),WCL(10)
      DATA TINY/1.E-10/
      WRITE(6,20)
      DO 10 I=1,NL
         CALL SUERRM(6.1,TYL(I),1.,20.,6.)
         WRITE(6,30)I,TYL(I),TKL(I)*.01,WCAD(I),WCWP(I),WCFC(I),WCST(I)
         IF(I.EQ.1) THEN
            ZL(I)=0.
         ELSE
            ZL(I)=ZL(I-1)+TKL(I-1)
         ENDIF
         MS    =AMAX1(0.,ZW-ZL(I)-0.5*TKL(I))
         CALL SUWCMS(I,2,WCL(I),MS)
         CALL SUERRM(6.2,WCL(I),WCAD(I),WCST(I),6.)
 10   CONTINUE
      ZLT   =ZL(NL)+TKL(NL)
 20   FORMAT(' SOIL CHARACTERISTICS PER COMPARTMENT: ',/,
     $' COMPARTMENT  TYPE NR   TKL    WCAD    WCWP    WCFC    WCST')
 30   FORMAT(3X,I4,8X,F5.1,3X,F5.3,3X,4(F5.4,3X))
      RETURN
      END

      SUBROUTINE SUSTCH(WCL,NL,INXSAT,JTOT,JJTOT)
C checks for presence of saturated layers. in L2SS. 9/87.
      IMPLICIT REAL (A-Z)
      INTEGER NL,INXSAT,I,J,JTOT,JJTOT,JJ
      COMMON /VOLWAT/ DUMMY1(20),WCST(10),DUMMY(10)
      DIMENSION WCL(10),INXSAT(10,10),JJTOT(10)
      DATA TINY/0.001/
      J     =0
      JJ    =0
      DO 10 I=1,NL
         IF(ABS(WCL(I)-WCST(I)).LT.TINY) THEN
            JJ    =JJ+1
            IF(JJ.NE.1) THEN
               J     =J+0
            ELSE
               J     =J+1
               JTOT  =J
            ENDIF
            INXSAT(J,JJ)=I
            JJTOT(J)=JJ
         ELSE
            JJ    =0
         ENDIF
10    CONTINUE
      RETURN
      END

      SUBROUTINE SUSTFL(NL,INXSAT,JTOT,JJTOT,TKL,DHH,
     $             FLXSTT,FLXSQ1,FLXSQ2)
C calculates tentative saturated fluxes. in L2SS. 04/88
      IMPLICIT REAL (A-Z)
```

```
      INTEGER I,J,N,NL,JTOT,JJTOT,INXSAT,INDX,IA,JA
      COMMON /HYDCON/ DUMMY(30),KST(10)
      DIMENSION A(11,11),TKL(10),INXSAT(10,10),JJTOT(10),B(11)
      DIMENSION DHH(10),FLXSTT(11),DIS(10),INDX(11),KS(10)
      DATA TINY/1.0E-10/, LARGE/-100./,DUMMY2/0./,DUMMY1/0./
      DO 190 J=1,JTOT
         DO 20 IA=1,NL+1
            DO 10 JA=1,NL+1
               A(IA,JA)=0.
               B(JA)=0.
10          CONTINUE
20       CONTINUE
         IF(ABS(DHH(J)).LT.TINY) THEN
            FLXSTT(J)=0.
         ELSE
            DO 30 IA=1,JJTOT(J)
               I=INXSAT(J,1)+IA-1
               KS(IA)=KST(I)
               DIS(IA)=TKL(I)
30          CONTINUE
            DO 40 IA=1,JJTOT(J)
               A(IA,JJTOT(J)+1)=-1.
               JA=IA
               A(IA,JA)=-KS(IA)/DIS(IA)
40          CONTINUE
            DO 50 JA=1,JJTOT(J)
               A(JJTOT(J)+1,JA)=+1.
50          CONTINUE
            A(JJTOT(J)+1,JJTOT(J)+1)=0.
            DO 60 JA=1,JJTOT(J)
               B(JA)=0.
60          CONTINUE
            B(JJTOT(J)+1)=DHH(J)
            N=JJTOT(J)+1
            CALL SUSTMD(A,N,INDX,D)
            CALL SUSTMS(A,N,INDX,B)
            FLXSTT(J)=B(N)
         ENDIF
         IF(DHH(J).GT.-TINY) THEN
            DO 80 IA=1,NL+1
               DO 70 JA=1,NL+1
                  A(IA,JA)=0.
                  B(JA)=0.
70             CONTINUE
80          CONTINUE
            IF(JJTOT(J).EQ.1) THEN
               FLXSQ1=LARGE
            ELSE
               DO 90 IA=1,JJTOT(J)-1
                  I=INXSAT(J,2)+IA-1
                  KS(IA)=KST(I)
                  DIS(IA)=TKL(I)
90             CONTINUE
               DO 100 IA=1,JJTOT(J)-1
                  A(IA,JJTOT(J))=-1.
                  JA=IA
                  A(IA,JA)=-KS(IA)/DIS(IA)
100            CONTINUE
```

```
                DO 110 JA=1,JJTOT(J)-1
                   A(JJTOT(J),JA)=+1.
 110            CONTINUE
                A(JJTOT(J),JJTOT(J))=0.
                DO 120 JA=1,JJTOT(J)-1
                   B(JA)=0.
 120            CONTINUE
                B(JJTOT(J))=DHH(J)+TKL(INXSAT(J,1))
                N=JJTOT(J)
                CALL SUSTMD(A,N,INDX,D)
                CALL SUSTMS(A,N,INDX,B)
                FLXSQ1=B(N)
             ENDIF
             DO 140 IA=1,NL+1
                DO 130 JA=1,NL+1
                   A(IA,JA)=0.
                   B(JA)=0.
 130            CONTINUE
 140         CONTINUE
             IF(INXSAT(J,1).EQ.1) THEN
                FLXSQ2=DUMMY2
             ELSE
                DO 150 IA=1,JJTOT(J)+1
                   I=(INXSAT(J,1)-1)+IA-1
                   KS(IA)=KST(I)
                   DIS(IA)=TKL(I)
 150            CONTINUE
                DO 160 IA=1,JJTOT(J)+1
                   A(IA,JJTOT(J)+2)=-1.
                   JA=IA
                   A(IA,JA)=-KS(IA)/DIS(IA)
 160            CONTINUE
                DO 170 JA=1,JJTOT(J)+1
                   A(JJTOT(J)+2,JA)=+1.
 170            CONTINUE
                A(JJTOT(J)+2,JJTOT(J)+2)=0.
                DO 180 JA=1,JJTOT(J)+1
                   B(JA)=0.
 180            CONTINUE
                B(JJTOT(J)+2)=DHH(J)-TKL(INXSAT(J,1)-1)
                N=JJTOT(J)+2
                CALL SUSTMD(A,N,INDX,D)
                CALL SUSTMS(A,N,INDX,B)
                FLXSQ2=B(N)
             ENDIF
          ENDIF
 190  CONTINUE
      RETURN
      END

      SUBROUTINE SUSTHH(INXSAT,JTOT,JJTOT,NL,HGT,HGB,
     $              WLO,WLOMX,HPBP,HPP,DHH)
C identifies hydraulic head across saturated layers. in L2SS. 04/88
      IMPLICIT REAL (A-Z)
      INTEGER I,NL,INXSAT,J,JTOT,JJTOT
      DIMENSION INXSAT(10,10),JJTOT(10),HHT(10),HHB(10),HGT(10),HGB(10)
      DIMENSION HPP(11),DHH(10)
      DATA TINY/1.E-5/
```

```
      HPTP  =AMIN1(WLO,WLOMX)
      DO 10 I=1,11
         HPP(I)=0.
 10   CONTINUE
      DO 20 J=1,JTOT
         EXTRAT=0.
         IF(INXSAT(J,1).EQ.1) EXTRAT=HPTP
         EXTRAB=0.
         IF(INXSAT(J,JJTOT(J)).EQ.NL) EXTRAB=HPBP
         HHT(J)=EXTRAT+HGT(INXSAT(J,1))
         HHB(J)=EXTRAB+HGB(INXSAT(J,JJTOT(J)))
         DHH(J)=HHB(J)-HHT(J)
         IF(DHH(J).GT.TINY) HPP(INXSAT(J,1))=DHH(J)
         IF(ABS(DHH(J)).LT.TINY) DHH(J)=0.
 20   CONTINUE
      RETURN
      END

      SUBROUTINE SUSTMD(A,N,INDX,D)
C decomposes matrix A. in L2SS. (PRESS etal, 1986.)
      IMPLICIT REAL (A-Z)
      INTEGER I,J,K,N,IMAX,INDX
      DIMENSION A(11,11),INDX(11),VV(11)
      DATA TINY/1.0E-10/
      D=1.
      DO 12 I=1,N
         AAMAX=0.
         DO 11 J=1,N
            IF(ABS(A(I,J)).GT.AAMAX) AAMAX=ABS(A(I,J))
 11      CONTINUE
         VV(I)=1./AAMAX
 12   CONTINUE
      DO 19 J=1,N
         IF(J.GT.1) THEN
            DO 14 I=1,J-1
               SUM=A(I,J)
               IF(I.GT.1) THEN
                  DO 13 K=1,I-1
                     SUM=SUM-A(I,K)*A(K,J)
 13               CONTINUE
                  A(I,J)=SUM
               ENDIF
 14         CONTINUE
         ENDIF
         AAMAX=0.
         DO 16 I=J,N
            SUM=A(I,J)
            IF(J.GT.1.) THEN
               DO 15 K=1,J-1
                  SUM=SUM-A(I,K)*A(K,J)
 15            CONTINUE
               A(I,J)=SUM
            ENDIF
            DUM=VV(I)*ABS(SUM)
            IF (DUM.GE.AAMAX) THEN
               IMAX=I
               AAMAX=DUM
            ENDIF
```

```
16        CONTINUE
          IF(J.NE.IMAX) THEN
             DO 17 K=1,N
                DUM=A(IMAX,K)
                A(IMAX,K)=A(J,K)
                A(J,K)=DUM
17           CONTINUE
             D=-D
             VV(IMAX)=VV(J)
          ENDIF
          INDX(J)=IMAX
          IF(J.NE.N) THEN
             IF(ABS(A(J,J)).LT.TINY) A(J,J)=TINY
             DUM=1./A(J,J)
             DO 18 I=J+1,N
                A(I,J)=A(I,J)*DUM
18           CONTINUE
          ENDIF
19    CONTINUE
      IF(ABS(A(N,N)).LT.TINY) A(N,N)=TINY
      RETURN
      END

      SUBROUTINE SUSTMS(A,N,INDX,B)
C solves a set of linear equations. in L2SS. (PRESS etal, 1986)
      IMPLICIT REAL (A-Z)
      INTEGER I,J,II,LL,N,INDX
      DIMENSION A(11,11),INDX(11),B(11)
      II=0
      DO 12 I=1,N
        LL=INDX(I)
        SUM=B(LL)
        B(LL)=B(I)
        IF(II.NE.0) THEN
          DO 11 J=II,I-1
            SUM=SUM-A(I,J)*B(J)
11        CONTINUE
        ELSE IF (SUM.NE.0.) THEN
          II=I
        ENDIF
        B(I)=SUM
12    CONTINUE
      DO 14 I=N,1,-1
        SUM=B(I)
        IF(I.LT.N) THEN
          DO 13 J=I+1,N
            SUM=SUM-A(I,J)*B(J)
13        CONTINUE
        ENDIF
        B(I)=SUM/A(I,I)
14    CONTINUE
      RETURN
      END

      SUBROUTINE SUUNST(SWICH3,SWICH4,WCL,TKL,NL,WLO,WLOMX,
     $            HPBP,MS,FLXUNT)
C calculates tentative fluxes of unsaturated layers; in L2SS; 04/88
      IMPLICIT REAL (A-Z)
```

```
INTEGER NL,I,SWICH3,SWICH4
COMMON /HYDCON/ DUMMY1(30),KST(10)
COMMON /VOLWAT/ WCAD(10),DUMMY2(10),WCST(10),DUMMY3(10)
DIMENSION WCL(10),TKL(10),DIS(11),FLXUNT(11),MS(10)
DIMENSION KMS(10),MFLP(10),MFLPQT(10)
DATA TINY1,TINY2/0.001,0.001/
HPTP  =AMIN1(WLO,WLOMX)
IF(WCST(1)-WCL(1).GT.TINY1) THEN
   CALL SUWCMS(1,SWICH4,WCL(1),MS(1))
   CALL SUMFLP(SWICH3,1,MS(1),MFLP(1))
   DIS(1)=0.5*TKL(1)
   IF(HPTP.GT.TINY1) THEN
      DZDH  =DIS(1)/(-MS(1)-HPTP)
      DMFLP =MFLP(1)-0.-KST(1)*HPTP
      FLXUNT(1)=(DZDH-1.)*DMFLP/DIS(1)
   ELSE
      FLXUNT(1)=0.
   ENDIF
ELSE
   FLXUNT(1)=0.
ENDIF
DO 10 I=2,NL
   IF(ABS(WCL(I)-WCST(I)).LT.TINY2) THEN
      IF(ABS(WCL(I-1)-WCST(I-1)).GT.TINY2) THEN
         DIS(I)=0.5*TKL(I-1)
         DZDH  =DIS(I)/(0.+MS(I-1))
         DMFLP =0.-MFLP(I-1)
         FLXUNT(I)=(DZDH-1.)*DMFLP/DIS(I)
      ELSE
         CONTINUE
      ENDIF
   ELSE
      CALL SUWCMS(I,SWICH4,WCL(I),MS(I))
      CALL SUMFLP(SWICH3,I,MS(I),MFLP(I))
      IF(ABS(WCL(I-1)-WCST(I-1)).LT.TINY2) THEN
         DIS(I)=0.5*TKL(I)
         DZDH  =DIS(I)/(-MS(I)-0.)
         DMFLP =MFLP(I)-0.
         FLXUNT(I)=(DZDH-1.)*DMFLP/DIS(I)
      ELSE
         CALL SUMFLP(SWICH3,I-1,MS(I),MFLPQT(I))
         CALL SUMFLP(SWICH3,I,MS(I-1),MFLPQT(I-1))
         DIS(I)=0.5*(TKL(I)+TKL(I-1))
         DMFLP1=MFLP(I)-MFLPQT(I-1)
         IF((MS(I)-MS(I-1))*(MFLP(I)-MFLPQT(I-1)).GT.-TINY1) THEN
            DMFLP1=0.
         ENDIF
         DMFLP2=MFLPQT(I)-MFLP(I-1)
         IF((MS(I)-MS(I-1))*(MFLPQT(I)-MFLP(I-1)).GT.-TINY1) THEN
            DMFLP2=0.
         ENDIF
         IF(ABS(DMFLP1).LT.TINY2.OR.ABS(DMFLP2).LT.TINY2) THEN
            DMFLP =0.
         ELSE
            SIGN  =DMFLP1/ABS(DMFLP1)
            DMFLP =SIGN*SQRT(DMFLP1*DMFLP2)
         ENDIF
         IF(MS(I).NE.MS(I-1)) THEN
```

```
                  DZDH   =DIS(I)/(-MS(I)+MS(I-1))
                  FLXUNT(I)=(DZDH-1.)*DMFLP/DIS(I)
               ELSE
                  CALL SUMSKM(SWICH3,I,MS(I),KMS(I))
                  CALL SUMSKM(SWICH3,I-1,MS(I-1),KMS(I-1))
                  KAV    =SQRT(KMS(I)*KMS(I-1))
                  FLXUNT(I)=+KAV
               ENDIF
            ENDIF
         ENDIF
 10   CONTINUE
      IF(ABS(WCL(NL)-WCST(NL)).LT.TINY2) THEN
         CONTINUE
      ELSE
         DIS(NL+1)=0.5*TKL(NL)
         CALL SUMFLP(SWICH3,NL,MS(NL),MFLP(NL))
         IF(HPBP.GT.TINY1) THEN
            DZDH   =DIS(NL+1)/(HPBP+MS(NL))
            DMFLP  =0.-MFLP(NL)+KST(NL)*(HPBP-0.)
            FLXUNT(NL+1)=(DZDH-1.)*DMFLP/DIS(NL+1)
         ELSE
            MSB    =-HPBP
            CALL SUMFLP(SWICH3,NL,MSB,MFLPB)
            IF(HPBP.NE.-MS(NL)) THEN
               DZDH   =DIS(NL+1)/(HPBP+MS(NL))
               DMFLP  =MFLPB-MFLP(NL)
               IF((MSB-MS(NL))*(MFLPB-MFLP(NL)).GT.-TINY1) DMFLP =0.
               FLXUNT(NL+1)=(DZDH-1.)*DMFLP/DIS(NL+1)
            ELSE
               CALL SUMSKM(SWICH3,NL,MS(NL),KMS(NL))
               FLXUNT(NL+1)=+KMS(NL)
            ENDIF
         ENDIF
      ENDIF
      RETURN
      END

      SUBROUTINE SUWCMS(I,SWICH4,WCL,MS)
C relates volumetric water content and suction; in L2SS, 03/87
      IMPLICIT REAL (A-Z)
      INTEGER I,SWICH4
      COMMON /VOLWAT/ WCAD(10),DUMMY1(10),WCST(10),DUMMY2(10)
      COMMON /PFCURV/ MSWCA(10)
      DATA TINY/0.001/
      IF(SWICH4.EQ.1) THEN
         CALL SUERRM(7.1,WCL,WCAD(I),WCST(I),6.)
         MS     =EXP(SQRT(-ALOG(AMAX1(WCAD(I),WCL)/WCST(I))/MSWCA(I)))-1.
      ELSE
         CALL SUERRM(7.2,MS,0.,1.E8,6.)
         WCL =AMAX1(TINY,WCST(I)*EXP(-MSWCA(I)*((ALOG(MS+1.))**2)))
      ENDIF
      RETURN
      END

      SUBROUTINE SUZECA(WCLQT1,EVSC,RAIN,FLX1,FLX2,WL0,
     $             DELT,ZE,AEXP,BEXP,C2)
C calculates the depth of the evaporation front; in L2SS; 04/88
      IMPLICIT REAL (A-Z)
```

```
      COMMON /VOLWAT/ WCAD(10),DUMMY1(10),WCST(10),DUMMY2(10)
      DATA TINY1,LARGE/1.E-3,10./
      IF(RAIN.GT.EVSC) THEN
         ZE    =0.
      ELSEIF(FLX2.LT.-TINY1.AND.FLX2.LT.FLX1) THEN
         ZE    =ZE-(FLX1-FLX2)/(0.5*WCST(1))
         IF(ZE.LE.TINY1) THEN
            ZE    =0.
         ENDIF
      ELSEIF(RAIN.GT.0.05) THEN
         CONTINUE
      ELSE
         WI    =WCLQT1/WCST(1)
         WTH   =WI-WCAD(1)/WCST(1)
         IF(WTH.LE.TINY1) THEN
            ZE    =LARGE
         ELSEIF(WL0.GT.TINY1.OR.ABS(WCLQT1-WCST(1)).LT.TINY1) THEN
            ZE    =0.
         ELSE
            C1    =AEXP*(EXP(AMAX1(0.,(WI-0.5)*BEXP))-1.)
            C3    =C2/(WTH+C1/C2)
            ZE    =SQRT(ZE*ZE+2.*C3*DELT)
         ENDIF
      ENDIF
      RETURN
      END
ENDJOB
```